Practical Spectroscopy of High-Frequency Discharges

PHYSICS OF ATOMS AND MOLECULES

Recent volumes in this series:

COINCIDENCE STUDIES OF ELECTRON AND PHOTON IMPACT IONIZATION
Edited by Colm T. Whelan and H. R. J. Walters

DENSITY MATRIX THEORY AND APPLICATIONS, Second Edition
Karl Blum

ELECTRON COLLISIONS WITH MOLECULES, CLUSTERS, AND SURFACES
Edited by H. Ehrhardt and L. A. Morgan

INTRODUCTION TO THE THEORY OF LASER–ATOM INTERACTIONS, Second Edition
Marvin H. Mittleman

INTRODUCTION TO THE THEORY OF X-RAY AND ELECTRONIC SPECTRA OF FREE ATOMS
Romas Karazija

PHOTON AND ELECTRON COLLISIONS WITH ATOMS AND MOLECULES
Edited by Philip G. Burke and Charles J. Joachain

POLARIZATION BREMSSTRAHLUNG
Edited by V. N. Tsytovich and I. M. Ojringel

POLARIZED ELECTRON/POLARIZED PHOTON PHYSICS
Edited by H. Kleinpoppen and W. R. Newell

PRACTICAL SPECTROSCOPY OF HIGH-FREQUENCY DISCHARGES
Sergei A. Kazantsev, Vyacheslav I. Khutorshchikov, Günter H. Guthöhrlein, and Laurentius Windholz

SELECTED TOPICS ON ELECTRON PHYSICS
Edited by D. Murray Campbell and Hans Kleinpoppen

THEORY OF ELECTRON–ATOM COLLISIONS, Part 1: Potential Scattering
Philip G. Burke and Charles J. Joachain

VUV AND SOFT X-RAY PHOTOIONIZATION
Edited by Uwe Becker and David A. Shirley

A Chronological Listing of Volumes in this series appears at the back of this volume.

A Continuation Order Plan is available for this series. A continuation order will bring delivery of each new volume immediately upon publication. Volumes are billed only upon actual shipment. For further information please contact the publisher.

Practical Spectroscopy of High-Frequency Discharges

Sergei A. Kazantsev
St. Petersburg State University
St. Petersburg, Russia, and
Paris Observatory
Meudon, France

Vyacheslav I. Khutorshchikov
Russian Institute for Radionavigation and Time
St. Petersburg, Russia

Günter H. Guthöhrlein
University of the Federal Armed Forces
Hamburg, Germany

and

Laurentius Windholz
Graz Technical University
Graz, Austria

Springer Science+Business Media, LLC

Library of Congress Cataloging in Publication Data

Practical spectroscopy of high-frequency discharges / Sergei A. Kazantsev . . . [et al.].
p. cm.—(Physics of atoms and molecules)
Includes bibliographical references and index.
ISBN 978-1-4419-3281-5 ISBN 978-1-4757-6229-7 (eBook)
DOI 10.1007/978-1-4757-6229-7
1. Radio frequency discharges. 2. Light sources. 3. Spectral analysis. 4. Electric lamps. 5. Quantum electronics. I. Kazantsev, Sergei A. II. Series.
QC711.8.R35P73 1998
537.5′34—dc21 97-40604
CIP

ISBN 978-1-4419-3281-5

Originally published by Plenum Press, New York in 1998

http://www.plenum.com

10 9 8 7 6 5 4 3 2 1

Preface

This monograph contains a thorough description of different applications of classical and novel techniques of optical spectroscopy for the investigation of light sources, operated on the basis of a high-frequency electrodeless discharge.

Such spectral sources have outstanding significance in fundamental spectroscopy, analytical spectroscopy, and quantum optics, not only as sources of line spectra of different chemical elements, but also as the substantial part of quantum electronics measuring devices, based on optical pumping and radio-optical double resonance.

Therefore, special attention is paid to high-frequency electrodeless discharges operated in a mixture of alkali metals (among these with high priority rubidium) and noble gases, which are directly used in quantum frequency standards, atomic clocks, and quantum magnetometers, and permit their high precision and sensitivity.

Besides characterization of high-frequency electrodeless discharges by their electrical quantities, spectroscopic methods are especially used to investigate the features of such discharges, plasma properties, plasma–surface interactions, collisional effects, intensity fluctuations, and long-term drifts. The results of these applications of practical spectroscopy are used to optimize the spectral parameters of the discharges and construct light sources, especially with respect to very long service lifetime.

The solution of problems related to the spectroscopic investigation of high-frequency discharges that are formulated and analyzed in this book should enable the improvement of different ground-based and satellite-borne navigation and time-support systems. Therefore this book will be of interest to readers not only among the wide community of researchers in industry and at universities, including Ph.D. students, concerned with optics and spectroscopy or gaseous discharge and plasma physics, but also among designers of light sources and quantum optics devices.

Contents

1
Introduction

1.1. High-Frequency Electrodeless Light Sources Emitting Line Spectra and Their Applications

The most widespread light sources that provide a line spectrum are different gaseous discharge devices. In high-frequency (hf) electrodeless spectral lamps (ELs) the discharge is excited by a hf field, created within the lamp with the help of electrodes or inductors located outside. The hf voltage, applied to initiate and sustain the discharge, is supplied to the inductor from a hf generator or a self-excited oscillator. The lamps, usually bulbs of spherical or cylindrical form, are filled with an inert gas or with an inert gas with additives of those elements which produce the desired radiation. The pressure of the gas is usually chosen between 0.7 and 20 mbar. The frequency of the exciting field is variable, depends on the size and shape of the glass bulb, and can be chosen within a range from several up to hundreds of MHz. The exciting electrodes are usually built as either an inducing coil (inductor) or a capacitor incorporated into the electric circuit of the hf field generator. When a proper discharge mode is chosen by setting the gas pressure (in case of vapors, via temperature), the power of the exciting field, as well as taking care of the design of the lamp, the emission of intense narrow spectral lines is observed.

The high intensity of optical radiation has brought about the widespread use of such light sources to generate emission spectra for various experiments, such as for the investigation of double resonance phenomena, sensitized fluorescence due to collision processes, shifts and broadening of spectral lines, and many other physical processes.

Technical applications of the hf ELs are quite broad, concentrating mostly on the field of spectral analysis and precision optical devices such as quantum frequency standards, goniometers, quantum magnetometers, and many other applications. We shall briefly discuss general requirements imposed upon spectral light sources in various spheres of their applications.

The general construction of a high-frequency electrodeless spectral source is shown in Figure 1.1: The lamp (1) is excited by an external unit (2) (normally

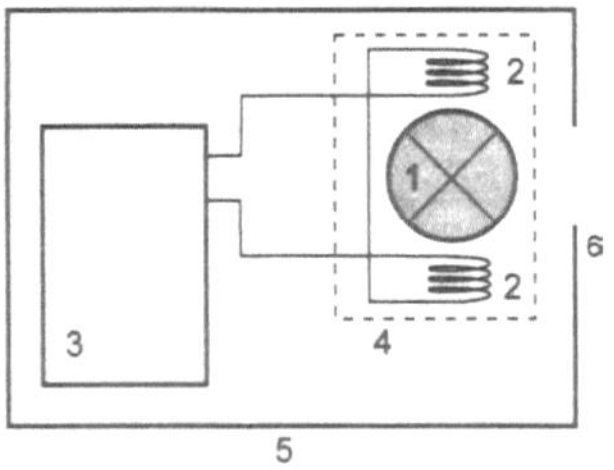

FIGURE 1.1. Main components of a high-frequency electrodeless spectral source: 1, electrodeless lamp; 2, inductors (coils or capacitor plates); 3, high frequency generator or self-excited oscillator; 4, thermostat; 5, housing; 6, exit orifice.

coils or capacitor plates; here, for example, coils) provided by a hf generator (3). Sometimes the temperature of the lamp is controlled by a thermostat (4). The whole equipment is placed inside a housing (5) with an orifice (6) for emergence of the radiation.

1.1.1. Analytical Spectroscopy: Atomic Absorption Techniques

The main requirement imposed on light sources for atomic absorption methods is the high brightness of radiation. If the intensity of the lamp is insufficient, the fluctuations of the measured signal A_I are not due to the internal noise of the lamp but to the shot noise of the photodetector. When increasing the light intensity the signal/noise ratio grows, as long as the linewidth of the irradiating light remains narrower than the absorption linewidth. This is the consequence of the fact that the signal A_I is proportional to the light intensity I, or more precisely to the integral

$$A_I \sim \int_0^\infty I(\nu)\exp(-k_\nu l)d\nu \tag{1.1}$$

or

$$A_I/A_0 \sim \int_0^\infty [I(\nu)/I_0]\exp(-k_\nu l)d\nu \tag{1.2}$$

where $I(\nu)$ is the shape of the radiation line, k_ν the factor of absorption at frequency ν, and l the thickness of the absorbing layer.

The noise factor is defined by the Schottky noise formula

$$\sqrt{\overline{i_{sh}^2}} = \sqrt{2ei\Delta\Omega} \tag{1.3}$$

(i is the photocurrent, $i \sim I_0$; e is the charge of the electron; $\Delta\Omega$ is the band within which the noise is measured), so noise is proportional to $I_0^{1/2}$.

The requirement for the stability of radiation in atomic absorption analyzers can be rather low when using two-channel recording systems, measuring the reference intensity I_0 (without absorption) and calculating $I(\nu)/I_0$. For one-beam devices highly stable light sources are indispensable.

Spectral light sources for absorption spectroscopy should emit a simple radiation spectrum and therefore allow the generation of intensive lines of a large number of elements including metals and chemically active species.

Special studies carried out elsewhere [1, 2] have shown that the above-stated general requirements are mostly fulfilled by hf ELs for the majority of elements.

1.1.2. Analytical Spectroscopy: Atomic-Fluorescence Techniques

In recent decades the atomic-fluorescence analysis [3], which is distinguished by higher sensitivity as compared with the atomic-absorption method, has gained wide application. For this method the main requirements regarding the light sources are again a high brightness of the emitted resonance lines, intensity stability, as well as simplicity and safety, long life period, and, if possible, low cost and availability. But there is a difference: in atomic-absorption analysis it is desirable to have the widths of the radiation lines considerably narrower than those of the absorption lines, while in the atomic-fluorescence analysis both widths should be comparable for the best sensitivity. These requirements are met to a high degree by hf ELs with metal vapors and halides of metals mixed with inert gases as well as by lamps with hollow cathodes [2].

During recent years lasers of various types [3] have been widely used for spectral analysis. The well-known merits of lasers extolled their importance in a number of applications, for example, in the analysis of trace quantities of substances as well as in other cases where their main advantage — the small spectral width of the emitted spectral lines — is necessary. However, for common applications they are rather expensive and cumbersome. Therefore, the development of laser analytical spectroscopy for special technical applications does not influence the majority of devices for spectral analysis which have reached a certain level of perfection and which are mainly based on the use of hf ELs. In view of the competition of classical discharge spectral sources with lasers, all the effort of developers and researchers of hf ELs for atomic absorption and fluorescence analysis as well as for other technical applications is directed to solving problems of longer durability and stability of radiation, multielement operation, etc.

A qualitative comparison of the main characteristics of various light sources for atomic-fluorescence analysis according to published data [3] is given in Table 1.1.

Table 1.1 shows the place occupied by hf ELs among other light sources used in physical experiments and engineering at the present time. The success in research and development of ELs (especially after publication of the work [3]) provided their relative position before the lamps with hollow cathode and continuous lasers in view of short-term stability, surpassing all the above-mentioned kinds of

TABLE 1.1. Comparison of Different Light Sources

Quantity	Comparison
Spectral brightness	IL $\gg$ CL $\gg$ EL, HCL $\gg$ SHP
Stability	*SHP* $\gg$ HCL, CL $\gg$ EL, IL
Ergometricity	*SHP* $\gg$ HCL, EL $\gg$ CL $\gg$ IL
Cost	IL, CL $\gg$ HCL, EL $\gg$ SHP
Number of analyzed lines	*SHP* $\gg$ EL, HCL $\gg$ IL, CL

[a] IL, impulse lasers; CL, continuous lasers; HCL, hollow cathode lamps; EL, hf electrodeless lamps; SHP, xenon arcs of superhigh pressure

light sources in long-term stability, reliability, and durability. The achievements listed above were attained by special studies concerning the use of hf ELs for quantum electronic devices with optical excitation. The applicability of such devices has been limited for a long time by the properties of the spectral light sources used.

1.1.3. Quantum Frequency Standards and Quantum Magnetometers

Quantum standards of frequency and quantum magnetometers are based on the stabilization of a quartz generator by locking its frequency to a hyperfine transition. In case of optical detection, it is necessary to create nonequilibrium populations of hyperfine sublevels in the ground state of the working atoms and then to record the change in the light intensity when scanning the frequency of the probing electromagnetic field around the transition frequency. The difference of populations created by optical pumping is defined by the optical pumping rate W_{ij} [4]:

$$W_{ij} = \int_0^\infty u(\nu)\left|w_{ij}\right|^2 d\nu \frac{\gamma/2}{[\nu-(\tilde{\nu}_0+\vec{k}_0\vec{v}/2\pi)]^2+(\gamma/2)^2} \tag{1.4}$$

where w_{ij} is the transition probability; i, j are the quantum numbers in the ground and excited state; $u(\nu)$ is the spectral density of optical radiation; $\gamma/2$ the half-width of absorption line in frequency units; $\vec{k}_0$ the wave vector of photons which are interacting with the atom; $\vec{v}$ the speed of atoms; ν_0 the unperturbed transition frequency between levels j and i; $\tilde{\nu}_0 = \nu_0 + \Delta E_0/h$, while ΔE_0 is the change of the energy difference between levels j and i by external effects.

It follows from Eq. (1.4) that the efficiency of excitation is at a maximum in case of overlapping of the shapes of absorption and radiation lines. For creating a population difference of the sublevels it is necessary to ensure different pumping speeds of each of the hyperfine sublevels. In case of frequency standards working with Rb, it is possible to select by means of a filter one of the hyperfine components, using the isotope ^{87}Rb or ^{85}Rb.

The sensitivity of radio-optical resonance methods is determined by the ratio of the amplitude of the resonance signal to the product of the width of the spectral line and the spectral density of noise. Therefore, the light source used requires a small spectral density of noise. To reduce the influence of low-frequency fluctuations of the lamp and the elements of the electrical circuit on the sensitivity, low-frequency modulation and synchronous detection are commonly used when observing the radio-optical resonance. Low-frequency fluctuations of the intensity of the pump source usually limit the sensitivity due to the effect of "light shifts." A "light shift" [4] is observed as a displacement of the absorption line in relation to the radiation line when the working atoms are perturbed by the light of excitation, resulting in a shift of the hyperfine levels of the ground state. The magnitude and sign of the light shift are defined by the intensity and line shape of the irradiating light as well as by its location in relation to the absorption lines. A change in light intensity of 1% results in a relative shift of the frequency of the hyperfine structure transition from 10^{-11} to 10^{-12}. Therefore the frequency stability of a quantum standard, based on a gas cell during the observation time $t \gg 10^3$–10^4 s, is mainly determined by intensity fluctuations of the light sources. Hence an increase in stability is possible when the magnitude of the light shift of the atomic levels, and therefore the fluctuations of the light source, are effectively reduced. The requirements for frequency standards are rather high and growing continuously, determining a high level of the light sources concerning:

(a) spectral resolution of the studied spectral lines;

(b) high spectral density of radiation (for different variants of quantum frequency standards the required intensity is different, but the width of the emission line should always be about the same as that of the absorption line);

(c) high stability of radiation intensity (not more than 1% per 24 h, 10% for the whole service life);

(d) long durability (continuous work for 10 years and more with a sufficient light intensity);

(e) nonreproducibility from switching-on to switching on should be not more than 1% for one device, from sample to sample not more than 10%;

(f) minimum consumed power, small size, reliability, noise immunity.

Similar requirements are imposed upon the lamps for quantum magnetometers. In quantum frequency standards mainly spectral lamps operating with rubidium vapor and inert gas, and in magnetometers mainly lamps operating with vapors of cesium and potassium and sometimes with silver and sodium vapors, are used. The development of frequency standards based on ion traps has lead to the creation of and research on lamps with mercury vapor possessing increased intensity of the ionic resonance lines [5]. In all these sources of light (except those working with rubidium and cesium vapors) the problem of increasing the stability and durability of lamps is considered to be the most difficult.

The light sources used at the present time meet mainly the above-stated requirements and quantum frequency standards are industrially available. However, the complex studies which have lead in many respects to solutions of the above-mentioned problems have not been sufficiently described in the literature. The aim of this book is to fill, at least partly, this gap.

1.1.4. Other Applications

One other practical application of spectral lamps is in goniometers. Here lamps filled with helium are used, and a difficult problem is to achieve high intensity of radiation with acceptable service lifetime of the spectral lamps. In recent years a suitable type of lamp with the required spectral parameters has been developed [5, 6].

Another wide field of applications comprises various physical experiments. A large manifold of designs of hf discharge lamps with radius from 3 to 30 mm, of various fillings and different forms such as spherical, cylindrical, cigar-shaped, toroidal, spherical with a concave surface at one of the walls and with filtering cells, have been elaborated and used. In every particular case the design was justified by optimizing the excitation of the required spectral lines (first or second resonance doublets, ionic lines) to achieve higher stability. All these questions and developments will be described in the corresponding sections of this book.

1.2. Historical Review of the Development of High-Frequency Electrodeless Spectral Lamps

The history of the scientific development and practical application of hf electrodeless discharges (EDs) as optical spectral sources goes back to the twenties. Jackson [7] began to use the radiation of a discharge, which was excited in a tube by a hf field with the help of electrodes placed outside the tube, to study the hyperfine structure of the rubidium lines. This tube was filled with rubidium vapor and a mixture of helium and neon at a total pressure of 0.7 mbar, and was heated to increase the intensity of the radiation of the rubidium lines [8]. Jackson [9] also studied the radiation of a hf ED in a mixture of cesium and helium vapors at a pressure of 2.7 mbar. These early works showed that the spectrum of this discharge source is characterized by an arc character and high spectral radiation density of the alkali resonance lines, which allowed one to record the hyperfine structure splitting. A similar design of the spectral source was used by different research groups until the beginning of the fifties to excite the spectra of various elements [10, 11]. The properties of such light sources in comparison with other types of spectral lamps were discussed in detail elsewhere [10]. A higher brightness of the radiation when decreasing the diameter of the middle part of the lamp down to several millimeters was detected, as well as a specific way of aging connected with the formation of a film of chemical compositions of the working element in the vicinity of the electrodes.

However, such light sources have not been widely used because of technical difficulties connected with the creation of the hf field, the lack of pure heavy noble gases, metals, and isotopes. At the same time intensive investigations of the electric characteristics of hf discharges have been stimulated by discussions between J. J. Thomson and J. Townsend on the nature of discharges (details of this discussion are given in [12]). As a result it was established that, depending on the type of generator, excitation of two principle kinds of electrodeless discharges is possible. A discharge excited by means of a coil was called an *H-discharge*, while a so-called *E-discharge* was excited within capacitor plates. The principal and more general difference between these two types of discharges, as shown in [13, 14], is that the H-discharge exists in the antinode of the magnetic field of an intensive discharge at rather high power, representing a significant load for a generator. In the antinode of the electrical field an E-discharge is excited with lower radiation intensity, being at the same time rather steady over wide ranges of pressure and power of the hf field oscillator.

In the fifties, hf EDs were investigated in pure buffer gases (Ar, Ne, N_2, Kr, Xe) in spherical bulbs at low and intermediate pressures and at frequencies between 4 and 8 MHz [15]. The excitation of two different kinds of discharge was observed. One of them was named *predischarge* and the other *ring discharge,* and they were connected to the potential and curling fields induced within the inductor.

At the beginning of the seventies hf EDs in spherical bulbs were studied in detail [16]. It was shown that the electron temperature is constant over the volume, if the electron concentration lies within the range 10^{11}–10^{12} cm^3. Further studies carried out in the eighties revealed some peculiarities in the distribution of the exciting fields and the concentration of electrons in the boundary area [17].

A discharge lamp filled with a mixture of a buffer gas and mercury vapor was proposed [18], and it was found that such a lamp radiates extremely narrow spectral lines, suitable for spectral analysis [11]. These studies, as well as further research [10, 19–21], made it possible to develop spectral light sources with alkali metal vapors and a number of other elements for purposes of optical pumping, as well as atomic-absorption and atomic-fluorescence analyses.

Interest in such light sources increased sharply after the publication of [18, 22, 23], where rather simple and reliable designs of light sources on the basis of hf EDs in mixtures of rubidium vapor and krypton or rubidium vapor and argon were proposed. In [22] a light source of 10 mm in diameter filled with rubidium and krypton vapor under a pressure of 2.1 mbar was described. The shot character of the noise, high working durability which exceeded 10000 h, and high density of optical radiation were noted. A flow of photons in the resonance lines of rubidium attaining 3×10^{18} photons/s was achieved by Brewer [23] with a lamp 30 mm in diameter. Similar results were received by Franz [24] using a hf EL with cesium vapor.

The results of earlier studies on hf ELs led to the conclusion that the high density of radiation, relative simplicity of design, and long work durability made these sources indispensable in optical pumping devices used in atomic-absorption analysis and several physical experiments, which required intensive and stable radiation of various atomic spectral lines. Therefore light sources on the basis of hf ELs became an object of intensive research. The significance of these spectral sources for problems of spectral analysis, and especially for atomic-absorption analysis, have been justified [1, 2, 25–27] and later ELs for such elements as Rb, K, Cs, Na, Zn, Cd, Hg, Se, and Tl began to be manufactured commercially. Suitable materials for cylindric lamps were chosen: alkali-proof glasses for alkali metals and quartz glass for lamps emitting ultraviolet radiation. Special studies of the spectral characteristics allowed one to specify a significant superiority of hf ELs over lamps with hollow cathodes. It appeared that for the cadmium lines hf lamps emit optical radiation 10 times more intense than hollow cathode lamps. Extremely high radiation intensity of hf lamps with mercury vapor was also noted.

During the sixties the study of hf discharge spectral sources for the purpose of optical pumping of atomic vapors was very intensive. Spectral lamps were studied extensively in Russia when creating quantum magnetometers in the Vavilov State Optical Institute (St.Petersburg), and quantum standards of frequency in the Gorky Device Designing Institute (Nizhny Novgorod) and the Scientific Research Radiotechnical Institute (St. Petersburg). The most systematic studies were conducted in the Institute of Radioelectronics of the Russian Academy of Science.

Detailed studies on the possibilities of utilizing such light sources in spectral analysis were carried out by different groups during the same period [1, 2, 5, 6, 10–12, 19–21, 25–51]. In course of these extensive studies the widths of spectral lines and the voltage–current characteristics of sources with various fillings of buffer gases were measured and the effects of the radiation output on the temperature of the thermostat were studied in detail. The relation of the radiation intensity to the mode of the lamp was established, and a limiting upper intensity of radiation of lamps with a particular diameter filled with a mixture of rubidium vapor and krypton as well as with cesium vapor and krypton was stated.

Careful studies of the structure of the rubidium absorption lines were carried out with the aid of a Fabry–Perot interferometer [52, 53]. Special processing allowed one to observe the splitting of the rubidium resonance line 794.76 nm due to the hyperfine structure in the 5 $^2P_{1/2}$ state, and to show that the line profile was dominated by the effect of Doppler broadening. Measurements of the widths of Rb spectral lines specially performed at the St. Petersburg State University with the use of a double Fabry–Perot interferometer allowed the broadening of the emission line profiles to be studied depending on the mode of the lamp, and even their shifts in relation to the lines of nonperturbed atoms were evaluated [54]. Newly developed spectroscopic methods, such as spectropolarimetric sensing, have also been applied to studies of hf discharge spectral sources. These methods allowed one to measure new characteristic parameters [34] and to establish many new results and conceptions of the physical processes in the plasma.

All this research promoted the development of the theory and practical applications of hf ELs for the optical pumping technique, the creation of quantum standards of frequency, and the perfection of methods of atomic-absorption analysis. Nevertheless the very rapid development of quantum electronics, and other fields where spectral lamps were applied, required an essential improvement in their characteristics, primarily, an increase in reliability and durability and a reduction in consuming power. The solution of these problems appeared to be possible only via a deeper understanding of the physics of spectral lamp operation. Therefore, later on in the seventies the improvement of hf spectral lamps continued in different works [49, 50, 52, 54–82]. Modeling of the processes in electrodeless discharges in mixtures of alkali metal vapors and inert gases [66, 69, 83] allowed one to explain all the main features in the formation of spectral line radiation. The basic model of this spectral source was proposed. The main concept of this model was that the alkali atoms being in the ground state, are concentrated in the vicinity of the lamp's surface, while excited atoms are predominately at some distance from it, so that a significant reabsorption occurs within the surface sheath [74, 84].

Disagreement between computed and experimental relationships when increasing the temperature or when decreasing the discharge power could be explained by radiation being reabsorbed within the wall sheath. The optical depth of this sheath of 0.1–5, estimated from the difference in the light intensities emitted from the center and border of the lamps, was close to that calculated in [69, 85]. It is shown [69, 85, 86] that the model of the discharge proposed in [66, 69] gives significant gradients of the density of atoms, resulting in a drift of atoms and, as a consequence, a shift between the lines of radiation and absorption, reaching 30 MHz [74].

In [69, 76, 87] self-absorption was used to explain the properties of the line shapes, including its asymmetry, over the whole range of the operating temperatures. Comparative studies of lamps filled with rubidium vapor and different gases — neon, argon, krypton, and xenon — showed that the change of characteristics as a function of the hf power is smoothest in lamps with krypton, which also provide a minimal linewidth [26, 54]. The intensity of the spectral lines was shown to depend on the type and pressure of the buffer gas. The intensity decreases in a row of buffer gases Ne, Ar, Kr, Xe and the optimal pressure for providing narrow lines was found to be 1.3–5.3 mbar. Optimization of hf ELs with Rb was the subject of a number of papers [22, 24, 26, 54, 72, 87]. Different effects such as relaxation, influence of the bulb dimensions, role of the buffer gas and excitation conditions have been covered. When choosing the gas, the intensity of the radiation of the spectral lines of the buffer gas appeared to worsen the signal-to-noise ratio. This effect is much higher in the case of Kr than for Ar or Xe [26].

All these effective studies resulted in the fact that, at the beginning of the seventies, the commercial production of lamps with various fillings and optical dimensions 10 and 20 mm in diameter was progressing, as well as lamps with improved operational characteristics, using a vacuum jacket [88]. Such spectral lamps filled with K, Na, Rb, Cs, Cd, Zn, Te, Sb, Sn, Tl, and Cu have been produced commercially since the middle of the seventies [31, 32]. However, as the service time and reliability of this generation of spectral lamps (especially for applications in quantum standards and other high precision devices) were low, the developers were compelled to continue basic research on the physical processes in the multicomponent plasmas affecting the spectral characteristics of the optical emission. As a result, spectral lamps originally designed for quantum frequency standards with minimum consumption of energy, were invented [89] and since then they have been fabricated, commercialized, and used in many other devices.

In the course of such scientific and engineering work the the spectrum of the lamp radiation as a function of the pressure and type of gas, the temperature of the thermostats, as well as the frequency of the exciting field has been analyzed [26, 27]. The role of the skin effect was studied [1–3, 22–25, 28–34, 36, 40, 46, 90–93] by special measurements in more detail than, for example, in [25], where the conductivity of the plasma was estimated experimentally with the help of Langmuir probes.

In the early seventies a model was suggested to explain the main features of lamps by the redistribution of atoms throughout the volume subject to condition of high electron concentration. A theory accounting for these effects was later developed [94–102].

Special types of cigar-shaped lamps filled with rubidium and krypton vapor were developed and investigated [56]. The lamps showed high stability: the intensity variations were lower than 8% during 14,000 h of continuous work at a level of about 5 mW per steradian [90].

In connection with the study of the sensitized fluorescence effect a large amount of original research and technological development has been undertaken [6, 37, 41, 103–105]. Spectral lamps with various fillings (more than half of the periodic table of elements) have been developed, as well as lamps with mixtures of vapors of two or three elements, lamps filled with helium showing high photon flux and long durability, lamps with rare isotopes, spectral light sources emitting mainly mercury ionic lines, and lamps emitting the normally suppressed spectrum of inert gases. Original methods of filling the lamps and dosing the working substance have been elaborated. Special attention has been paid to the creation of lamps on the basis of helium and isotopes of mercury [42]. Helium lamps, owing to their original technology, have a durability (operating period) of more than 1000 h with a very rich spectrum: helium lines originating up to the level $n = 20$ have been excited [43]. It was found and studied as an important effect for applications in which adding rubidium to mercury vapor lamps reduces the radiation of krypton [106, 107].

A significant number of studies has been undertaken for the development of quantum magnetometers [52]. The requirement of high radiation intensity in connection with high level of stability is extremely important in this case.

Systematic studies of the durability and reliability of lamps were performed in the seventies [80]. The operating time before failure for these lamps appeared to be several thousands of hours. The duration of the service life τ depends on the power W of the discharge as $\tau \sim W^a$, where $a = 2/3$. In [80] a conclusion was reached as to the possibility of increasing the durability of lamps up to 10–20 years when decreasing the discharge power. Studies of the aging of lamps by the method of accelerated tests [108, 109], in a real time scale [68, 69] and theoretically [97, 99], showed a weaker dependence on the power of the discharge and a relation between the speed of aging and the design and mode of the discharge lamp. A new kind of failure, connected with the reduction in power caused by metallization of the lamp walls and shielding of internal areas of the discharge, was found [110]. The process of film formation [10] is most intensive in locations which are in direct contact with the hf-discharge plasma. It results in a change of the working impedance of the lamp, and causes an increase in the breakdown potential of the hf discharge and a reduction in the intensity of the radiation spectrum. In [57], as elsewhere [63, 67, 69], the service life of the lamp was observed to depend on the pressure and was studied in detail. In order to solve the problem of increasing the durability and reliability, methods of controlling the parameters of the spectral lamps were developed [67–69, 108, 111] including the filling process [112]. The theory governing processes within the spectral lamps [69, 99] was experimentally tested [98, 113] and special attention paid to factors which determine the aging of the lamps [69, 109, 114]. As a result, special modes which ensure a durability of 50,000–100,000 h were chosen and the technology of filling and controlling the parameters was improved. Lamps with increased pressure that provide highest durability at high radiation intensity were especially studied [115–118]. With the proper manufacturing technology and the use of specially processed alkali-proof glass, the low-power mode allowed one to obtain practically unlimited service life of the lamps in the E-discharge mode [58–60, 69, 89]. But in some cases spectral lamps made of sapphire, which is extremely alkali-proof, had a considerably improved durability and reliability [119]. However, spectral lamps used in atomic-absorption analysis nowadays possess rather low reliability and a durability of about 300–500 h, and even shorter in the case of filling with alkali metals [40].

The stability of ELs is limited by the transfer and condensation of metal, but can be eliminated by selection of a temperature gradient between the area of discharge and the glass surface. For example, in lamps working with sodium vapor (diameter 30 mm, frequency 27 MHz, power 100 W) the metal was placed in a long tank connected to the lamp, in this way suppressing fluctuations. To reduce the low-frequency fluctuations it was also proposed that a rod made of ferromagnetic material be placed inside the lamp [26]. This rod, being a heat conductor, reduces simultaneously the dissipation of the hf field, and hence the consumed power within the bulb of the lamp becomes lower.

As a result a radiation stability of better than 1–2% per 1–6 h and about 5% per 20–200 h was attained [90, 120]. Such spectral lamps lose out in comparison with lamps excited by microwaves, among which there are spectral lamps with stability up to 1% or even up to 0.1% [81]. In [56, 90] cigar-shaped lamps were investigated. In [22] it was stated that with the temperature stabilization of the lamp and a rational choice of the working conditions of the generator, the radiation fluctuations of the lamp can be reduced up to a limit determined by the shot fluctuations of the light detector. Experimental measurements of the noise of the lamp, described in [23], showed that the observable level of noise is close to the theoretical one and does not depend on the frequency of the exciting hf circuit.

Measurements of the radiation stability of the Cs I line $\lambda = 455.5$ nm showed [2] that, down to light flows which correspond to a photocurrent of 4×10^{-11} A, the fluctuations do not exceed the level of the shot noise, making up to 0.03% of the signal (at a time constant of the recording device of 0.2 s). In [40] it was found that the intensity instability during 1 min usually amounts to 1%, and during 30 min to 10%.

However, systematic research of the fluctuation characteristics of lamps was undertaken in a series of investigations in the eighties [121–125]. In [124], the spectral density of the intensity fluctuations was studied as a function of the frequency. An approximation of this dependency on the measurement period for the H-discharge has been proposed [125]. Following these studies, the durability of spectral lamps depending on the amount of metal introduced into the lamp and on the sort of the bulb glass was studied in detail [68, 69, 126, 127].

The role of an external magnetic field at low intensities was discussed [128, 129]. Magnetic fields of less than 1 mT destroy the alignment of excited atoms in these lamps, which results in the angular redistribution of intensities. The effect of weak fields ($\ll 0.1$ mT) on the radiation intensity should be taken into account when designing and constructing light sources. According to [130, 131] the reabsorption of radiation results in a freezing of the ground state of metal atoms. This effect, as well as the redistribution of the intensity of the hyperfine components for hf modulation of the radiation [132], has a universal character.

The most substantial studies of light sources based on hf EDs have been carried out in connection with the development of the rubidium atomic clock. Therefore spectral lamps with rubidium vapor have been more intensively investigated than lamps filled with cesium and potassium vapors. Far fewer studies have been devoted to lamps with mercury vapor, and there are only few publications on other lamp fillings. In this connection the main attention in this book is given to lamps with fillings of those elements which have the most extended field of applications and for which much experimental material has been accumulated. Nevertheless, most of the results obtained for spectral lamps with a mixture of rubidium vapor and noble gases are rather universal and can be applied to many other cases.

In conclusion, the improvement of known methods of practical spectroscopy, the development of new techniques, and the comprehensive application of all these methods to high-frequency discharge lamps have allowed the parameters of these spectral sources to be optimized. A large number of design and engineering problems were solved in order to achieve an improved level of sensitivity, stability, longevity, and reliability for many devices of quantum metrology, navigation, time-scale support, analytical optics, and many fundamental physical and technical applications.

2

General Characteristics of High-Frequency Electrodeless Spectral Lamps

2.1. Designs of Light Sources and Types of Gas Discharges Used for the Excitation of a Spectrum

2.1.1. Features of Discharges in Electrodeless Spectral Lamps

At least two kinds of discharges have been known since the 1930s. Following [13] they were named E- and H-discharges. As mentioned in the introduction, the E-discharge is excited in the antinodes of the electric field, and the H-discharge in the antinodes of the magnetic field. Accordingly, in the E-discharge the breakdown and maintenance of the discharge are insured by an electric potential field while in the H-discharge by a curling electric field. Different types of hf discharges have been discussed in detail for discharge parameter variation over a wide range [15, 133–137]. In reality, the discharge is excited and supported by an electromagnetic field, which is described by Maxwell's equations:

$$\begin{aligned} &\operatorname{rot}\vec{H} = \vec{j} + \frac{\partial \vec{D}}{\partial t}, \qquad && \operatorname{rot}\vec{E} = -\frac{\partial \vec{B}}{\partial t} \\ &\operatorname{div}\vec{B} = 0, && \operatorname{div}\vec{D} = \rho \end{aligned} \tag{2.1}$$

where $\vec{E}$ is the electric field strength, $\vec{H}$ the strength of the magnetizing field, $\vec{D}$ the induction created by the electric field, $\vec{B}$ the magnetic field strength, $\vec{j}$ the density of electric current created by free charges, and ρ the charge density of the free charges.

In order to take into account the electromagnetic waves moving in matter, we have only to incorporate the effects of electrical polarization $\vec{P}$ and magnetization $\vec{M}$ into the relations between $\vec{D}$ and $\vec{E}$ on one side and $\vec{B}$ and $\vec{H}$ on the other. This is done in the following steps:

$$\vec{j} = \sigma\vec{E}, \qquad \vec{P} = \varepsilon_0\chi\vec{E}, \qquad \vec{M} = \kappa\vec{H} \tag{2.2}$$

where $\vec{P}$ and $\vec{M}$ are vectors of electrical polarization and magnetization per unit volume of the medium, σ is the electric conductivity of the medium, χ the electric susceptibility, κ the magnetic susceptibility. Instead of χ and κ, the permittivity $\varepsilon = \varepsilon_r\varepsilon_0 = (1+\chi)\varepsilon_0$ and permeability $\mu = \mu_r\mu_0 = (1+\kappa)\mu_0$ are introduced and, as a result, we get the equations

$$\vec{j} = \sigma\vec{E}, \qquad \vec{D} = \varepsilon\vec{E}, \qquad \vec{B} = \mu\vec{H} \tag{2.3}$$

which, together with (2.1) — unchanged — and initial and boundary conditions, form a closed system which describes the electromagnetic field in matter (as well as in the plasma of the discharge).

The electric field which arises as a result of the change in the magnetic field has a curling character and causes a corresponding current when influencing the plasma. Besides, a so-called current of displacement exists and is induced by the polarization of the plasma when an external potential field is applied. Thus, the breakdown and the discharge itself may be conditioned by electric fields of different origin, and it is possible to expect differences in their behavior for different discharge modes.

At first we shall consider the case of an E-discharge. If the voltage U is applied to the lamp (between capacitor plates spaced l apart) we can write

$$U = 2U_w + U_p = 2E_w l_w + E_p l_p \tag{2.4}$$

where l_w is the thickness of the wall of the bulb, l_p the thickness of the plasma layer, U_w, U_p, E_w, E_p are the voltages and electric field strength in the materials of the wall and inside the bulb, respectively.

As shown elsewhere [134], the current flow in the case under consideration is described by the following system of equations:

$$\begin{aligned} &\varepsilon E_w = q, && E - \varepsilon E_w = -q_p \\ &U = 2\frac{q}{\varepsilon}l_w + \frac{j_p}{\sigma}l_p, && j_p = \sigma(q - q_p) \\ &\frac{d\rho}{dt} = j, && \frac{d\rho_p}{dt} = j_p \end{aligned} \tag{2.5}$$

i.e., before the breakdown, the conductivity of the plasma $\sigma = 0$ and all the voltage is applied to the gas in the volume of the lamp sphere.

In this case the strength of the electric field is high ($\sim 10^2$–3×10^2 V/cm), and therefore "on a free run" an electron can accumulate an amount of energy high enough for the excitation and ionization of atoms. Hence, fluorescence will be observed in the locations of the highest voltages of the hf field. This behavior occurs even in the absence of alkali metal vapors and for not too high frequency of the exciting field.

While the conductivity is growing, the power failure on the walls of the cylinder will increase and the voltage of the electric field applied to the discharge interval appears to be less than U by the value $2E_\delta l_\delta$. This fact is the reason for the hindered initiation of the E-discharge in bulbs with thick walls. Since, as will be shown further, the H-discharge arises usually after the excitation of the E-discharge it is desirable in this case, to use lamp cylinders with thin walls. When increasing the power of the discharge, the power falling on the walls of the lamp grows as well. The growth of energy being transmitted to the discharge is also hindered by the polarization of the plasma, resulting in a shielding of its internal volume.

In fact, the Debye shielding results in a reduction of the external electric field E subject to the law

$$E \sim E_p \exp(-x/r_D) \tag{2.6}$$

where r_D is the length of Debye shielding given by

$$r_D = \left[\frac{\varepsilon_0 k_B T_e T_a}{e^2\left(n_e T_a + n_i z^2 T_e\right)}\right]^{1/2} \tag{2.7}$$

T_e and T_a are the temperatures of electrons and atoms, respectively, n_e and n_i the concentrations of the electrons and ions; z is the degree of ionization of the atoms. Hence, when $n_e = 10^{10}$ cm^{-3} and $T_e = 10{,}000$ K, $r_D \sim 1.4 \times 10^{-3}$ cm and the external field practically does not penetrate into the plasma. Since the applied field is variable in direction, for an evaluation of its efficiency it is necessary to take into account the relaxation time τ of the plasma [65]: $\tau = r_D/\overline{v}$, where $\overline{v}$ stands for the average velocity of the electrons. As a result one obtains $\tau \gg 1/f$ at $n_e = 10^7$ cm^{-3} and $\tau \ll 1/f$ at $n_e = 10^{12}$ cm^{-3}, where f is the frequency of the hf field [in Hz].

Thus, the radiation intensity in the E-discharge at first grows quickly with increasing power, and then its growth is slowed down because of the decrease in the discharge efficiency. The magnetic field penetrates into the volume of the plasma practically unimpeded and, at high powers, the H-discharge appears much more effective than the E-discharge. The skin effect is a restriction on the power being induced, because it sharply decreases the plasma volume, in which the hf energy is effectively induced. The thickness δ of the skin layer is equal to

$$\delta = 1/\sqrt{\pi\sigma\mu f} \tag{2.8}$$

where μ is the permeability of the plasma. In practice the restriction often appears earlier, caused by the final value of the power of the hf generator.

It should be noted that for a long time it was assumed that the skin effect predetermines the high values of spectral density of radiation of such light sources. However, it is not difficult to be convinced that at $\sigma = 2\Omega^{-1}\text{cm}^{-1}$ and $f = 50$ MHz [70], $\delta \simeq 5$ mm and that at the highest measured conductivities of a discharge in a mixture of krypton and rubidium [138] the thickness of the skin layer is no larger than $d = 2$ mm. For characteristic temperatures and pressures in the lamp the optical thickness of the skin-layer is much higher than unity [85]. It is important to note that the space heterogeneity of introducing the hf power into the discharge has no effect upon the temperature of the electronic cloud. In reality, the thermal flow q_e being carried by electrons will be expressed as

$$q_e = -\lambda_e \frac{dT_e}{dx} \tag{2.9}$$

with

$$\lambda_e = \frac{5}{2} \frac{n_e T_e k_B^2}{m_e n_a \sigma_{ea} v_e} \tag{2.10}$$

where λ_e is the factor of heat conductivity by electrons, m_e and v_e are the mass and speed of the electrons; σ_{ea} the cross section of elastic collisions of electrons with atoms; n_a the concentration of atoms, and T_e the electron temperature.

Hence, it is possible to obtain the following expression (x is a spatial coordinate):

$$T_e(x) = T_0 - \frac{q_e}{\lambda_e} x \tag{2.11}$$

On substituting the values $T_0 = 10^4$ K, $n_a = 10^{17}$ cm^{-3}, $\sigma_{ea} = 2 \times 10^{-20}$ m^2, $v_e \simeq 6 \times 10^7$ cm/s, and $q_e \sim 1$ W/cm^2, we obtain

$$T_e = T_0 - \frac{5 \times 10^{15}}{n_e} x \tag{2.12}$$

It is obvious that for the size of a noncurrent zone of about several millimeters a heterogeneity in temperature of order 10% appears when $n_e > 10^{12}$ cm^{-3}. Actually the heterogeneity will be much less, as the losses of heat conductivity are less than the energy applied. Therefore at $p < 13$ mbar and $n_e > 10^{12}$ cm^{-3}, the distribution of T can be considered as uniform throughout the volume of the lamp. Thus, the spectral characteristics of the hf ED cannot be defined by the skin effect, though the electrical parameters are dependent on it. So, from the viewpoint of the conductivity σ [Ω^{-1} m^{-1}] of the plasma, the criterion for the most effective input of the hf radiation into the discharge [85, 139] is

$$\frac{0.5 \times 10^9}{\sigma^2 (2r_0)^2} < f[\text{Hz}] < \frac{1.5 \times 10^9}{\sigma^2 (2r_0)^2} \tag{2.13}$$

where r_0 is the radius of the lamp (cm) and f the frequency of the self-excited oscillator.

However, the peculiarities of the discharge are not caused by how the energy is supplied, but by the physical processes in the lamp and, primarily, of all, by the mechanisms of ionization and recombination. We have shown above that in an E-discharge direct ionization is possible, but when the frequency of the field is increased or if easily ionizable components (e.g., alkali metals) are used, a transition to a discharge with the symmetry of the bulb and with considerably higher brightness of radiation can be observed. According to its external attributes such as an intensive luminescence, a significant load on the generator, and a high conductivity, the E-discharge can be compared with the positive column of a glow discharge and, even more correct, with an arc discharge. A representative example is a discharge in neon. Here a high-intensity emission in the red part of the spectrum is accompanied by an increase of loads on the generator (growth of the current through the generator and strong warming is observed). The probable reason for this behavior is the transition to an arc discharge with a stepwise excitation and an electron emission under the action of ionic bombardment of the lamp walls. The role of the stepwise excitation can be evaluated if we use the criterion [140]

$$n_e \geq \sum_j \left[\frac{\left(U_i + \frac{2kT_e}{e}\right) \exp\left(-\frac{eU_{a_j}}{kT_e}\right)}{\left(U_i - U_{a_j} + \frac{2kT_e}{e}\right) \overline{\sigma_j} \overline{v} \tau_{a_j}} \right] \tag{2.14}$$

where the summation is carried out over all excited levels j, U_i is the potential of ionization of the atom, U_{a_j} the potential of excitation of j level, T_e the electronic temperature in the discharge, v the mean electron velocity; $\overline{\sigma_j}$ the excitation cross section, τ_{a_j} the lifetime of the atom in the excited level j.

We will simplify the expression by assuming that the main contribution to the formation of ions is made by the processes of ionization from metastable states. It is evident that the estimated concentration of electrons, for which stepwise ionization predominates, is then slightly overestimated. In this case

$$n_e \geq \frac{\left(U_i + \frac{2kT_e}{e}\right) \exp\left(-\frac{eU_a}{kT_e}\right)}{\left(U_i - U_a + \frac{2kT_e}{e}\right) \overline{\sigma_a}\, \overline{v} \tau_a} \tag{2.15}$$

We will evaluate the concentration of electrons for the case of a discharge in argon at a pressure of 7 mbar, assuming a Maxwellian velocity distribution of electrons and with the Fabricant approximation for the dependence of the excitation cross sections on the speed of the electrons [140]. It turns out that in a lamp of radius 6.5 mm it is necessary to have a concentration of electrons $n_e > 10^{11}$ cm^{-3} for the stepwise ionization to predominate. It is essential that the lifetime of ions decreases inversely with the radius of the lamp and, consequently, for a radius of 6 mm a concentration of electrons $n_e > 4 \times 10^{11}$ cm^{-3} will be required, and for a radius of 4 mm, $n_e > 10^{12}$ cm^{-3}. Hence, if the sizes of the lamps are smaller, the transition to stepwise excitation will occur at higher concentrations of electrons and the excitation of an intensive discharge, when reducing the size of the bulb, requires higher specific energy (Chapter 3). At the same time, when increasing the frequency, the concentration of electrons, under which transition to the stepwise ionization is possible, will be decreased owing to the growth of intensity of the curling field.

It should be noted that so-called α- and γ-forms of E- and H-discharges are described in the literature [133, 134]: where α is the weak-current E-discharge with current density 1 mA/cm^2 or less; γ is the high-current E-discharge with current density higher than 10 mA/cm^2, which is usually excited when the walls of the bulb cylinder or of the electrodes are atomized. In hf ELs these versions of the E-discharge, which differ in density of the current and symmetry, are also observed and they can be compared with the α- and γ-discharges. So, at low temperature, the discharge is usually concentrated at the walls of the lamp near the electrodes. When warming-up the lamp, the discharge begins to fill the volume while continuously reducing the luminescence near the walls of the lamp. A similar phenomenon can be observed in the H-discharge. When increasing the power of the discharge a change in the distribution of radiation is possible, together with a considerably greater brightness of emission. As this effect is observed for the rubidium filling, it is supposed to correspond to with the emission of electrons by electron bombardment or the photoeffect. This situation corresponds to the γ-discharge, which is accompanied by the reorganization of all processes in the lamp and is characterized by another spectrum of radiation and transition from direct ionization to stepwise ionization.

In the field of pressures and discharge modes in use, ambipolar diffusion with recombination on the wall predominates [66, 140]. However, when increasing the pressure the volume recombination begins to play an essential role [79, 105] via the influence on the spectral structure of emitted light.

2.1.2. Classification of Modes of Operation of Electrodeless Spectral Lamps

The discharges in the hf spectral lamps differ regarding the techniques of excitation, mechanisms of ionization and recombination, thermal modes, and density of the current. We will consider the main kinds of discharges and their classification.

(a) In terms of excitation (the inductor is outside the lamp) it is possible to distinguish between capacitive discharge (E-discharge), which is excited in the antinode of the hf potential field, and inductive discharge (H-discharge), which is excited by the curling electric field arising in the antinode of the magnetic hf field.

One must bear in mind that, for a hf discharge in electrodeless spectral lamps, it is difficult to operate a pure E- or H-discharge. For example, when placing the lamp in the field of inductance between the ends of the inductor, a potential field, which is superior in value or comparable with the curling field, will also be formed. In practice, at a 10^2 V voltage of the inductor, a 10 mm distance between its ends, and a 100 MHz frequency of the field, the strength of the curling field will be 1–5 V/cm, but the strength of the potential field will be in the range of 50 up to more than 200 V/cm. Thus, depending on the design of the inductor, the ratio of the strengths of the curling and potential fields lies within 0.005–0.1. The relative proportion of the fields in the discharge is determined by the parameters of the plasma, i.e., conductivity and polarization.

When a spectral lamp is placed in the field of a capacitor, a curling component always exists. Actually [134], the strength of the curling field is $E_c \sim fBr_0/c \sim fr_0^2 j/c^2$, and of the potential field is $E_p \sim \rho r_0$ and $f\rho \sim j/r_0$, therefore $E_p \sim j/f$. Hence, $E_c/E_p \sim [fr_0/c]^2 = (r_0/\lambda)^2$. In case of a lamp of diameter 10 mm, driven with a frequency of 150 MHz, the ratio $E_c/E_p \sim 0.038$, which is close to the value of the ratio of the intensities of curling and potential fields in the inductive discharge. When changing the frequency of excitation from 10 up to 200 MHz and the diameter of the lamp from 50 up to 10 mm, the intensity ratio between the curling and potential fields varies from 10^{-7} up to 4×10^{-2}. Thus, at rather high frequencies the curling field is comparable with the fields created in an inductance, while it is insignificant in lamps with diameter > 20 mm at frequencies below 60 MHz. Hence, as the frequency increases, the difference between the E- and H-discharges disappears.

In the subsequent discussion we will return to these questions. However, we note here that these fields always coexist. The availability of a potential field is necessary for excitation of the H-discharge. In its turn the curling field, when warming-up the lamp, begins to determine the processes in the lamps containing vapors of rubidium or other alkaline metals. Practically, this manifests itself in the change of the structure of the discharge, i.e., in the transition from a predominant emission from locations adjacent to the electrodes, to an emission which eventually fills the volume. This is especially well detectable for a nonuniform potential field.

(b) Classification of lamps according to the mechanism of ionization:

(1) Direct ionization of atoms is observed in the E-discharge ($n_e < 10^{11}$ cm^{-3}).

(2) Step by step excitation predominates at $n_e > 1 \times 10^{11}$ cm^{-3}, especially in the H-discharge.

(3) Electronic emission from the internal surface of the lamp cylinder under the action of the flow of ions accelerated in the plasma field and photoemission

conditioned by a flow of ultraviolet photons, which corresponds to resonant transitions in atoms, are observed at a concentration of electrons $n_e > 5 \times 10^{13}$ cm^{-3}. This can be seen visually because a transition from a discharge mode with minimum radiation at the walls of the lamp (in a mode of ambipolar discharge) to a mode with maximum excitation at the walls takes place. This transition is accompanied by a further increase in conductivity of the plasma and a growth of radiation intensity. These processes are externally similar to the so-called α- and γ-discharges in the hf electrodeless E-discharge, described, for example, in [136]. However, in the modes and designs usually used, photoemission is observed not in the E- but in the H-discharge. Under these conditions the spectrum of radiation becomes similar to that of an arc discharge.

(c) Classification of lamps according to the mechanism of recombination:

(1) The mode of ambipolar diffusion, where recombination occurs on the wall of the lamp after the joint diffusion of electrons and ions to this location. It is characterized by a bell-like distribution of electrons and emitting atoms of the inert gas throughout the volume of the lamp.

(2) The mode with volume recombination, when electrons and ions recombine in the volume of the hf spectral lamps. It is realized at a high pressure of the gas (according to [141], in xenon with $p > 4$ mbar) and increased density of the rubidium vapor. This effect results in a broadening of the spectral lines.

(d) Classification of lamps according to the thermal conditions, which strongly influence the character of the discharge, in case of the use of saturated vapors as working substance:

(1) The temperature of the tank with a condensed element is chosen to obtain a vapor pressure smaller than 10^{-6} mbar (to be more precise, it is necessary to consider the fulfillment of the condition $\nu_{ea} \ll \nu_{eb}$, where ν_{ea}, ν_{eb} are frequencies of collisions of electrons with atoms of the vapor and atoms of the buffer gas). In this case the parameters of the discharge are completely defined by the processes in the buffer gas. The radiation of the gas eventually fills the volume of the lamp.

(2) The temperature t corresponds to the vapor pressure of the saturated vapor $10^{-6} \ll p_a(t) \ll 10^{-3}$ mbar. In this case a joint discharge in the saturated vapor and in the buffer gas is observed. These modes are characterized by the maximum density of optical radiation and they are usually of highest interest for various applications.

(3) Temperatures where $p_a > 10^{-3}$ mbar. In this case a discharge is only observed in the vapor of the substance, which determines all processes in the lamp and the parameters of radiation.

The specified boundaries do not exclude each other, and therefore all three forms of discharge practically always coexist, but there exist combinations of substances (e.g., rubidium and neon) where mode (2) is unstable. When reducing the power of the discharge and the concentration of electrons, the boundaries are shifted into the area of the working substance with smaller concentration.

(e) The construction of the source influences the thermal working order (thermal mode) significantly. The choice of the thermal connection of the lamp with the thermostat predetermines the time at which the final working temperature is reached, the power consumed, long-term stability, and serviceability.

(1) Low thermal connection (heat conduction of 10^{-3} W/K).
This mode is realized in a design with an evacuated lamp housing or with a lamp mounted on a heat isolating holder. Thanks to small heat capacity the desired thermal mode can be attained rapidly, but a stationary mode requires increased stability of the hf power supply and careful dosing of the substances in the lamp.

(2) Strong thermal connection (heat conduction > 1 W/K).
In this case the requirements with respect to the stability of the source of the hf energy are lower, but the accuracy of the thermostatic control of the lamp temperature should be high enough, which is not difficult to ensure. This mode is realized in lamp designs, where the major part of the condensed substance is in good thermal contact with the thermostatically controlled surface or connected to it by means of a rod with good heat conductivity.

(3) Intermediate thermal connection.
This is characterized by heat conductivity from 10^{-3} up to 1 W/K and is realized in a design of the light source where the lamp is connected with a thermostat-heat conductor, but the metal is simultaneously under the influence of the plasma of the discharge.

(f) The types of discharge can be classified by the density of the current flowing through the plasma. In the case of a hf discharge such a classification is to some extent arbitrary, as the current is indirectly evaluated. For the glowing discharge, current densities from 10^{-4} to 10^{-1} A/cm^2 are characteristic and for the arc discharge approximately 1 A/cm^2 and higher [134]. The density of the current through the plasma in hf ELs can be evaluated by measuring the heat and intensity of the field in the discharge. The results of relevant measurements are indicated in Section 2.2.2. They permit us to evaluate the current density < 0.2 A/cm^2 in the E-discharge and > 1 A/cm^2 in the H-discharge. Thus, by this parameter, it is possible to compare the E-discharge to the positive column of a glow discharge and the H-discharge to the positive column of an arc discharge. However, when increasing the frequency of the field and the size of the lamp, and when using an easily ionizable element as a working substance, these distinctions become negligible.

All these forms of discharges are transformed into one another either smoothly or with discontinuities in the voltage–current characteristic. A low current discharge, which is concentrated close to the electrodes, is usually observed when initiating the discharge. As the discharge power or the frequency increases, or when the lamp is placed into the antinode of the magnetic field, a sharp transition to a discharge, characterized by larger radiation intensity and larger currents, occurs. This discharge is frequently identified with the H-discharge, but it can be obtained in the E-discharge as well, at least at rather high frequencies. For example, in a lamp 10 mm in diameter at a krypton pressure of 2.5 mbar, such a discharge is observed at frequencies above 260 MHz. The probable reason for such a discharge mode is transition from direct to stepwise excitation, accompanied by a growth of the concentration of electrons and increasing power of the discharge. The growing conductivity leads to an increase in the polarization of the plasma and a decrease in the efficiency of the effective field. However, if a rather strong curling electric field exists, it is capable of supporting the discharge. Accompanying effects are transition to the mode of ambipolar diffusion and Maxwellization of the velocity distribution function. This distribution is also determined by the volume of the lamps and leads to a characteristic shape of the radiation, with concentration of the brightest part in the center of the lamp volume and reduction of the intensity to zero in the vicinity of its walls. The distribution of the radiation intensity reflects the distribution of electrons throughout the volume of the lamp. As the power of the discharge increases, the role of processes on the wall — photoelectric emission and emission under the metastable Kr, Kr^+, Rb^+ collisions with the wall — increases. In this case a bright luminescence occurs near the wall of the lamp and the radiation is characterized by especially high optical density, i.e., by narrow and intensive resonance lines. Such modes are similar to the arc discharge. In the case of increasing discharge power in this mode, intensive atomization of the wall material begins and eventually destruction of the lamp is possible.

In conclusion we will compare the hf ED with the dc discharge. Most essential is the fact that the electron temperature T_e for the hf discharge is less than that for the dc discharge. Hence, to receive the same value of the electron temperature T_e in a hf discharge, large fields must be applied. At the same time, if we only operate by internal parameters of the plasma, the properties of different discharges will be about the same if the values of n_e, T_e, n_a, R_e, etc., are similar. This is an important result which enables us to apply results obtained for a dc discharge, with similar filling and sizes, to the analysis of processes in the hf ED.

2.1.3. Designs of the Light Sources

A variety of designs of light sources based on ED are described in the literature. The dimensions vary within a significant range — from less than 2 to more than 1000 cm^3. The main components of a hf light source are a spectral lamp filled with an inert gas or with a mixture of an inert gas and saturated vapor of the element under study (sometimes several elements), an inductor of excitation located near the lamp, and a generator of the hf field (Figure 1.1). The lamp itself may have various shapes: spherical (most widespread), cylindrical, cigar-shaped, disk-like and more "exotic" designs (see Chapter 9).

Lamps with saturated vapors at a certain temperature are operated employing thermostats of various design. At the beginning of the sixties thermostats with a bimetallic plate, which provided an accuracy in temperature stability up to 1 °C, were frequently used. Subsequently, they were substituted by electronic stabilizers of various designs. The accuracy of sustaining the temperature is in the order of 0.01 °C. The real accuracy is usually significantly lower and is determined by the design of the source and, primarily, by the lamp design (the method of its fastening, mode of discharge, etc.) [5, 75, 81]. We will return to these questions when analyzing the features of the thermal mode of the lamp.

In order to reduce the consumed power and decrease the dependence of the intensity on the temperature of the environment, lamps in evacuated housings are employed. Such lamps, containing saturated vapors of different elements, are operated in both the H- [32] and E-discharge modes [89]. In both cases the time taken to establish the stationary mode may be rather long due to the long self-warming-up time. Various electric circuits of generators for excitation of hf EDs are possible. The following are commonly in use:

(1) Clapp self-excited oscillators [142–144] (Figure 2.1).

(2) Self-excited oscillators, usually two-step, which ensure high efficiency and easily allow one to realize modes which are similar to the arc discharge [2, 145, 146].

(3) Oscillators with external excitation, which permit one to attain a higher efficiency than in the self-excited oscillators and, when using quartz stabilized generators for exciting, make it possible to obtain a stable frequency of excitation of the discharge. The latter is important for increasing the stability of the light source.

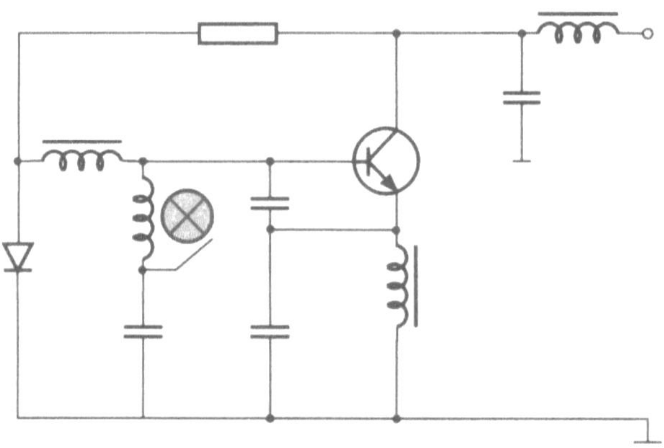

FIGURE 2.1. Electrical circuit of the Clapp self-excited oscillator for excitation of the H-discharge.

There is no basic difference between the characteristics of lamps when using different oscillators.

Possible instabilities of the light sources determined by the oscillator, such as "relaxation," blinking, etc., may be eliminated by decreasing the coupling between the hf oscillator and the lamp. The features of the self-excited oscillator for hf ELs will be considered in detail in Chapter 8.

An important problem when creating an effective source is the correct choice of the design of the excitation inductor. For the excitation of a H-discharge, inductors such as an inductor coil "reversed" in a manner so that potential turns are close to each other in order to ensure the steady ignition of discharge [22], are used. Besides such conventional inductors, solenoids were also applied for cigar-shaped lamps [56]. Such inductors constructed as printed coils, which ensure high reproducibility of lamp parameters, are very effective. For the E-discharge two kinds of constructive condensers are applied. They may be formed by the fastening elements of the lamp and the system of thermostat control [58, 60] or by condensers sprayed on the surface of the lamp [89]. In the former case the simple and reliable design allows one to create a compact light source, but significant losses of hf power connected with initiation of Foucault currents are often observed. In the latter case it is possible to ensure that the light source works at a very low power of the hf self-excited oscillator. The majority of research on E-discharges was carried out using the first variant of constructive condensers, which allow the simple change of lamps and variation in the modes of operation.

Let us compare characteristics of different inductor designs used in experimental and industrial light sources. Figure 2.2 shows the designs of typical inductors. As shown in [65], the excitation of the H-discharge in a spectral lamp

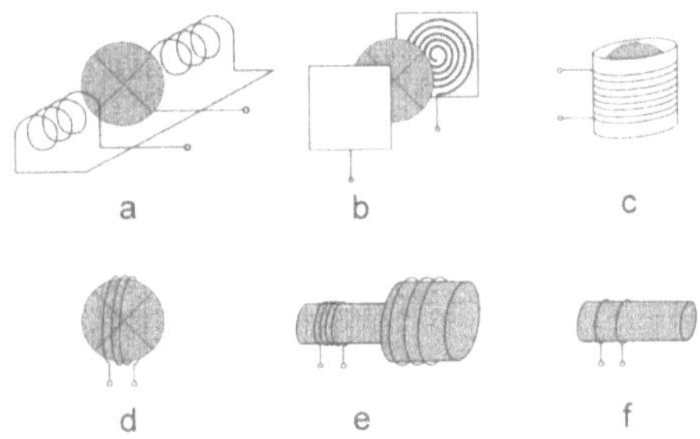

FIGURE 2.2. Designs of inductors used for exciting electrodeless discharges: (a) two external coils; (b) printed circuits; (c) long coil; (d) short coil. (e) coil with additional shortcut coil; (f) two rings.

is carried out by the electric curling field

$$\vec{E} = -\frac{\partial \vec{A}}{\partial t} \tag{2.16}$$

after a preliminary excitation of the E-discharge by a potential field

$$\vec{E}_p = -\operatorname{grad} U \tag{2.17}$$

The inductor has to ensure the creation of high intensity of both curling and potential fields.

When the self-excited oscillator is operating in a linear mode the electromotive force U_B of the curling electric field, within the volume of the lamp with surface S, will be

$$U_R = -\frac{d\Phi}{dt} = -\frac{d}{dt}\oint B_z dS \tag{2.18}$$

where Φ is the flow of magnetic induction while B_z is the projection of the magnetic component of the electromagnetic field on the axis of symmetry of the spectral lamp–inductor system.

When evaluating the distribution of the magnetic field the following assumptions were made:

(1) the inductors consist of individual circular turns with a current;

(2) the current is constant (quasi-stationary approximation);

(3) dielectrics and conductors inside and outside the inductors do not exist.

Results for the strength of the magnetizing field are shown in Table 2.1, where $\kappa_z = (\pi/H_{000})H_z$, H_z is the longitudinal component of the magnetizing field and $H_{000} = i/(2a)$ is the field created by the current i at the center of a turn of radius a.

TABLE 2.1. Relative Values of the Magnetic Field Strength κ_z for Inductors of Different Design[a]

Measured at point no.	Type of inductors a	b	c	d
1	5.26	4.54	18.8	7.62
2	2.30	1.36	6.6	3.02
3	0.50	0.32	3.6	0.00
4	7.36	3.41	7.2	7.26
5	7.33	16.39	36.4	7.00

[a] Location of the points where κ_z has been determined: 1, center of lamp; 2, lamp envelope opposite the reservoir; 3, reservoir top; 4, lamp envelope at 45° (from center) relative to the axes of symmetry of the conductors; 5, on the symmetry axes at $0.75r_0$.

In the general case the hf current i can be represented by

$$i = \sum_{k=1}^{\infty} i_{0k} e^{j\omega kt} \tag{2.19}$$

Usually, the first harmonic $i_1 = i_{01} \cdot e^{j\omega t}$ is used. In Eq. (2.19), the imaginary unit is denoted by j in order to avoid confusion with the current i.

If we measure the current $i_y = i_{01}/\sqrt{2}$ and investigate it at the moment of ignition in the working mode, it is possible to evaluate the efficiency of an inductor. However, it is difficult to do this without violating the conditions of the hf discharge. The power $W = U_0 i_0$ consumed by the oscillator (U_0 is the power supply voltage of the self-excited oscillator and i_0 is the constant part of the collector current for the transistor self-excited oscillator) was measured in the experiments. The first harmonic of the collector current is connected with the resonant value of the current i in the coil through the quality factor Q of the oscillatory circuit: $i_0 = Qi$ [147].

The losses in the oscillatory circuit when igniting the discharge are determined by losses in the connecting wires, in the capacitors of the oscillatory circuit, and in the inductor coil. All these values are approximately of the same order. Therefore, in order to reduce the consumed power when igniting the discharge, it is desirable to have an inductor with a high quality factor, i.e., for the transistor oscillator not less than 100–150, and for the lamp self-excited oscillator not less than 150–200.

After ignition of the discharge in the spectral lamp the losses in the lamp are much higher than in the other elements of the circuit. So the quality factor of the inductor is no longer significantly high (it should not be less than 75–85 when the discharge is burning).

TABLE 2.2. Evaluation of the Efficiency of Different Inductors[a]

Type of inductors (Figure 2.2)	Q	W_{ig} [W]	W_{ex} [W]
a	300	1.0	1.0
b	150	1.3	1.4
c	120	1.5	1.5
d	160	0.9	0.7

[a] Q, quality factor; W_{ig}, power of ignition; W_{ex}, power of extinguishment of the discharge.

Studies have been carried out at a frequency $f = (90 \pm 10)$ MHz. Since the characteristics of a self-excited oscillator depend on the value of the inductance, devices with inductance of $(3 \pm 1) \times 10^{-7}$ H, which is the optimum from the standpoint of adjusting the self-excited oscillator and transfer of energy into the discharge, have been chosen for definiteness when comparing inductors.

Thus, by measuring the power of ignition W_{ig} and extinguishment W_{ex} of the discharge in terms of adjustment and inductance, it is possible to evaluate the efficiency of inductors of different designs. The results are given in Table 2.2. The measurements were performed with a transistor self-excited oscillator (assembled according to the Clapp circuit) for inductors of Figure 2.2a and d, and with a lamp self-excited oscillator for inductors of Figure 2.2b and c. It is evident, that the most effective design, for the excitation of the discharge and transfer of energy from the self-excited oscillator into the discharge is the design presented in Figure 2.2d.

In the absence of restrictions on the power consumed by the light source, it is expedient to use in series production an inductor shown in Figure 2.2c distinguished by its adaptability to manufacturing processes and reproduction of parameters.

Constructive condensers are employed for excitation of an E-discharge. They have better reproducibility of parameters than the inductance ones and higher quality factors up to 250. Since they are designed very differently for certain types of lamps, a systematic comparison has not been carried out.

2.2. *Electrical Characteristics of High-Frequency Electrodeless Lamps*

2.2.1. *Voltage of Ignition of the Discharge*

As mentioned above, many different spectral light sources have been constructed so far. The general laws of ignition of a hf ED will be considered for the example of a lamp of spherical form, which is the most widespread

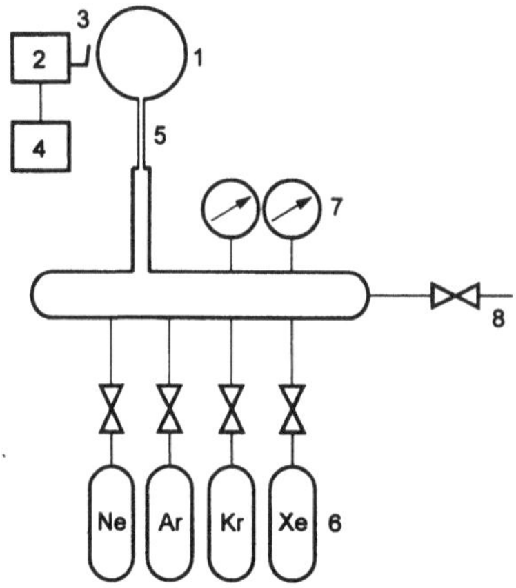

FIGURE 2.3. Device for studying the ignition of the spectral lamp: 1, spectral lamp; 2, generator for excitation of the discharge in the lamp; 3, inductor; 4, power supply 5, capillary tube; 6, vacuum system with gases under study; 7, absolute pressure gauges; 8, pumping line.

variant of a hf EL.

The design of a typical experimental setup is shown in Figure 2.3. The internal cavity of the lamp, after being evacuated to 10^{-6} mbar, is filled by the gas under study. The pressure and sort of the gas can be changed, while an invariable position of the inductor of excitation and the envelope of the lamp is fixed. For usual experiments, when lamps are changed during the experiment, the irreproducibility of the source parameters makes the reliability of the results quite a complicated problem.

The lamp under investigation had an inner diameter of 12.2 mm and was connected through a capillary of 0.8-mm diameter and 20-mm length with the vacuum vessel. The excitation of the discharge was carried out with the help of a self-excited oscillator (Figure 2.4).

A two-section screw coil was used as an inductor of excitation, those turns

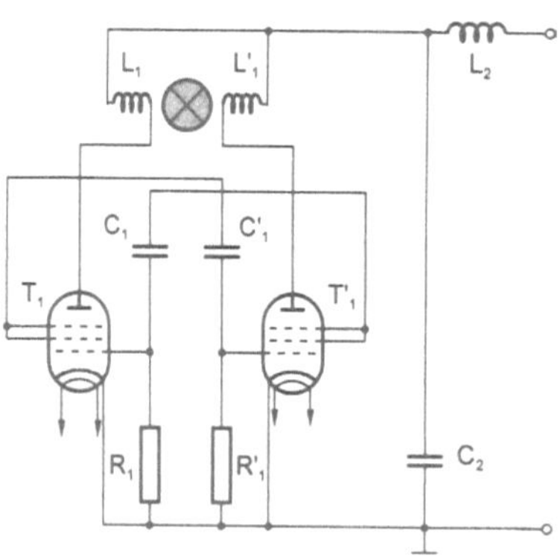

FIGURE 2.4. The circuit design of the self-excited hf oscillator: T_1, T_1', pentode vacuum tubes; C, capacitors; R, resistors; L, inductivities.

which were directly connected to the oscillator being inverted to the lamp, (Figure 2.2a). A similar oscillator was used when measuring the frequency dependence of the discharge parameters. Its readjustment to another frequency was carried out with the help of a variable capacitor in the oscillatory circuit. The hf voltage was measured by a voltmeter connected through a capacitor divider to reduce the influence of the generator on the circuit.

The installation described was employed to investigate experimentally the dependence of the hf voltage at the inductor, causing the ignition of the discharge in the lamp, on different fillings (Ne, Ar, Kr, Xe, and a mixture of Ar + Ne with pressures from 0.3 to 13 mbar). In the case of argon the ignition of the discharge as a function of frequency (between 40 and 100 MHz) was specially investigated.

Before the measurements the sphere of the lamp was evacuated down to 1×10^{-6} mbar. Afterward the sphere and the connected glass tubes were heated to a temperature of (320 ± 20) °C for 20–30 h. After cooling, the pressure of the residual gases lay between 10^{-7} and 7×10^{-7} mbar. With a getter a pressure of less than 1×10^{-6} mbar could be sustained in the volume for some hours without further pumping. Finally, the sphere in which the measurements were intended to be performed was cleaned in an atmosphere of the same gas.

The pressure of the gas under study was controlled by a calibrated manometer. The time to establish the pressure in the lamp, evaluated with an accuracy of 1% from the formula for the gas viscosity, was 30 s in pure gases and 2.5 h in a mixture of gases with allowance for the mutual diffusion. Therefore measurements with pure gases were carried out 5 min after filling, and with gas mixtures 3 h after filling. Spectrally pure gases Ne, Ar and specially cleaned Kr, Xe were used for such measurements.

The ignition of E and H-discharges (Figure 2.5) was studied. In the present experiment the emission of E- and H-discharges differed considerably: for E-discharge the optical radiation was concentrated close to the electrodes and the generator was loaded insignificantly; for H-discharge the optical radiation was spherically symmetric and, when being ignited, the voltage across the coil of excitation decreased sharply.

The ignition of the H-discharge was only possible together with the ignition of the initial E-discharge. Therefore the observable minimum of the ignition voltage for the H-discharge is caused by the dependence of the ignition voltage on the pressure of both the H-discharge and E-discharge.

The curves of Figure 2.5 and the data given in Table 2.3 show that the hf ignition voltage of the E- and H-discharges (U_E and U_H) and the optimum pressure p_E and p_H grow when the atomic mass of the gas is reduced.

Figure 2.6 illustrates the dependence of the ignition voltage of the discharge on the pressure of Ar and on the frequency of the field. When the frequency of the field was increased the ignition voltage of the H-discharge remained constant and the ignition voltage of the E-discharge decreased.

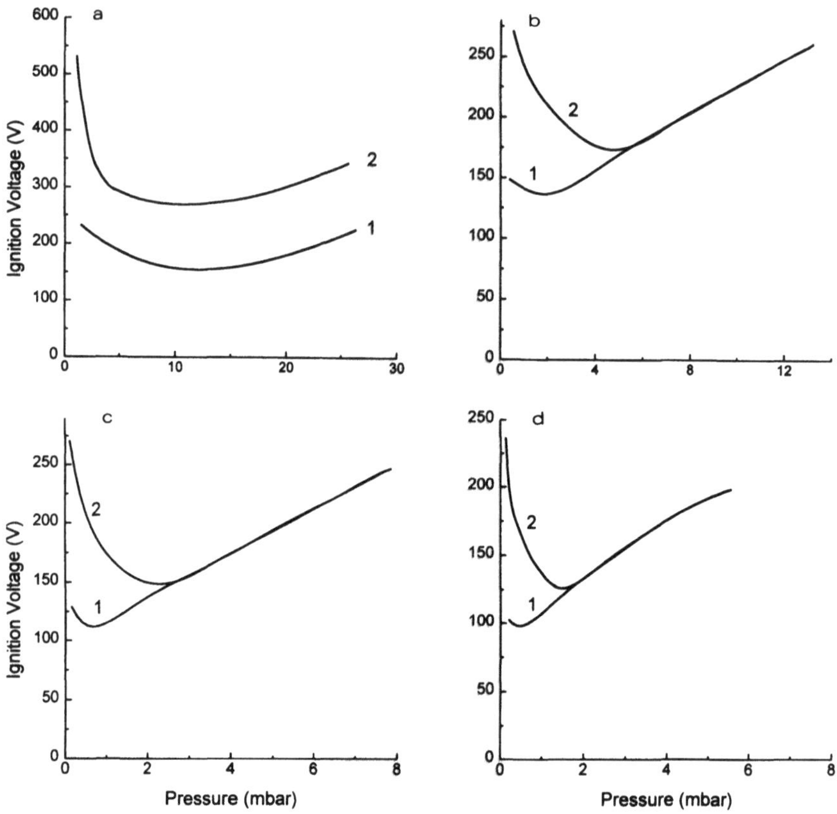

FIGURE 2.5. Ignition voltage of E- and H-discharges in spectral lamps as dependent on the gas pressure for different noble gases: 1, ignition voltage of the E-discharge; 2, ignition voltage of the H-discharge; (a) Ne; (b) Ar; (c) Kr; (d) Xe.

TABLE 2.3. Minimal Ignition Voltage U_E of the E-Discharge and U_H of the H-Discharge and Corresponding Gas Pressures p_E and p_H

	Gas				
Characteristics	Ne	Ar	Kr	Xe	90%Ne + 10%Ar
U_E [V]	250	120	90	70	110
U_H [V]	380	180	160	110	200
p_E [mbar]	1.5–15	0.8	0.5	0.4	1.5–15
p_H [mbar]	1.5–15	3	1.8	1.2	2.5–8

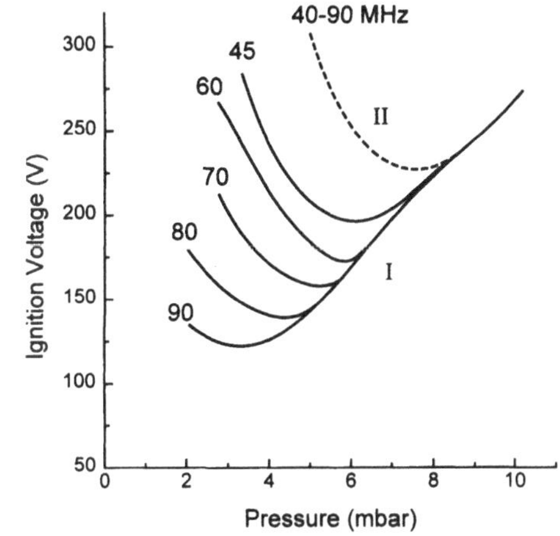

FIGURE 2.6. Dependence of the ignition voltage on the Ar pressure and on the frequency of the exciting field: I, E-discharge; II, H-discharge.

Figure 2.7 illustrates the dependence of the ignition voltage on the mixture Ne + Ar. We note that from 20 to 100% Ar the ignition voltage depends only weakly on the argon concentration. An insignificant reduction of the ignition voltage in comparison with the expected value (the Penning effect) is likely to be caused by the time of diffusion drift of the metastable atoms to the walls, which is comparable to the lifetime of the excited state owing to the small dimensions of the lamps. The difference of the ignition voltages for the E- and H-discharges is caused by the accelerated deactivation of metastable atoms Ne* by collisions with the electrons. This effect is 10^2–10^3 times larger in H-discharges than in E-discharges.

Thus, the voltage of ignition grows in the row of Xe, Kr, Ne + Ar, Ar,

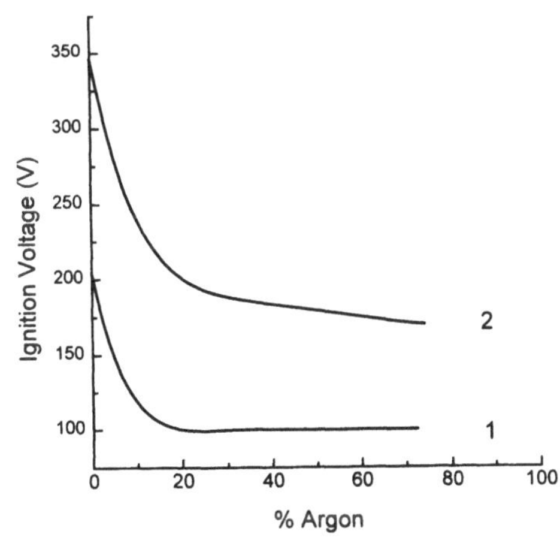

FIGURE 2.7. Dependence of the ignition voltage on the Ar percent in a Ne–Ar-mixture. 1, E-discharge; 2 H-discharge.

TABLE 2.4. Optimal Gas Pressure as Dependent on the Electron–Atom Collision Frequency ν_{ea}

Gas	ν_{ea} [10^9 s^{-1} mbar^{-1}]	p_E [mbar]
He	2.0	2.2
Ne	1.2	4.2
Ar	5.3	0.7
Kr	7.5	0.5
Xe	11	0.3

Ne. The ignition voltage of the E-discharge decreases in the range of frequencies 40–100 MHz, while in the H-discharge it remains constant.

The comparison of typical dependences [134, 148] shows a qualitative correspondence with results of experiments carried out by different authors over the range of frequencies 10–100 MHz where the discharges were excited in vessels with characteristic sizes 2–100 mm. The reduction in the optimum pressure and ignition voltage with increasing frequency under these conditions can be explained as follows: the left-hand branch of the ignition curve is caused by the growth of losses of electrons on the wall of the bulb because the mean free path and the characteristic size of the vessel are comparable [148]. The increase in frequency reduces the drift of the electrons to the walls of the bulb, and therefore the optimum pressure is reduced and the required ignition voltage decreases. By setting the amplitude of oscillations of the electrons in the hf field equal to the characteristic size Λ of the volume under study, we obtain the formula [148]

$$\frac{eE_p}{m_e f \nu_{ea}} = \Lambda \tag{2.20}$$

On substituting the values of collision frequencies ν_{ea} (for He, Ne, Ar from [136], for Kr, Xe from [148]), we determine the optimum ignition pressure for an E-discharge (Table 2.4). Within the error bars these data given by the approximate formula are in good agreement with the data given in Table 2.3.

On the other hand, if we assume that for the discharge breakdown an electron has to accumulate energy in the electric field that is sufficient for the ionization of atoms, and we take into account that $\nu_{ea} \gg f$, it is possible to write

$$\frac{m_e v^2}{2} = \frac{e^2 E_p^2}{m_e \nu_{ea}^2} \sin^2(2\pi f t) \simeq \frac{e^2 E_p^2 f^2}{m_e \nu_{ea}^4} \sim eU_i \tag{2.21}$$

Thus upon increasing the frequency a smaller amplitude of the field is required, since $E_p \sim 1/f$ (see Figure 2.6).

As a result, the above mechanism describes correctly the observable ignition features of the E-discharge in inert gases under conditions which are characteristic for ELs. This means that we can use this model to evaluate the ignition field strength for an E-discharge in lamps of different sizes and with different pressures (see Chapter 9).

In case of an H-discharge the ignition voltage does not depend on the frequency of the exciting field. The point is that the effective voltage depends on the frequency as follows [134]:

$$U \sim U_0 \sqrt{(\nu_{ea}^2 + f^2) / \nu_{ea}^2} \tag{2.22}$$

When the frequency f is reduced from 90 MHz down to 45 MHz, the ignition voltage grows only by 0.2%. Analysis of the dependencies of the ignition voltage of the H-discharge upon the conditions of this experiment should take into account that under these circumstances the right branch of the ignition curve was mainly determined by the E-discharge.

2.2.2. Voltage–Current Characteristics of the Discharge

Voltage–current characteristics of the discharge in hf ELs are shown in Figure 2.8. When increasing the voltage, an increase in current approximately proportional to the voltage is observed until the ignition of the discharge.

Initial ignition of the H-discharge is possible depending on the design of the inductor, the adjustment of the generator, or filling of the lamp. Sometimes an initial ignition of a low-power E-discharge takes place. Accordingly, the voltage–current characteristics of the light sources are distinguished: if initially an H-discharge appears, a significant jump of the current caused by the essential change of load on the oscillatory circuit takes place, and in the second case only an insignificant change of parameters is observed. However, further increase in the energy transmitted to the discharge leads to ignition of the discharge and a corresponding alteration in the input resistance. The different character of the voltage–current characteristics is determined by a different relationship between the curling and potential fields for various designs of the inductor and magnitudes of the curling and potential fields, necessary for the breakdown as a function of the pressure and type of gas. Hysteresis is observed for all types of voltage–current characteristics. This phenomenon is determined by the fact that a higher ignition power is needed as compared to the power of extinguishment.

The ignition current through the generator for a H-discharge can grow or decrease depending on the relation between the internal resistance and the resistance induced by the discharge into the oscillatory circuit. In the case of spectral lamps, in which the density of the vapor is determined by the power of the discharge, a falling voltage–current characteristic with stabilization in the mode of the discharge in the vapors of the easily ionizable component is possible. Usually it is possible to apply a stabilized voltage in order to stabilize a certain

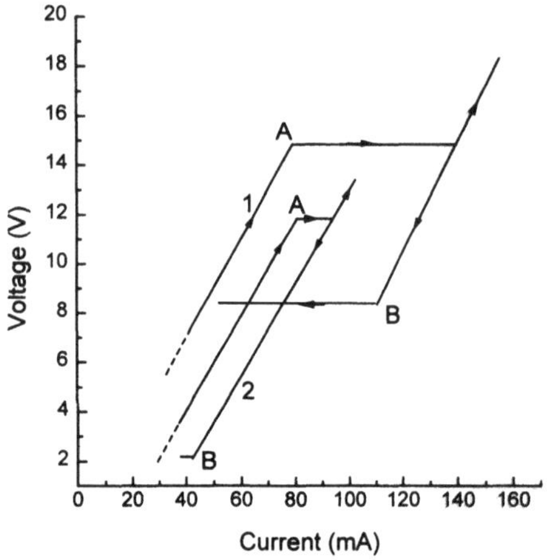

FIGURE 2.8. Typical voltage–current characteristics of discharges in hf ELs: 1, H-discharge; 2, E-discharge. A, ignition of the discharge; B, extinguishment of the discharge.

mode, but in case of a falling characteristic one has to use a current stabilizer. It is not difficult to evaluate the requirements for stabilization for the known voltage–current characteristic of a lamp. The requirements for the fluctuation parameters will be considered in Chapter 5.

The equivalent electrical circuit of the H-discharge and its connection with the elements of an oscillatory circuit is given in Figure 2.9a. The discharge is represented as a short-circuit winding of the transformer with inductivity L_p and resistance R_p, related to the oscillatory circuit by the mutual induction M. If there is a discharge in the lamp, the inductivity of the oscillatory circuit decreases by ΔL_k and the active resistance increases by ΔR_k:

$$\Delta L_k = L_p \frac{\omega^2 M^2}{R_p^2 + \omega^2 L_p^2}, \qquad \Delta R_k = \frac{R_p \omega^2 M^2}{R_p^2 + \omega^2 L_p^2} \tag{2.23}$$

where ω is the frequency of fluctuations of the generator.

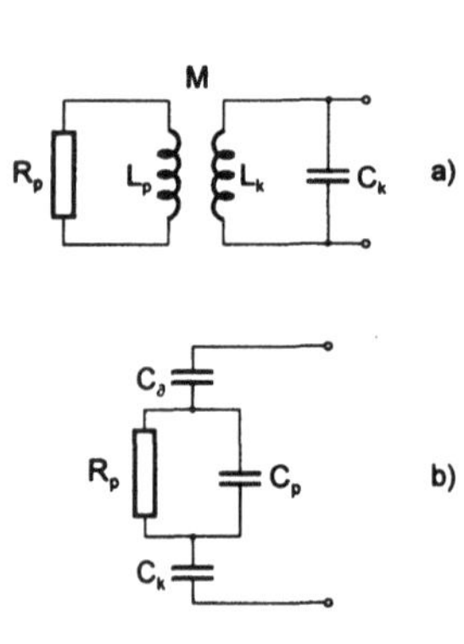

FIGURE 2.9. Equivalent electric circuit. (a) H-discharge; (b) E-discharge. R, L, C equivalent resistor, inductivity, capacity; p lamp plasma; k generator circuit; M mutual induction; for other designations, see text.

The change ΔL_k can be evaluated within 10^{-8} H. The variation in the quality factor of the oscillatory circuit before and after ignition of the discharge permits one to find ΔR_k, which appears to be in the range from 10 to 25 Ω.

In the E-discharge the conductive circuit $(C_p||R_p)$ is contained in the oscillatory circuit. After ignition, the overall capacity grows and reduces the frequency of the oscillatory circuit. For example, when igniting an E-discharge in a lamp filled with argon at a pressure of 7 mbar, the frequency changes from 81.5 MHz to 80.4 MHz.

Figure 2.9b shows the equivalent circuit of the E-discharge in the lamp following [134]. Here R_p and C_p are the resistance and capacitance of the discharge, C_∂ the dielectric capacitance of the lamp envelope, and C_k the capacity of the condenser of the oscillatory circuit. With the values of f and Δf it is possible to evaluate $C_k = 5.6 \times 10^{-13}$ F and $C_p = 3 \times 10^{-14}$ F. Using the data of [136] one can define the resistance of the discharge, which (at a power of 0.1–0.2 W) appears to be in the range from 150 to 200 Ω, and the conductivity, which amounts to 0.015–0.020 $\Omega^{-1}\,\text{cm}^{-1}$.

Ignition of the E-discharge increases the active resistance and capacity of the oscillatory circuit by the following amount:

$$\Delta R_k = R_p \frac{(C_\partial/2)^2}{(C_k + C_\partial/2)^2 + \omega^2 C_k^2 R_p^2 (C_\partial/2)^2}, \qquad \Delta C_k \cong C_\partial/2 \tag{2.24}$$

In the case considered $\Delta R_k < 1\Omega$ and $\Delta C_k \sim 3 \times 10^{-13}$ F.

When driving a hf ED the self-excited oscillator is more often assembled with partial inclusion of an oscillatory circuit in the collector circuit (the Clapp circuit). Such an electric scheme can be applied to increase the hf voltage and is useful for the ignition process and to sustain the discharge. In case of the interface with the supply unit, it is usually required to ensure a certain value of the input resistance of the generator. By tuning the elements of the oscillatory circuit this resistance can be varied within some limits without any decrease in the generator efficiency.

Actually, the internal resistance R_{in} of the self-excited oscillator of the light source equals

$$R_{in} = (U + i_0 R)/i_0 \sim U/i_0 \tag{2.25}$$

where U is the value of the constant voltage on the collector, i_0 the amplitude of the constant collector current, and R the resistance of the oscillator circuit.

Using the expression for the constant component and the first harmonic of the collector current $i = S_0\gamma_0 U_{be}$ and $i_1 = S_0\gamma_1 U_{be}$, where S_0 is the slope of the linearized characteristic of the collector current, U_{be} the base–emitter voltage, and γ_0, γ_1 the zero and first expansion coefficients of the pulse of the collector current, it can be shown that

$$R_{in} = U/(S_0\gamma_0 U_{be}) = U\gamma_1/i_1\gamma_0 = U\gamma_1 R_H/(U_1\gamma_0) \tag{2.26}$$

where U_1 is the amplitude of the voltage of the first harmonic on the collector and R_H the resistance of the oscillatory circuit in relation to the collector–emitter points. In the Clapp circuit the resistance of the circuit R_H at resonance is $R_H = C/C_2R_k^2$, where R_k is the resistance of the oscillatory circuit at the resonance, while C and C_2 are the capacitors in the collector–emitter and base–collector circuits. Therefore

$$R_{in} = U\gamma_1 C^2\omega^2 L^2/\left(U_1\gamma_0 C_2^2 R\right) \tag{2.27}$$

Thus R_{in} depends on active losses of the oscillatory circuit through U_1 and γ, and variation in the internal resistance of the self-excited oscillator is possible.

2.2.3. Influence of Change of the Pressure upon the Electrical Characteristics of the Discharge

With increasing operating time of the light source, the number density of the atoms of the inert gas in the spectral lamp decreases due to several processes [68]. Since the lamp is a nonlinear complex resistance for the self-excited oscillator, the reduction in pressure of the inert gas can influence the electrical characteristics of the self-excited oscillator and hence the radiation intensity.

Lamps filled with krypton were investigated at a pressure of 1.5–12 mbar, where the H-discharge is supported by the azimuthal component of the electric field E_b, which in the absence of the skin effect is directly related to the voltage across the coil U_c:

$$E_b = \frac{\mu\mu_0}{2}\frac{N}{L}k(l,z,r)\,U_c \tag{2.28}$$

where N is the number of inductor coil turns per unit length, and k the form factor which connects the magnetic field of the real inductor coil B_p with the magnetic field of the long solenoid B_z:

$$B_p = k(l,r,z)B_z \tag{2.29}$$

From Eq. (2.28) it can be seen that the dependence of E_b on the parameters of the lamp are mirrored by the dependence of U_c on these characteristics. The experimental dependence of U_c on p_{Kr} is shown in Figure 2.10 (curve 1). Within the range of the krypton pressure (3–11 mbar), the voltage required to sustain the discharge varies poorly (we note that the data in Figure 2.10 were obtained for a sinusoidal voltage in the circuit of the self-excited oscillator and a fixed power consumption of the latter). At other values of the consumption power of the self-excited oscillator the character of $U_c(p_{\mathrm{Kr}})$ remains unchanged.

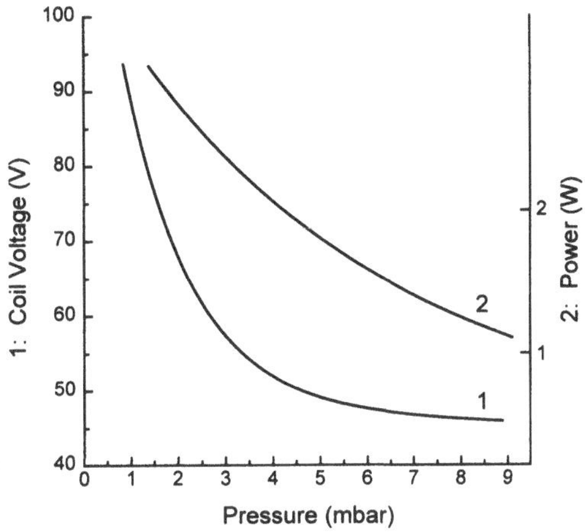

FIGURE 2.10. 1, coil voltage U_c; 2, discharge power as dependent on krypton pressure.

The obtained function $U_c(p_{\text{Kr}})$ shows that in the specified range of the krypton pressure the resistance induced by the plasma in the oscillatory circuit of the generator depends poorly on p_{Kr}, despite the increase in the concentration of electrons n_e with increasing p_{Kr} [64]. Due to the fact that the plasma induces resistance in the oscillatory circuit, the resistance is proportional to the conductivity σ of the plasma for $f^2 \ll \nu_{ea}^2$ and one kind of gas atom:

$$\sigma \sim e^2 n_e/(m_e \nu_{ea}) \tag{2.30}$$

where ν_{ea} is the frequency of collisions of the electrons with the atoms and f is the frequency of the hf field. The growth of n_e with increase in $p \sim n_e$ is accompanied by an increase of ν_{ea}, therefore σ is kept approximately constant. When reducing the krypton pressure to less than 3 mbar the voltage across the inductor grows rapidly, which implies the change in σ.

The function represented in Figure 2.10 (curve 1) reflects conceptually the well known fact that it is necessary to increase the intensity of the field in order to sustain the discharge when the pressure of the gas is reduced. It is evident that in the investigated lamps the rapid growth of intensity of the field begins at a pressure below 3 mbar (for krypton). The reduction in krypton pressure below 3 mbar during long operation of the light sources results in a change of the oscillator characteristic.

The most important characteristic of the light source is the specific energy W_l, which is transmitted by the oscillator to the spectral lamp. Measurements of W_l were carried out with the help of a calorimetric method [69] at constant power of consumption of the self-excited oscillator. It was found, that within 10–15% experimental error, the value W_l is practically independent of krypton pressure.

The minimum energy necessary for sustaining the discharge depends on the pressure of the inert gas (see Figure 2.10, curve 2) and on the pressure of the saturated rubidium vapor (i.e., on the temperature of the spectral-lamp thermostat). It is conditioned by the reduction in the intensity of the electric field necessary to sustain the H-discharge in the lamp with growth of p_{Kr}, and the reduction of p_{Rb} in the considered areas of pressure of the inert gas and of the vapor pressure of the alkali metal.

Thus, within the krypton pressure range of 3–11 mbar, the change of the gas pressure in the lamp does not influence the electrical characteristics of the self-excited oscillator of the light source. At a krypton pressure less than 3 mbar, the reduction of the pressure in the lamp influences the characteristics of the generator, which means in particular that a large field is required to sustain the discharge. An increase in krypton pressure within a range of 1–11 mbar results in a reduction of minimum energy necessary for sustaining the H-discharge. An increase in the density of the rubidium vapor results in a growth of the minimum energy.

2.2.4. Concentration and Average Energy of Electrons

All the main parameters of the hf EL are defined by the concentration and distribution of electrons throughout the energy levels. The first results of research on these parameters were given in [25] where, similar to the discharge driven by a constant current, the concentration of electrons between 5×10^{11} and 5×10^{12} cm^{-3} was evaluated. In [70] the data of electron concentration measurements in a lamp of spherical form are given.

To obtain formulas which relate the impedance characteristic of the oscillatory circuit with the parameters of the plasma, we will go back to the expression for the susceptibility of a conducting ball in a homogeneous rapidly varying magnetic field [70]:

$$\kappa = \kappa_r + i\kappa_i = -\frac{3}{2}\left[1 - \frac{3}{r_p^2 m^2} + \frac{3}{r_p m}\cot r_p m\right] \tag{2.31}$$

$$m = K\exp(i\theta), \qquad K = \{\omega\mu_0(\sigma_r^2 + \omega^2\varepsilon_0^2)^{1/2}\}^{1/2}$$

$$\theta = \frac{1}{2}\arctan\left(-\frac{\sigma_r}{\sigma_i + \omega\varepsilon_0}\right), \qquad \sigma = \sigma_r + i\sigma_i$$

where r_p is the radius of the conducting ball; σ, σ_r, σ_i are the total, real, and imaginary components of the conductivity of the medium, $\omega = 2\pi f$.

Taking into account the well-known relation of the magnetic penetration μ with the susceptibility as well as with the magnetic flows without plasma (ϕ_0) and with plasma (ϕ), namely $\phi/\phi_0 = \mu$, it is possible to write an expression for these flows in terms of (2.31):

$$\frac{\phi}{\phi_0} = 1 - \frac{3}{2}\left[1 - \frac{3}{r_p^2 m^2} + \frac{3}{r_p m}\cot r_p m\right] \tag{2.32}$$

When $\cot r_p m$ is expanded in a series in terms of Bernoulli numbers and the first three terms of the series are retained (if $r_p m < 1$, the error in terminating of the series is not more than 10%, if $r_p m < 0.5$ it is no more than 2.5%, and if $r_p m < 0.2$ it is no more than 0.5%), it is possible to obtain from (2.32)

$$\frac{\phi}{\phi_0} = 1 - \frac{1}{10} r_p^2 m^2 \tag{2.33}$$

Since $\phi/\phi_0 = Z/Z_0$, where $Z = R + iX$ is the impedance of the coil with plasma and $Z_0 = iX_0$ is the impedance of the coil, and if we set $\alpha = R/X_0$ and $\beta = X/X_0$, then

$$\frac{1}{10}k^2 r_p^2 \cos 2\theta = 1 - \beta; \qquad \frac{1}{10}K^2 r_p^2 \sin 2\theta = \alpha; \qquad \tan 2\theta = \frac{\alpha}{1-\beta} \tag{2.34}$$

The quantities α and β are expressed in terms of characteristic parameters of the oscillatory circuit, usually measured in the experiment, as follows [70]:

$$\alpha = \left[\frac{C_0}{C_p}\frac{1}{Q} - \frac{1}{Q_0}\right]\rho^{-2}, \qquad \beta = 1 - \left[1 - \frac{C_0}{C_p}\right]\rho^{-2} \tag{2.35}$$

where C_p and C_0 are the capacitance of the measuring circuit with and without plasma, $\rho = r_0/r_k$, while r_0 and r_k are the radii of the lamp and probing coil, respectively.

Thus, having measured the change of capacity and the quality factor of the oscillatory circuit after inducing the plasma, and having calculated α and β by means of (2.35), we determine K and θ and the conductivity σ from (2.31). From the known conductances it is possible to determine the frequency of collisions ν_{ea} and the concentration of electrons n_e from the relations

$$\sigma_r = \frac{n_e e^2 \nu_{ea}}{m_e(\omega^2 + \nu_{ea}^2)} K_{\sigma_r}\left(\frac{\nu_{ea}}{\omega}\right), \qquad \sigma_i = \frac{n_e e^2 \omega}{m_e(\omega^2 + \nu_{ea}^2)} K_{\sigma_i}\left(\frac{\nu_{ea}}{\omega}\right) \tag{2.36}$$

where K_{σ_r} and K_{σ_i} are dimensionless factors which depend on the relation between the collision frequency and the frequency of the probing hf field.

The experimental setup is shown in Figure 2.11. The probe coil had an inner diameter of 13.8 ± 0.2 mm, with four windings of copper wire (0.6-mm diameter). To reduce the electrostatic connection with the plasma, it was placed into a grounded screen made of aluminum foil of 0.02-mm thickness. The excitation coil (11-mm diameter) consisted of two sections spaced 12 mm apart. The

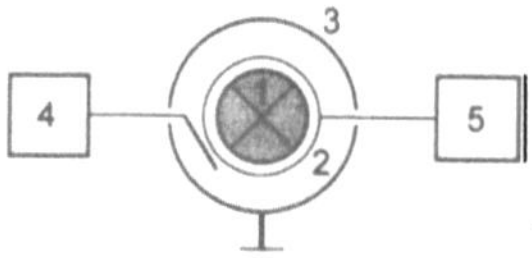

FIGURE 2.11. Experimental setup for measuring the conductivity of the hf electrodeless plasma: 1, lamp under study; 2, connection coil of the q-meter; 3, screening box; 4, inductor for discharge excitation, 5, q-meter.

lamp (12.8-mm diameter) was filled with krypton at a pressure of 3 mbar. For a reliable measurement of the frequency of collisions the frequency of probing must satisfy the condition [139] $0.1\nu_{ea} < f < 10\nu_{ea}$ and, when using the formula for determining the conductivity, the following condition must be fulfilled:

$$\frac{r_0}{\delta} < 0.7 \tag{2.37}$$

where δ is the thickness of the skin layer.

When measuring the conductivity the probing was carried out at a frequency of 8.9 GHz, while when measuring the frequency of collisions a frequency of 35 GHz was used. Calculation of the frequency of collisions and the concentration of electrons was performed at those conductivities for which condition (2.37) was fulfilled.

The circuit parameters were determined with the help of a quality factor meter, the error in measuring the quality factor was less than 5%, and the frequency shift was about 10–20%. The results of measuring the conductivity and computing the concentration of electrons are given in Figure 2.12. In the experiment

$$\overline{\sigma} = (1/v) \int \sigma_r dv \tag{2.38}$$

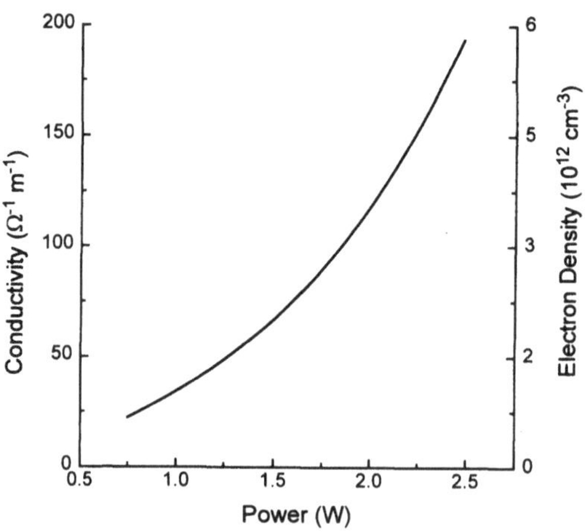

FIGURE 2.12. The plasma characteristics as a function of the discharge power determined in krypton.

was determined, where $\sigma_r = \sigma_0 \cos \pi r/(2r_0)$ and σ_0 is the conductivity of the plasma at the center of the lamp.

When calculating the concentration of electrons the experimentally determined frequency of collisions was $\nu_{ea} = (1.06 \pm 0.3) \times 10^9$ s^{-1} [$T_l = (1.4 \pm 0.6) \times 10^4$ K for a Maxwellian distribution of electron velocities], and ν_{ea} did not depend on the power of the discharge within the error bars. The power of the discharge plotted on the abscissa axis was measured by the calorimeter method with an error of 20%. The error in determining the conductivity (theoretical) was caused by the imperfectness of the measurement of the diameter of the discharge plasma and the diameter of the probe and was about 30%, the error of reproducibility being no more than 10%.

For the earlier considerations pertaining to expression (2.35), uniformity of the probing field was assumed. For the estimation of the systematic error connected with this assumption, the conductivity was measured when the spectral lamps were replaced by bulbs of the same size filled with calibrated solutions of sulfuric acid. We note that the conductivities obtained by the probe are somewhat smaller than those presented elsewhere [149]. In this context a correction factor of 1.5 was introduced into the final result (see Figure 2.12).

The determination of the concentration of electrons in an E-discharge by the given method involves difficulties arising from the small change of the quality factor. The evaluation gives a change of conductivity from 0.4 up to 8 (Ω^{-1} m^{-1}) corresponding to the change of the electron concentration from 2×10^9 to 4×10^{11} cm^{-3} with variation in the discharge power.

As the concentration of electrons in the E-discharge is much smaller than in the H-discharge, it cannot be measured reliably by the radio-frequency method used before. The concentration of electrons in the E-discharge was measured by means of the cavity method, one of the microwave diagnostic techniques. In this method the plasma under study is placed in a resonator and the variation in the eigenfrequency f_r and Q-factor of the resonator is measured. If the condition $\nu_{ea}^2 \ll f_r^2$ is fulfilled, the plasma introduces only a weak perturbation into the resonator, and it is possible to obtain the relationship between the parameters of the discharge (concentration of electrons n_e and frequency of collisions ν_{ea}) and of the resonator:

$$\frac{\Delta f_r}{f_r} = \frac{1}{2} C_v \frac{\bar{n}}{n_{cr}} \frac{v_p}{v_r}, \qquad \Delta\left(\frac{1}{Q}\right) = B_v \frac{\bar{n}}{n_{cr}} \frac{\nu_{ea}}{f_r} \frac{V_p}{V_r} \tag{2.39}$$

where n_{cr} is the critical concentration given by

$$n_{cr} = m_e \varepsilon_0^2 f_r^2 / e^2 \approx 1.24 \times 10^{10} f_p^2 \tag{2.40}$$

$\bar{n}$ is the concentration of the electrons averaged over the volume of the lamp, V_r the volume of the resonator, V_p the volume of the plasma in the resonator, f_p the plasma frequency, while C_v and B_v determine the distribution of the field and the plasma throughout the volume of the resonator:

$$C_v = \frac{V_r \int\limits_{V_r} |E|^2 \, dV}{V_p \int\limits_{V_p} |E|^2 \, dV} \tag{2.41}$$

For our measurements we used the change of frequency of the resonator, hence the expression for B_v is not given. We will only note that $B_v = \text{const}$ when the frequency of collisions is independent of the radius of the lamp.

Hence the concentration of the electrons in the plasma can be determined by the system (2.39), measuring the change of the resonator frequency when igniting the discharge, and then used to calculate the factor C_v. The highest concentration of the electrons that can be measured is the critical concentration n_{cr} which corresponds to the plasma frequency f_p. Therefore, it is necessary to use as high resonator frequencies as possible to expand the range of measurable concentrations. At the frequency $f = 8.8$ GHz, n_{cr} is 10^{12} cm^{-3}. The use of higher frequencies is rather difficult because of the necessity to have the lowest type of oscillations. The ratio d/L, where d is the diameter of resonator and L its length, is chosen to ensure the excitation of the TEM_{010} mode and to exclude the occurrence of other modes at the given frequency. The limits of reduction in d and L are restricted by the sizes of the spectral lamp and of the inductor as well as by the decrease in the quality factor of the resonator and thus by the decrease in the accuracy of the method.

In view of these arguments a cylindrical resonator of internal diameter 26 mm and variable length from 19 to 23 mm was made. Its quality factor Q appeared to equal 800. The excitation coil (diameter 21 mm) had three windings made of a wire (1.2 mm in diameter) and was placed axially in the resonator. To reduce losses and ensure excitation of the discharge in the resonator, some slots were made in its side surface. The resonance frequency of the resonator with a lamp (without discharge) and with the excitation coil was 8.88 GHz. The resonator was connected to the waveguide by a coupling probe. In the case of best coupling between the frequency of the oscillator and the resonant frequency, the reflected signal observed on the screen of the oscilloscope was minimal. The frequency shift was measured with the aid of a frequency meter with an error of 1.5 MHz, which corresponds to a 10^8 cm^{-3} error in the measurement of $\bar{n}_e$. The factor C_v was calculated by Eq. (2.41), assuming a homogenous distribution of electrons, as well as a series representation of the Bessel function in the expression for the distribution of the field in the resonator. The value of C_v was found to be 9.0 ± 1.3. The error in calculating C_v was the main error source when determining the concentration of the electrons. The overall accuracy was estimated to be 20%.

As in [70] a spectral lamp of diameter 13 mm and filled with krypton at a pressure of 2 mbar was investigated. The concentration of electrons in the E-discharge varied from less than 1×10^8 cm^{-3} to more than 3×10^{10} cm^{-3}, when the voltage on the connection coil, placed outside the resonator, was varied between 110 and 780 V. The E-discharge was only ignited at 350 V; the ignition voltage for an H-discharge was 750 V. When igniting the H-discharge, the quality factor Q dropped sharply and the measurement of the frequency shift failed. From a measurement of the quality factor, it is possible to evaluate the concentration of the electrons in the H-discharge at larger than 5×10^{11} cm^{-3}.

It is noteworthy that the measurements of the electron concentration with the aid of the resonator method became possible only in the absence of rubidium in the spectral lamp. Otherwise, the quality factor of the resonator was insufficient to carry out a reliable measurement of the resonance frequency shift.

In a mixture of rubidium and inert gas vapors, the electron concentrations were measured by the probe technique [150]. The concentration of the electrons appeared to be in a range between 1×10^{13} to 2×10^{14} cm^{-3}; the mode of the discharge corresponded to an H-discharge.

2.2.5. Thermal Mode of High-Frequency Electrodeless Lamps

When saturated vapors of metals are used as working substances, the thermal working condition (thermal mode), characterized by the working temperature, defines completely both the value and the stability of the radiation intensity of the hf EL. This behavior is caused by the exponential dependence of the vapor concentration on the temperature, described by a factor of the form $\exp(A + B/T)$, where A is a constant, B a constant proportional to the work for evaporation of an atom, and T the temperature of the lamp; A and B are usually determined experimentally. Besides, the thermal mode defines the power of the discharge, the time to establish constant intensity, and the working range of temperatures within which the light source can operate.

Let us consider the thermal balance of the lamp. The hf light source consists of the EL itself, a thermostat which stabilizes the temperature T_{th} of the lamp environment, a holder for the lamp, and an outlet for the optical radiation. The energy consumed by the lamp in the stationary mode is defined by the heat emission from the surface of the lamp. It comprises the heat removal from the lamp through convection (P_c), heat removal through thermal radiation (P_{th}), heat removal through the inductor due to convection and conduction (P_c and P_{cdi}), heat removal from the surfaces of the holder through convection, thermal radiation and conduction (P_{ch}, P_{thh}, and P_{cdh}). A sketch of the lamp under consideration was given in Figure 1.1. The temperature of the thermostat in different designs varies from 50 up to 130 °C in the case of lamps filled with rubidium, cesium, mercury, and potassium, and more than 700 °C in the case of lamps with silver vapor. In

TABLE 2.5. Temperatures of Different Parts of the Lamp[a]

$2r_0$ [mm]	W_{gen} [W]	W [W]	Material of holder	Temperature [°C] thermostat	surface of lamp	basement of holder
				H-discharge		
10	2.5	1.2	textolite	75	145 ± 8	110
13	3.6	1.9	textolite	105	250 ± 5	120
13	2.8	1.5	brass	115	170 ± 10	120
				E-discharge		
13	1.2	0.3	brass	95	108 ± 5	95
13	0.7	0.2	brass	95	100 ± 5	95

[a] r_0, radius of the spherical lamp bulb; W_{gen}, power of hf generator; W, power consumed by the lamp.

the following we will consider a lamp operating with rubidium vapor.

The temperature of the lamp surface as dependent on the power W of the discharge is given in Table 2.5. Depending on the power of the discharge, the temperature in the lamp is seen to vary in the H-discharge mode from less than 145 °C up to more than 250 °C. In the E-discharge, the temperature of the lamp surface exceeds moderately the temperature of the thermostat and the thermal mode of the lamp is defined by the design of the thermostat.

First we will calculate the losses of an H-discharge through convection, assuming that the lamp is in a spherical thermostat. Then

$$P_c = \lambda_{eq} \frac{2(T_l - T_{th})}{(d_1 - d_2)d_1/d_2} S_1 \tag{2.42}$$

where T_l is the temperature of the surface of the lamp, T_{th} the temperature of the thermostat, d_1 and d_2 the internal and external diameters of the spherical internal layer, S_1 the area of the internal surface of the spherical internal layer, and λ_{eq} the equivalent factor of heat conduction.

In case of a cylindrical internal layer we obtain

$$P_c = \lambda_{eq} \frac{2(T_l - T_{th})}{d_1 \ln(d_1/d_2)} S_1 \tag{2.43}$$

where d_1 and d_2 are the internal and external diameters of the cylindrical internal layer while S_1 is the area of the internal surface of the cylindrical intermediate layer.

The results of computing the convection power of the lamps are shown in Figure 2.13 (curves I). The power of the thermal radiation of the lamp, P_{th}, was calculated for the model of a spherical bulb in a cylindrical thermostat by the

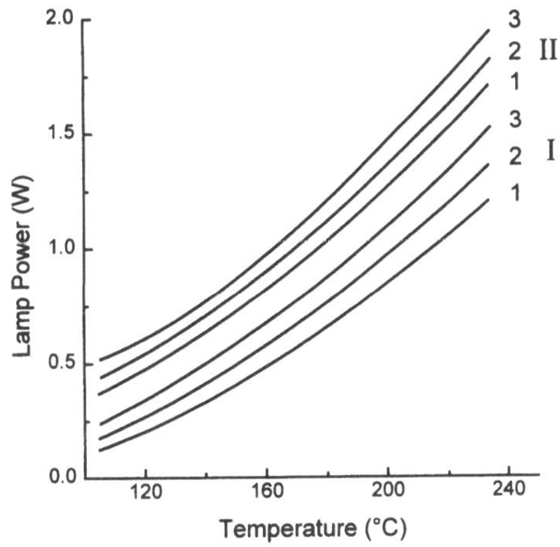

FIGURE 2.13. Lamp power as a function of the lamp surface temperature: I, convective power; II, heat radiating power. Thermostat temperature: 1, 110 °C; 2, 90 °C; 3, 70 °C.

formula

$$P_{th} = \frac{\sigma_0 F_1}{\frac{1}{\varepsilon_1} + \frac{F_1}{F_2}\left(\frac{1}{\varepsilon_2} - 1\right)} \left[\left(\frac{T_1}{100}\right)^4 - \left(\frac{T_2}{100}\right)^4\right] \tag{2.44}$$

where F_1 is the radiating surface and F_2 the surface of the thermostat; emission factors $\varepsilon_1 = 0.89$ (glass) and $\varepsilon_2 = 0.85$ (thermostat), while $T_1 = 453$ K and $T_2 = 373$ K, σ_0 being the Stefan–Boltzmann constant. The results of computing the radiation power of the lamp are given in Figure 2.13 (curves II).

The power of the optical radiation passing through the window of the area F_1 in the thermostat can be found by the formula

$$P_{opt} = \sigma_0 (T_1^4 - T_2^4) F_1 \varphi_{F_1 - F_2} \tag{2.45}$$

where

$$\varphi_{F_1 - F_2} = \frac{1}{2}\left[1 - \frac{1}{\sqrt{1 + (a/k)^2}}\right] \tag{2.46}$$

a is the radius of the exit window, k the distance between the center of the lamp and the exit window, F_1 the radiating surface, and F_2 the surface of the window. At the parameters chosen this power appeared to be P_{opt}=0.08 W.

The conductive capacity of the inductor P_{cdi} may be calculated by the formula

$$P_{cdi} = 2\Delta T \lambda S / L \tag{2.47}$$

where ΔT is the difference of temperature at both ends in the inductor; λ the factor of heat conduction of the inductor material, S the cross-section area of the inductor, L the inductor length; while the factor 2 takes into account the availability of two inductors (in case of a two-section coil). For the case considered, $P_{cdi} = 0.47$ W.

The convective power of the inductor P_{ci} is given by

$$P_{ci} = \sum_i P_i = \sum 2\alpha_i(T_i - T_{ei})S_i \tag{2.48}$$

where $i = 1,2,3$ denotes the section number with corresponding temperature 133–100 °C, 133–107 °C, and 107–100 °C; the convection factor for thin conductors $\alpha_i = 25$–30 W/(m^2 K), T_{ei} is the environment temperature of the ith section, and S_i - surface of heat exchange of the ith section. In our case, $P_{ci} = 0.11$ W.

The convective power of the holder P_{cd} depends on the orientation of the holder. It achieves the maximum value at its vertical position. In a first approximation we assume natural convection effects from a cylindrical surface into unlimited space. If the thermostat diameter D and temperature drop $T - T_C$ are not too large, the movement of air is laminar and

$$\alpha = A_2\left(\frac{T - T_c}{d}\right)^{1/4} \approx 11.8\ \text{W m}^{-2}\text{K}^{-1}$$

[$A_2 = 0.54(\beta g P_r)^{1/4}(\lambda/\nu^{1/2})$, where β is the volume extension temperature coefficient, P_r the Prandtl criterium factor, λ the thermoconductivity, ν the kinematic viscosity coefficient, and g the free fall acceleration]. Using the formula $P_{cd} = \alpha(T_A - T_C)S$ and supposing natural convection within the cylinder interlayer, we obtain for the convective power of the holder

$$P_{cd} = \lambda_a \frac{2(T_A - T_{th})}{d_1 \ln(d_1/d_2)} S_1 \tag{2.49}$$

where λ_a is the heat conduction factor of air.

The power of the thermal radiation of the holder P_{thh} can be obtained from formula (2.44), when quantities ε_1 and F_1, related to the parameters of the holder will be used.

The heat conductive power of the holder P_{cdh} was calculated using the basic model of the limited rod from the side surface, where heat emission occurs subject to Newton's law and by radiation; the temperature of the end face of the rod is stabilized by the thermostat. At the initial moment the temperature is constant along the rod and at this moment the energy, which is transferred conductively to the rod, is applied to one of the end faces. An equation describing heat transport along the holder has the form

$$c\gamma \frac{\partial T(x,\tau)}{\partial \tau} = \lambda \frac{\partial^2 T(x,\tau)}{\partial x^2} - w; \qquad \tau > 0, \qquad 0 < x < 1 \tag{2.50}$$

where c, γ, and λ are the specific heat capacity, density, and heat conduction factor of the rod; w is the quantity of heat transported by a unit of rod volume per unit of time τ into the environment. It is possible to set

$$w = \frac{\alpha}{S\Delta l}\left[T(x,\tau) - T_{th}\right]C\Delta l = \frac{\alpha}{h}\left[T(x,\tau) - T_{th}\right] \tag{2.51}$$

where $h = S/C$, S being the cross-sectional area of the rod, C the perimeter of the rod, Δl the length of a small section of rod. Consequently Eq. (2.50) can be rewritten as

$$\frac{\partial T(x,\tau)}{\partial \tau} = a\frac{\partial^2 T(x,\tau)}{\partial x^2} - \frac{\alpha}{c\gamma h}\left[T(l,\tau) - T_{th}\right] \tag{2.52}$$

where a is the temperature conductivity with initial and boundary conditions:

$$T(x,\tau = 0) = T_l = \text{const}; \qquad T(l,\tau) = \text{const}; \qquad -\lambda S\frac{\partial T(0,\tau)}{\partial x} = P$$

(the discharge power P is introduced into the rod at $x = 0$).

The solution of the resulting equation received with the help of a Laplace transform has the form

$$T(x,\tau) - T_{th} = \frac{P}{\sqrt{\lambda\alpha SC}}\frac{\sinh\left[\sqrt{B_i}(l-x)/h\right]}{\cosh\left[\sqrt{B_i}(l-x)/h\right]} + \sum_{n=1}^{\infty}(-1)^n\frac{2Pl\sin(\mu_n(l-x)/l)}{\lambda S(\mu_n^2 + B_iD^2/h^2)}\exp\left(-\frac{\mu_n^2 + B_il^2}{h^2}F_0\right) \tag{2.53}$$

Here λ is the heat conductivity of the rod, $B_i = \alpha h/\lambda$ [93], $F_0 = a\tau/l^2$ is the Fourier number, τ is the time; $\mu_n = \pi(2n-1)/2$, $n = 1,2,\ldots$.

In the stationary case the temperature field along the length of the rod has the following form:

$$\Theta(x) = T(x) - T_{th} = \frac{P}{\sqrt{\lambda\alpha SC}}\frac{\sinh\left[\sqrt{B_i}(l-x)/h\right]}{\cosh\left[\sqrt{B_i}(l-x)/h\right]} \tag{2.54}$$

Analysis of the solutions shows that lamp holders with minimum heat conduction should be used for faster attainment of the operating mode. When the thermostat temperature is increased, a smaller power is required to sustain the discharge of a chosen type. When reducing the power, the gradual change of parameters can result in a qualitative change of the discharge. At the same time the maximum temperature and correspondingly the maximum intensity and density of radiation, is reduced. In this case, the requirements on the stability of the power in the discharge and on the external temperature conditions are growing. Table

TABLE 2.6. Thermostat Parameter Requirements for Intensity Stability of 1%[a]

Material	Temperature [°C]						δW
	T_l	T_t	T_w	δT_t	δT_w	δT_{bh}	[W]
Textolite	206	102	98	0.23	16	6.2	0.23
Brass	177	104	100	0.28	0.8	7.5	0.39
Brass[b]	—	—	—	0.72	0.5	19.5	0.93

[a] T_l, lamp temperature; T_t, thermostat temperature; T_w, temperature of outlet window; δW, power stability of the discharge; δT_{bh}, temperature change in the base of the lamp holder.

[b] Constructed for negligible heat exchange between the holder and the condensed Rb inside the lamp.

2.6 indicates the requirements on the stability of power in the discharge δW, on the temperatures of the thermostat δT_t, on the temperature of the basis of the lamp holder δT_{bh}, and on the temperature of the outlet window for the optical radiation δT_w when using a synthetic material (textolite) and brass as the material of the holder. These requirements must be fulfilled to maintain the level of instability of the intensity lower than 1% [93].

In case of a metal holder, a significant temperature gradient on the surface of the lamp appears, but overall, the requirements on the stability of the power and the environment temperature are reduced. All of the considered channels of power loss are equally important and none can be neglected.

3

Modeling of Processes in the Plasma of High-Frequency Light Sources

The present chapter deals with questions concerning the modeling of the physical processes in the electrodeless discharge of a lamp and the formation of radiation. Methods and results of experimental studies of the distribution of atoms in the discharge plasma are discussed in Chapter 4. When modeling processes in a spectral lamp, cases with and without taking into account the excitation of atoms of the buffer gas are analyzed separately. Though in the first case the results of the excitation of atoms of the buffer gas modeling are influenced by the significant simplification that the excitation of atoms of the buffer gas is neglected, this approximation permits one to carry out a comprehensive, computerized, and operative analysis.

3.1. Basic Equations

The power density of the radiation of atoms is described by the formula of A. Einstein:

$$I_{ki} = h\nu_{ki}A_{ki}n_k \tag{3.1}$$

where h is Planck's constant, ν_{ki} the transition frequency ($\omega_{ki} = 2\pi\nu_{ki}$), A_{ki} the probability of transition from state k to state i, n_k the number density of atoms in state k having energy E_k. The concentration n_k is defined by the condition governing the stationary state [151]:

$$\sum_{r>k} \Delta n_{rk} = \sum_{i<k} \Delta n_{ki} \tag{3.2}$$

where on the left side of the equation the sum of all radiative and collisional transitions between level k and levels r having higher energy ($E_r > E_k$), resulting in the population (or depopulation) of level k, is given and the sum over all transitions between level k and levels i having lower energy which depopulate (or populate) k is on the right. In addition, the condition

$$\sum_{k=0}^{\infty} n_k + n_{ion} = n \tag{3.3}$$

should be fulfilled in the stationary mode, i.e., the sum of the number densities of atoms in all possible states and the number density of the ions should equal the total number n of nuclei of the given kind per unit volume. The number of collisionally induced transitions from all possible states i into state k will be

$$\Delta n_k = n_0 n_e \int_0^{\infty} \sum_{i=1, i\neq k}^{m+1} \sigma_{ik}(v) f(v) v^2 dv \tag{3.4}$$

where n_0 is the concentration of neutral atoms, v the electron velocity, $\sigma_{ik}(v)$ the cross section of the process of excitation of atoms from state i into state k, $f(v)$ the function of the velocity distribution of electrons, and m the number of collisionally coupled states.

Neglecting collisions between atoms and ions and assuming only excitation by electron impact from the ground level as well as cascade processes, the population of the emitting level k is given by

$$n_k = \frac{n_0 n_e \int_0^{\infty} \sigma_{0k}(v) f(v) v^2 dv}{\sum_{r=0}^{k-1} A_{kr}} + \frac{\sum_{i=k+1}^{m+1} n_i A_{ik}}{\sum_{r=0}^{k-1} A_{kr}} \tag{3.5}$$

The first term on the right side of Eq. (3.5), which depends on the excitation cross section σ_{0k}, takes into account the role of direct excitation modes. The second term considers the role of cascade transitions, where n_i is determined from the corresponding cross sections σ_{0i}.

When cascade transitions are neglected, the formula for the population is further simplified and the intensity of radiation from the given state, I_{ki}, is determined by the cross section of the direct excitation process:

$$I_{ki} = h\nu_{ki} A_{ki} \frac{n_0 n_e \int_0^{\infty} \sigma_{0k}(v) f(v) v^2 dv}{\sum_{r=0}^{k-1} A_{kr}} \tag{3.6}$$

This formula for the population of the excited state applies to low density of the discharge current in a single atomic gas and in the absence of reabsorption of radiation. The cross sections of the processes of excitation depend in a complicated manner on the energy of the electrons ε. Therefore various approximations, such as the following, are used in calculations:

$$\sigma_{0k}(\varepsilon) = b_{0k}[1 - (U_0 - U_k)/\varepsilon] \tag{3.7}$$

where b_{0k} is tabulated for a series of elements; see for example, [152]. Here $\varepsilon \geq U_0 - U_k$ is the energy difference between the ground (0) and excited (k) states.

The velocity distribution function of the electrons is the solution of the Boltzmann equation, which is a consequence of the more general Liouville equation, describing the movement of an ensemble of atoms in the phase space. If the rate of inelastic collisions ν_m is much smaller than the rates of elastic collisions ν_{ee} and ν_{ea}, the velocity distribution differs only slightly from the Maxwellian distribution function. So we can regard the distribution of the electrons as a Maxwellian distribution for energies lower than the excitation or ionization energy of an atom. For large energies the number of electrons decreases sharply, and we will assume that their velocity distribution corresponds to another Maxwellian distribution with a lower value of the mean energy.

Having calculated the electron distribution, it is possible to determine the distribution of atoms over the quantum states and the local spectrum of the radiation of the atoms in the plasma with the aid of Eqs. (3.5) and (3.6). The emitted spectral line has a finite width determined by the natural lifetime τ of the atoms:

$$I(\nu) = \frac{I_0}{2\pi} \frac{\gamma}{(\nu - \nu_{ki})^2 + (\gamma/2)^2} \tag{3.8}$$

where $\gamma = 1/\tau$ is the natural width of the radiating level (in frequency units).

If the lifetime is reduced by collisions, the shape remains the same, but now $\gamma = 1/\tau + 1/\tau_{col}$. Additionally, a shift $\delta\nu$ of the line appears, which is proportional to the concentration of the perturbing atoms:

$$I(\nu) = \frac{I_0}{2\pi} \frac{\gamma}{(\nu - \nu_{ki} - \delta\nu)^2 + (\gamma/2)^2} \tag{3.9}$$

An exact theory [151] gives

$$\gamma = 8.16 C_6 v^{2/5} n^{3/5}, \qquad \delta\nu = 0.36\gamma \tag{3.10}$$

where C_6 is the Lennard-Jones constant. In general, the line shape becomes more complex, i.e. asymmetry and occurrence of satellites is possible, but at pressures which are characteristic for our light sources ($p < 20$ mbar) these effects are negligibly small.

The spectral line shifts and broadenings determined by electrons (the contribution of ions is much smaller) are represented by the Weisskopf–Lindholm formulas and are

$$\gamma_e = 1.82 C_4 v^{2/3} n_e^{1/3}; \qquad \delta\nu_e = 1.56 C_4 v^{2/3} n_e^{-1/3} \tag{3.11}$$

where C_4 is the constant of interaction.

The velocity distribution of the emitting atoms leads to a corresponding frequency distribution of the radiation intensity, the Gaussian profile:

$$I(\nu) = I_0 \exp\left[-4(\ln 2)\,(\nu-\nu_0)^2/\Delta\nu_D^2\right] \tag{3.12}$$

where ν_0 is the emission frequency of an atom at rest (without line shift $\nu_0 = \nu_{ki}$). The full width of the Doppler profile, $\Delta\nu_D$, at half maximum (FWHM) is given by

$$\Delta\nu_D = \frac{\nu_0}{c}\sqrt{8k_BT(\ln 2)/m} \tag{3.13}$$

where k_B is the Boltzmann constant, T the absolute temperature, and m the atom mass.

The superposition of Gaussian and Lorentzian profiles gives a Voigt profile:

$$I(\omega) = I_0\frac{a}{\pi}\int_0^{\infty}\frac{e^{-y^2}dy}{a^2+(w-y)^2} \tag{3.14}$$

where

$$a = \frac{\sqrt{\ln 2}\,\Delta\nu_L}{\Delta\nu_D}, \qquad w = \frac{2\sqrt{\ln 2}\,(\nu-\nu_0)}{\Delta\nu_D}, \qquad y = \frac{2\sqrt{\ln 2}\,(\nu'-\nu_0)}{\Delta\nu_D}$$

and $\Delta\nu_L$ is the FWHM of the Lorentzian profile; ν' runs during the integration from ν_0 (lower bound, $y = 0$) to ∞ (upper bound, $y = \infty$) [178].

The profile is tabulated for example, in Ref. [151]. However, the shape of an emitted spectral line may differ from Eq. (3.14) due to reabsorption of radiation in the volume of the plasma before the radiation can escape from the emitting volume. The general approach is to use the equation of transfer of Milne–Eddington form, which can be written as [153]

$$\begin{aligned}\cos\theta\frac{dI(\nu,\theta,z)}{k_0dz} &= -P(\nu)I(\nu,\theta,z)\\ &\quad +4\pi\lambda\int_{4\pi}\int_{-\infty}^{+\infty}I(\nu',\theta',z)R(\nu',\vec{n}';\nu,\vec{n})d\nu'd\Omega' + \frac{\varepsilon_0}{k_0}\end{aligned} \tag{3.15}$$

where $R(\nu',\vec{n}',\nu,\vec{n})$ is the probability of absorption of a photon with frequency ν' moving in a direction $\vec{n}'$ which holds for an atom emitting a photon with frequency ν in direction $\vec{n}$, z is a geometrical coordinate orthogonal to the optically dense layer, θ the inclination of the light beam relative to this normal, λ the escape factor, ε_0 the real intensity output per length, and

$$P(\nu) = \int_{-\infty}^{\infty}R(\nu,\nu')d\nu' \tag{3.16}$$

When the radiation is examined within a small solid angle ("one-beam approximation") it is possible to write expression (3.15) in the more convenient form:

$$\frac{dI(\nu)}{d\tau} = -I(\nu) + \varphi(\tau)B(\nu_0, T) \tag{3.17}$$

where $B(\nu_0, T)$ is a temperature-dependent factor,

$$\tau = s\int_0^l P_a(v, l')f_a(l')dl'; \qquad \varphi(\tau) = \frac{f_e(l)P_e(v, l)}{f_a(l)P_a(v, l)} \tag{3.18}$$

Here, P_e and P_a are line shapes of radiation and absorption, respectively, where

$$f_e(l) = \frac{n_e(l)}{\int_0^{l_0} n_e(l)dl}; \qquad f_a(l) = \frac{n_a(l)}{\int_0^{l_0} n_a(l)dl}; \qquad s = \frac{\pi e^2}{mc} f_{ik}\bar{n}_a \tag{3.19}$$

and $\bar{n}_a$ is the averaged concentration of absorbing atoms over length l. For calculations employing the formulas (3.17)–(3.19) it is necessary to find the initial distribution of corresponding atoms in the emitting volume. This is possible by solving the appropriate equation of continuity:

$$\frac{\partial n}{\partial t} + \operatorname{div}(n\vec{v}) = \frac{\delta n}{\delta t} \tag{3.20}$$

where v is the diffusion velocity of particles of the given type. Subsequently it is important to regard correctly the processes in the plasma primarily, the recombination and ionization paths. We note here that a series of other processes, such as photoionization and volume recombination (especially thc associative process) are accounted for in the balance equations. The parameters of the discharge are in many respects determined by the balance of power.

The kinetic equation can be used to derive an expression for the balance of power by introducing a new variable $mv^2/2$.

Then, for the expectation value of the kinetic energy, $\langle E_k \rangle = \langle mv^2/2 \rangle$, we will have

$$\frac{\partial}{\partial t}(n\langle E_k \rangle) = -\operatorname{div}\vec{W} + ZenvE + \frac{\delta}{\delta t}(n\langle E_k \rangle) \tag{3.21}$$

where Z is the charge number, e the electron charge, v the drift velocity, E the electric field strength. From this equation it is evident that the change in kinetic energy ($\frac{\partial}{\partial t}(n\langle E_k \rangle)$) of particles of any type is caused by transfer of kinetic energy ($\operatorname{div}\vec{W}$), heating ($ZenvE$), and collisions. Following [154], the equation of energy balance in case of a weakly ionized plasma can be written as

$$\frac{\partial T_a}{\partial t} + \vec{v}_a \operatorname{grad} T_a + \frac{2}{3} T_a \operatorname{div} \vec{v}_a + \frac{2}{3} \operatorname{div} \vec{q}_a = -\kappa_{ab} \nu_{ab} (T_a - T_b) + \frac{2}{3} \nu_{ab} \frac{m_a^2 m_b}{(m_a + m_b)^2} v_a^2 \quad (3.22)$$

Here

$$\vec{v}_e = -\frac{1}{n} \operatorname{grad}(D_a n) - b_e E$$

where $D_a = (T_e + T_i)/(\mu_{ia} \nu_{ia})$ and $b_e = e/(m_e \nu_{ea})$; $\vec{q}_a$ is the heat flow transported by atoms, κ_{ab} the energy transfer coefficient for collisions between particles a and b, D_a the ambipolar diffusion coefficient, D_e the diffusion coefficient for electrons, b_e the electron mobility, and μ the reduced mass. Then, the heat flow transported by electrons will be equal to $\vec{q}_e = -\frac{5}{2} D_e \operatorname{grad} T_e$.

Finally, all processes are determined by the energy introduced into the discharge

$$\frac{\partial}{\partial t}\left[\frac{\varepsilon \varepsilon_0 E^2}{2} + \frac{B^2}{2\mu\mu_0} + \sum E_k\right] - \int q_w df - W_{opt} = -\int q df \quad (3.23)$$

where $\vec{E}$ and $\vec{B}$ are the strengths of the electrical and magnetic component of the hf field, E_k is the kinetic energy, q_w and q are flows of power through the surface of the lamp, W_{opt} is the power of the optical radiation.

Besides

$$\begin{aligned} \operatorname{rot} \vec{H} &= \sigma \vec{E} + i f \varepsilon \vec{E}, \qquad & \operatorname{div} \vec{E} &= 0 \\ \operatorname{rot} \vec{E} &= -i\omega\mu \vec{H}, \qquad & \operatorname{div} \vec{H} &= 0 \end{aligned} \quad (3.24)$$

where σ is the conductivity of the plasma for frequency f. The solution of these equations in the quasi-stationary approximation gives the field on the surface of the lamp as

$$\left|\vec{E}\right| \sim \sqrt{\frac{f\mu}{\sigma}} \left|\vec{H}\right| \quad (3.25)$$

and the energy flow into the lamp

$$q \sim \vec{H}\vec{E} \sim \sqrt{\frac{f\mu}{\sigma}} H_0^2 \quad (3.26)$$

The quantity of heat being released in unit volume of the plasma is expressed by

$$\frac{W}{V} = \frac{jf}{k^2} \left\{ \operatorname{Im} \varepsilon^l(f,\vec{k}) \left| \left(\vec{k}\vec{E}\right) \right|^2 + \operatorname{Im} \varepsilon^{tr}(f,\vec{k}) \left| \left[\vec{k}\vec{E}\right] \right|^2 \right\} \quad (3.27)$$

where $\vec{k}$ is the wave vector while ε^l and ε^{tr} are the longitudinal and transverse components of the dielectric permittivity:

$$\varepsilon^{l,tr}(f,\vec{k}) = 1 + \frac{i}{2\pi f}\sigma^{l,tr}(f,\vec{k}) \tag{3.28}$$

with σ^l and σ^{tr} the longitudinal and transverse conductivity of the plasma. The first component in Eq. (3.27) determines the absorption of the longitudinal (uncurling) electrical field, the second that of the transverse (curling) electrical field. The quantity of induced hf energy is proportional to the concentration of electrons in the volume of the lamp, at least at not too high conductivity of the plasma. With growing hf energy the skin effect begins to appear and the released energy is concentrated in the surface layers of the discharge plasma in the lamp. However, due to the high thermal conductivity of the electrons, their temperature at $n_e > 10^{11}$ cm^{-3} is about constant over the volume of the lamp. For smaller concentrations the temperature of the electrons is not homogeneous. This fact can be seen visually, as there is a substantially brighter area in the antinode of the E-field.

3.2. *Mixture of an Alkaline Metal Vapor and an Inert Gas*

3.2.1. *A Simple Model of the Discharge in a Two-Component Mixture Taking into Account Only the Excitation of Alkali Metal Atoms*

It is expedient to begin the study of processes in the lamps with the primary excitation of the alkali metal atoms. Subsequently the processes in the lamp are described by a simpler system of equations. Experimentally, modes close to those under this simplified approach are often realized for low-power discharges, especially for E-discharges. In such cases a thermal regime of operation is established, ensuring the highest intensity of radiation of the resonance lines of the alkali metal. As shown in Chapter 4, the processes in the alkaline metal vapor prevail under such conditions. When using lamps in the modes of the highest intensity of the H-discharge, conditions close to those considered in the present section may also be realized. For all such cases a boundary problem which describes the discharge in the lamps is proposed and justified, and the main features of the solution are analyzed in detail.

We consider a spherical volume filled with a mixture of krypton and rubidium vapor. The pressure of krypton varies from 0.65 up to 13 mbar, the temperature from 84 up to 140 °C, and the radius of the sphere from 6.5 to 25 mm.

When calculating the process the following assumptions were made:

(1) The ionization of atoms occurs in the volume when electrons collide with the rubidium atoms, and recombination occurs on the wall of the sphere of the lamp (ambipolar diffusion mode).

(2) The atoms of the metal are described by a three-level model (ground, excited, and ionized states).

(3) The processes of excitation and ionization of the buffer gas (krypton) may be neglected in the balance of energy and particles.

(4) The temperature of the atoms and ions is equal to the temperature of the wall of the sphere, which determines the density of the rubidium vapor at the wall.

(5) The temperature of the electrons is identical at all points of the lamp.

(6) The velocity distribution function of the electrons is approximated by a two-temperature function. The temperature for electrons with energy lower than the first excitation potential is T_e and for more energetic electrons T_{et}.

The first condition is rather well satisfied in the investigated range of pressure. Only for low pressure should an appreciable contribution of dissociative recombination be expected. Therefore dissociative recombination was not taken into account. The third condition means that we limit the calculation to modes in which the radiation of spectral lines of the inert gas is not observed. Experimentally, such modes will be easily realized both in the E- and in the H-discharges for a rather high temperature of the wall of the lamp (high density of the vapor of the metal). Condition (4), that the temperatures of the atoms and ions are both equal to the temperature of the wall, is satisfied for a discharge power lower than 200 mW. The fact that the potential of the plasma of the ED relative to the bulb walls can reach 100 V was ignored [12]. This voltage difference causes additional luminescence in the bulk of the lamp and is one of the reasons for intensity instabilities. These instabilities are due to the bombardment of metal drops (sitting on the wall) by ions, accelerated in the electrical field between the volume of the plasma and the drops of metal (when assuming the alkali metal vapor pressure as equilibrium pressure above the liquid metal at wall temperature). This effect influences also the speed of aging of the lamp. Despite the important role of these effects, we cannot take them into account, as they depend on particular conditions of the experiment, such as the arrangement of conductors near the sphere or the position of the inductor in relation to the lamp. This can be the reason for the disagreement of particular experiments with the results of calculations.

When fulfilling the initial assumptions, the density of electrons $n(r)$, the density of metal atoms in the ground $n_1(r)$ and excited $n_2(r)$ states are, in the stationary case, the solutions of the following boundary-value problem:

$$D_a \Delta_r n + K(1,3) n_1 n + K(2,3) n_2 n = 0 \tag{3.29}$$

$$D_1\Delta_r n_1 - [K(1,3)+K(2,1)]n_2n_1 + K(1,2)n_2n + \frac{n_2}{\tau_{eff}} = 0 \tag{3.30}$$

$$K(1,2)n_1n - \left[K(2,1)n + K(2,3)n + \frac{1}{\tau_{eff}}\right]n_2 = 0 \tag{3.31}$$

$$W = -\frac{5}{2}k_BT_eD_a4\pi r_0^2\left.\frac{dn}{dr}\right|_{r=r_0} + \int_0^{r_0}[P_{el}+P_{inel}]n(r)4\pi r^2dr \tag{3.32}$$

$$\left.\frac{dn_1}{dr}\right|_{r=r_0} = 0, \qquad \left.\frac{dn}{dr}\right|_{r=r_0} = 0 \tag{3.33}$$
$$n_1(r=r_0) = n_w, \qquad n(r=r_0) = 0$$

$$n_w = 133.322\frac{p}{k_BT_a} \tag{3.34}$$

Here (3.29) is the equation governing the balance of electrons; (3.30) and (3.31) are equations of the balance of metal atoms in the ground and excited states; (3.32) is the equation of the balance of power; D_a, D_1, and D_2 are factors of the ambipolar diffusion and diffusion of atoms and ions (D_2 will be used later); T_e and T_a are temperature of electrons and atoms (ions) in K; n_w is the concentration of metal atoms at the wall of the lamp, and τ_{eff} the effective lifetime of a photon in the volume of the lamp, which is determined by the concentration of atoms and their distribution in the volume; $K(i,j)$ is the transition rate of atoms from state i to state j; P_{el} and P_{inel} are the rate of elastic and inelastic collisions, respectively, Δ_r is the Laplace operator with respect to the radial variable, and r_0 the radius of the lamp.

The vapor pressure of rubidium is given by $p = 10^{A+B/T_a+CT_a+D\ln(T_a/1\mathrm{K})}$, with $A = 15.88$, $B = -4529.61$ K, $C = 5.8$ K^{-1}, and $D = -2.997$ (empirical coefficients from [155]).

The expressions for τ_{eff}, $K(i,j)$, P_{el}, and P_{inel} are the same as in [156], where a dc discharge in vapors of sodium and cesium was studied:

$$P_{el} = \left(1-\frac{T_w}{T_e}\right)\sum_j\frac{2m_e}{m_j}\left(\frac{m_e}{2\pi k_BT_e}\right)^{3/2}\frac{8\pi}{m_e^2}\int_0^\infty E^2\exp\left(-\frac{E}{k_BT_e}\right)n_j\sigma_jdE \tag{3.35}$$

where σ_j is the cross section of elastic collisions with atoms j, n_j the concentration of atoms, m_j the mass of the atom j; m_e and E_e the mass and energy of an electron, and T_w the temperature of the wall.

The inelastic losses of the electrons can be calculated by the formula

$$\begin{aligned} P_{inel} = & n_1 \left[K(1,3)(U_3 - U_1) + K(1,2)(U_2 - U_1)\right] \\ & + n_2 \left[K(2,3)(U_3 - U_2) - K(2,1)(U_2 - U_1)\right] \end{aligned} \tag{3.36}$$

where U_j and n_j are the energy and concentration of atoms in the jth level.

Quantity τ_{eff} is defined by

$$\tau_{eff} = \left(\frac{1}{3\tau_{eff,1/2}} + \frac{2}{3\tau_{eff,3/2}}\right)^{-1} \tag{3.37}$$

where $\tau_{eff,1/2} = \tau_{nat,1/2}\langle N_{1/2}\rangle$, $\tau_{eff,3/2} = \tau_{nat,3/2}\langle N_{3/2}\rangle$, $\langle N_j\rangle$ is the average number of photons scattered into state j, while $\tau_{nat,1/2}$ and $\tau_{nat,3/2}$ are the lifetimes of the $5\,^2P_{1/2}$ and $5\,^2P_{3/2}$ levels. Values for the constants and the factors of diffusion can be found in [157], for the cross sections for excitation and ionization processes in [152], and for the energy splitting in [158].

If some cumbersome transformations of Eqs. (3.29)–(3.33) are conducted, it is possible to obtain the connection between n and n_1 in the form

$$n_1 = n_w - \frac{D_a}{D_1} n \tag{3.38}$$

which permits one to replace boundary problem (3.29)–(3.33) by boundary problem (3.29), (3.31)–(3.33), (3.38) on reducing the order of the system of differential equations to second order. The velocity distribution of the electrons was assumed to be Maxwellian, leading to the energy distribution

$$f(E) = \frac{2}{\sqrt{\pi}} \frac{\sqrt{E}}{(k_B T)^{3/2}} \exp\left(-\frac{E}{k_B T}\right) \tag{3.39}$$

where T is the temperature of the main part of electrons T_e or of the tail of the distribution T_{et} [159].

The cross sections of inelastic collisions are approximated by the expressions

$$\begin{aligned} & \sigma_{ij} = b_{ij}\left(1 - \left|U_i - U_j\right|/E\right) && \text{for } E \geq \left|U_i - U_j\right| \\ & \sigma_{ij} = 0, && \text{for } E < \left|U_i - U_j\right| \\ & K(i,j) = b_{ij}\left(\frac{8k_B T}{\pi m}\right)^{1/2} \exp\frac{-\left|U_i - U_j\right|}{k_B T} && \end{aligned} \tag{3.40}$$

The values of the constants are given in Table 3.1.

The system of equations (3.29), (3.31)–(3.33), (3.38) was solved numerically by the "method of adjustment" [varying the initial value $n(0)$ and temperature T

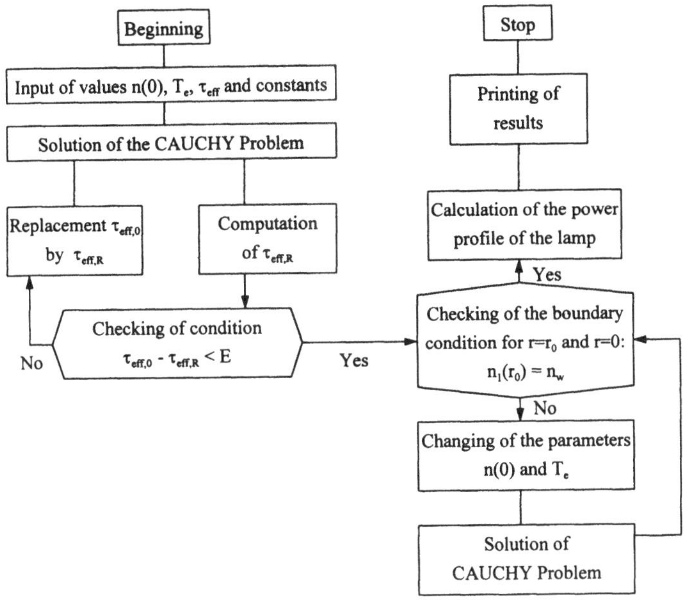

FIGURE 3.1. Flow chart of the algorithm for the numerical solution of Eqs. (3.29), (3.31)–(3.33), (3.42) by the "method of adjustment."

until the conditions are satisfied at the point $r = r_0$]. The block diagram of the algorithm for solving the problem is shown in Figure 3.1.

The term dn/dr in expression (3.32), which causes uncertainty at the point $r = 0$, was varied from 0 up to 10^9 without essential influence on the solution. A numerical value was assigned to τ_{eff} first and then made more precise with each solution of the Cauchy problem, irrespective of the fulfillment or default of conditions at the point $r = r_0$. The series appeared to converge rapidly to some constant value and did not vary further. According to a criterion presented in [162] a one-temperature function for the electron distribution was used.

The solutions of the system (3.29), (3.31)–(3.33) do not exist for all values of the parameters. The zones of existence of the solutions are situated to the right of the curves given in Figure 3.2.

A physical reason for the nonexistence of the solution, when the temperature of the metal tank was reduced or the vapor density was decreased, is the nonfulfillment of the initial assumption for this model concerning the decisive role of the metal vapors. When reducing the temperature, the role of the gas increases steeply due to the growth in the temperature of the electrons, and it is then the buffer gas which determines all processes in the discharge. Thus, the absence of

TABLE 3.1. Atomic Constants Used in the Calculations[a]

Element	Parameter	Amount or formula	Ref.
^{23}Na	σ_{12} [m^2]	$53\times10^{-20}(1-20/E)$	[156]
	σ_{13} [m^2]	$11\times10^{-20}(1-5.14/E)$	
	σ_{23} [m^2]	$0.39\times10^{-20}(1-3.04/E)$	
	σ_L(Ne) [m^2]	119×10^{-20}	
	σ_L(Ar) [m^2]	252×10^{-20}	
	σ_{el}(Na) [m^2]	$0.88\times10^{-22}(490-123.8E+19.479E^2-1.4615E^3+0.040246E^4)$	
	$\sigma_{el}(Na^+)$ [m^2]	$26\times10^{-18}\ln\Lambda/(2E)^2$	
	D_1(Ne)	$5\times10^{-5}(T_g/425)^{1/2}$	
	D_1(Ar)	$3\times10^{-5}(T_g/273)^{1/2}$	
	D_3(Ne)	$2.2\times10^{-5}(T_g/273)^{2/3}$	
	D_3(Ar)	$0.81\times10^{-5}(T_g/291)^{2/3}$	
	$\lambda(3^2P_{1/2}\leftrightarrow3^2S_{1/2})D_1$ – line [nm]	589.592	
	$\lambda(3^2P_{3/2}\leftrightarrow3^2S_{1/2})D_2$ – line [nm]	588.995	
	$f(3^2P_{1/2}\leftrightarrow3^2S_{1/2})$	0.327	[158]
	$f(3^2P_{3/2}\leftrightarrow3^2S_{1/2})$	0.655	
	I	3/2	
	$A(3^2S_{1/2})$ [MHz]	885.8	
	$A(3^2P_{1/2})$ [MHz]	94.3	
	$A(3^2P_{3/2})$ [MHz]	18.7	
	$B(3^2P_{3/2})$ [MHz]	2.9	
Rb	σ_{12} [m^2]	4.4×10^{-19} (maximal value)	[160]
	σ_{13} [m^2]	$(2.0\text{–}10)\times10^{-20}(1-4.2/E)$	[138]
	σ_{23} [m^2]	$(12.4\pm4)\times10^{-20}(1-1.6/E)$	
	D_1(Ne)	$3.1\times10^{-5}(T_g/273)^{1/2}$	[161]
	D_1(Ar)	$2.4\times10^{-5}(T_g/273)^{1/2}$	
	D_1(Kr)	$1.6\times10^{-5}(T_g/273)^{1/2}$	
	D_1(Xe)	$1.3\times10^{-5}(T_g/273)^{1/2}$	
	D_3(Kr)	$3.3\times10^{-5}(T_g/273)^2 p$	[161]
	D_3(Kr – Kr)	$3.76\times10^{-5}(T_g/273)^2 p$	
	$\lambda(5^2P_{1/2}\leftrightarrow5^2S_{1/2})D_1$ – line [nm]	794.760	
	$\lambda(5^2P_{3/2}\leftrightarrow5^2S_{1/2})D_2$ – line [nm]	780.023	

TABLE 3.1. *Continued*

Element	Parameter	Amount or Formula	Ref.
Rb	$f(5^2P_{1/2} \leftrightarrow 5^2S_{1/2})$	0.4	[151]
	$f(5^2P_{3/2} \leftrightarrow 5^2S_{1/2})$	0.8	
	natural abundance ^{87}Rb[%]	27.8	
	natural abundance ^{85}Rb[%]	72.2	
	$I(^{87}\mathrm{Rb})$	3/2	
	$I(^{85}\mathrm{Rb})$	5/2	
	$A^{85}\mathrm{Rb}(5^2S_{1/2})$ [MHz]	1011.9	[158]
	$A^{85}\mathrm{Rb}(5^2P_{1/2})$ [MHz]	120.7	
	$A^{85}\mathrm{Rb}(5^2P_{3/2})$ [MHz]	25.0	
	$B^{85}\mathrm{Rb}(5^2P_{3/2})$ [MHz]	25.9	
	$A^{87}\mathrm{Rb}(5^2S_{1/2})$ [MHz]	3417.3	
	$A^{87}\mathrm{Rb}(5^2P_{1/2})$ [MHz]	406.2	
	$A^{87}\mathrm{Rb}(5^2P_{3/2})$ [MHz]	84.8	
	$B^{87}\mathrm{Rb}(5^2P_{3/2})$ [MHz]	12.5	
^{133}Cs	σ_{12} [m^2]	$160 \times 10^{-20}(1-4.2/E)$	[156]
	σ_{13} [m^2]	$11 \times 10^{-20}(1-3.89/E)$	
	σ_{23} [m^2]	$0.58 \times 10^{-20}(1-2.47/E)$	
	σ_L [m^2]	157×10^{-20}	
	$\sigma_{el}(\mathrm{Cs})$ [m^2]	$10^{-18}(2.8-2.1\,\lg E)$	
	$\sigma_{el}(\mathrm{Cs}^+)$ [m^2]	$26 \times 10^{-18}\ln\Lambda/(2E)^2$	
	$D_1(\mathrm{Ar})$	$1.34 \times 10^{-5}(T_g/273)^{1/2}$	
	$D_3(\mathrm{Ar})$	$0.56 \times 10^{-5}(T_g/273)^{1/2}$	
	$D_3(\mathrm{Kr-Kr})$	$3.76 \times 10^{-5}(T_g/273)^2 p$	
	$\lambda(6^2P_{1/2} \leftrightarrow 6^2S_{1/2})D_1$ – line [nm]	894.350	
	$\lambda(6^2P_{3/2} \leftrightarrow 6^2S_{1/2})D_2$ – line [nm]	852.110	
	$f(6^2P_{1/2} \leftrightarrow 6^2S_{1/2})$	0.394	[158]
	$f(6^2P_{3/2} \leftrightarrow 6^2S_{1/2})$	0.814	
	I	7/2	
	$A(6^2S_{1/2})$ [MHz]	2298.2	
	$A(6^2P_{1/2})$ [MHz]	291.9	
	$A(6^2P_{3/2})$ [MHz]	50.3	
	$B(6^2P_{3/2}$ [MHz]	−0.38	

[a] E, energy (in eV); T_g, temperature of ground state atoms (in K); $\ln\Lambda$, coulomb logarithm: $\ln\Lambda = \ln\left(\frac{12\pi}{e^3}\frac{(\varepsilon_0 k_B T_e)^{3/2}}{n^{1/2}}\right)$ (where n is electron number density; and T_e, electron temperature).

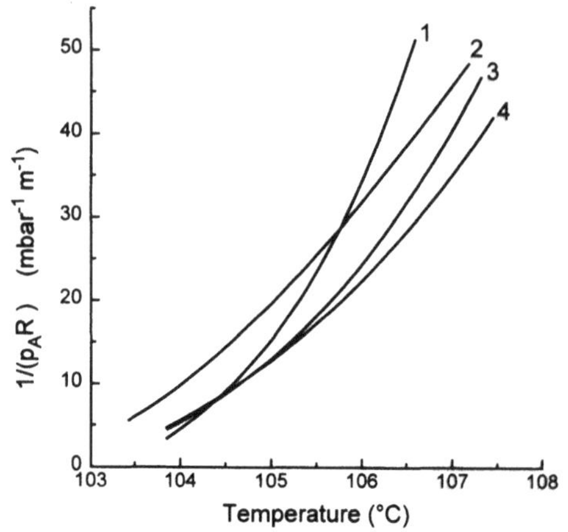

FIGURE 3.2. Zones of existence of the solutions (to the right of the curves). 1, $n_e = 10^{11}$ cm^{-3}, $r_0 = 20$ mm; 2, $n_e = 10^{13}$ cm^{-3}, $r_0 = 15$ mm: 3, $n_e = 10^{11}$ cm^{-3}, $r_0 = 15$ mm; 4, $n_e = 10^{12}$ cm^{-3}, $r_0 = 20$ mm.

a solution in this case restricts the model. The decrease of pressure also results in the disappearance of the solution of the set of equations. The condition $\nu_i \geq \nu_r$ (ν_i and ν_r are the frequency of ionization and recombination, respectively) is not fulfilled in this case. During experiment, also a growth in the intensity of the lines of the inert gas is usually observed. The interesting fact of a disappearing solution upon increasing the temperature of the electrons is connected with the nonfulfillment of the condition $\nu \geq \nu_r$, as the average energy of electrons exceeds the energy of ionization. It is evident that under these conditions an oscillatory process can arise. In order to eliminate this process, the temperature of the electrons must be decreased, i.e., by reducing the power of the discharge, increasing the pressure of the gas, or increasing the diameter of the lamp. Solutions exist at electron temperatures from 8000 up to 45,000 K. Correspondingly, the electron temperature grows, when decreasing the pressure of the metal vapor, the pressure of the inert gas, or the size of the lamp.

Under typical conditions — an alkali vapor pressure of 10^{-4} mbar, lamp diameter 13 mm, and 2.4 mbar pressure of the inert gas — the electron temperature appeared to lie within 20,000–25,000 K. In general, this result is in agreement with the results of data extrapolation [163, 164]. Under these conditions the power of the discharge varies between 0.005 W and 0.1 W, which is much lower than the power of the H-discharge, but it lies within the characteristic limits for the E-discharge.

Figure 3.3 shows the dependence of the electron temperature T_e on the temperature T of the rubidium vapor (i.e., on the rubidium number density). When increasing the density of the vapor, the electron temperature decreases according to a hyperbolic law. When reducing the buffer gas pressure (at a fixed temperature T) the electron temperature grows quickly. For example, at $T = 375$ K, reducing the pressure from 10.7 to 0.7 mbar results in a growth of T_e from

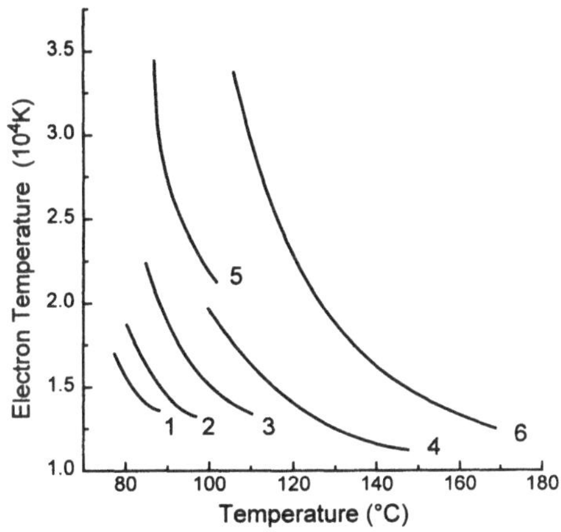

FIGURE 3.3. Dependence of the electronic temperatures T_e on the temperature of the saturated rubidium vapor. The electron density is around $n = 10^{11}$ cm^{-3}, the pressure p of krypton: 1, $p = 13$ mbar; 2, $p = 10.7$ mbar; 3, $p = 5.3$ mbar; 4, $p = 2.4$ mbar; 5, $p = 2$ mbar; 6, $p = 0.7$ mbar.

1.2×10^4 up to 4×10^4 K. This type of dependence can also be observed in other versions of an electronic discharge.

When decreasing the lamp radius (Figure 3.4) the temperature also grows steeply, from 10^4 K at $r_0 = 25$ mm up to more than 3×10^4 K at $r_0 = 6.5$ mm. A study of the results indicates that for some modes a rather high electronic temperature, which determines the excitation and ionization of krypton atoms in general, may violate the initial assumptions of the model used. The excitation and ionization of krypton atoms results in an increase of electron concentration and increase of power released in the discharge, and also leads to reduction in the temperature of the "tail" of the velocity distribution function of the electrons.

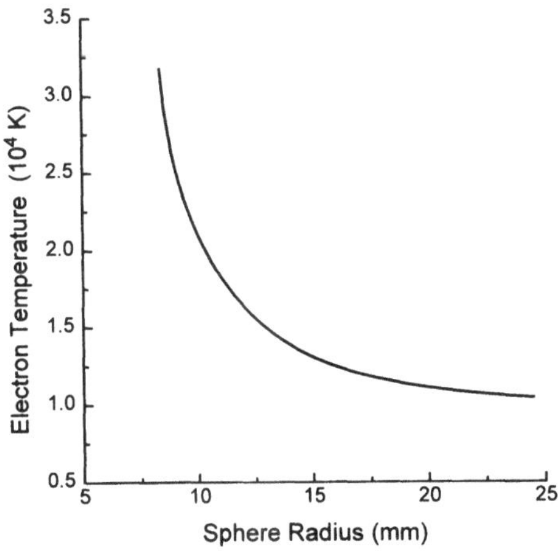

FIGURE 3.4. Dependence of the electron temperature on the lamp radius r_0.

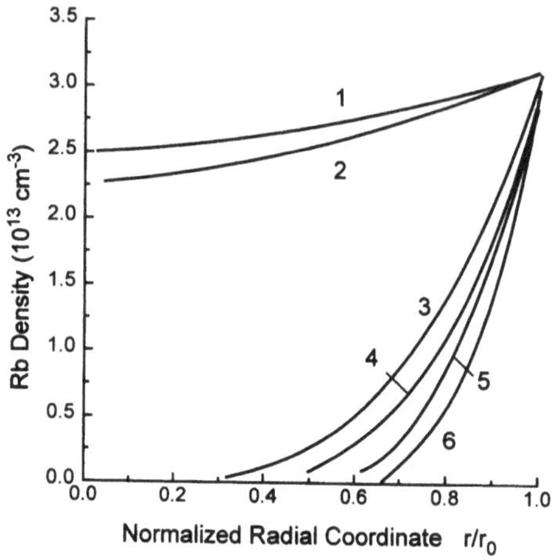

FIGURE 3.5. Calculated radial distribution of rubidium atoms (ground state) in the lamp: 1, power of discharge $W = 0.08$ W; 2, 0.1 W; 3, 0.3 W; 4, 0.6 W; 5, 1.2 W; 6, 1.7 W.

However, the influence of these processes on the total balance of power requires additional modeling, taking into account the excitation of krypton atoms. The spatial distribution of excited rubidium atoms is determined by the concentration and temperature of those electrons which possess an energy smaller than the first excitation potential of the krypton atom. Nevertheless, the characteristic dependences should not differ significantly from the calculated ones for the presence of excitation processes of krypton atoms.

Figure 3.5 shows calculated distributions of rubidium atoms in the ground state throughout the volume of the lamp. A decrease takes place in the concentration of atoms toward the center of the lamp. This effect is more pronounced for higher electron concentrations. At a low power of the discharge, the maximum concentration of excited atoms is at the center of the lamp. When increasing the power, the maximum of the excited atom density is shifted to the walls of the lamp and a minimum appears in the center. This distribution differs strongly from other theoretical predictions and also from experimentally observed distributions for a high discharge power due to the excitation of inert gas atoms (see Section 3.2.2).

The spatial distribution of atoms for the E-discharge has not been studied sufficiently in experiments, but the analysis of preliminary results on the distribution of concentration shows that they behave as predicted in the computed relationships when changing the modes.

3.2.2. The General Model of Processes of Excitation in a Two-Component Mixture

A drawback of the theory presented in Section 3.2.1 is neglect of the excitation of the atoms of the inert gas. At modes with maximum density of optical radiation, in E- as well as in H-discharges, simultaneous excitation of atoms of the inert gas and alkali metal atoms usually occurs. Calculation of the discharge parameters also shows that in certain operating modes the electron temperature is rather high and therefore excitation of atoms of the inert gas should be observed.

In this connection a system of balance equations for the concentration of electrons, n_e, rubidium atoms in ground and excited states, n_1 and n_2, and krypton atoms in ground and excited states, n_4 and n_5, as well as the equation for the balance of power W were set up to describe processes in ELs containing metal vapor:

$$D_a\Delta_r n_e + K(1,3)n_1n_e + K(2,3)n_2n_e + K(4,6)n_4n_e + K(5,6)n_5 = 0 \tag{3.41}$$

$$D_1\Delta_r n_e + K(1,2)n_1n_e - K(1,3)n_1n_e - K(2,1)n_2n_e + \frac{n_2}{\tau_{eff}} = 0 \tag{3.42}$$

$$K(1,2)n_1n_e - K(2,1)n_2n_e + K(2,3)n_2n_e - \frac{n_2}{\tau_{eff2}} = 0 \tag{3.43}$$

$$D_4\Delta_r n_4 - K(4,5)n_4n_e + K(5,4)n_5n_e + \frac{n_5}{\tau_{eff5}} + K(4,6)n_en_4 = 0 \tag{3.44}$$

$$D_4\Delta_r n_5 + K(4,5)n_4n_e - K(5,4)n_5n_e + K(5,6)n_5n_e - \frac{n_5}{\tau_{eff}} = 0 \tag{3.45}$$

$$W = -\frac{5}{2}kT_1\int_0^{r_0}[K(1,3)n_1 + K(2,3)n_2 + K(4,6)n_4 + K(5,6)n_5]4\pi r^2 dr + \int_0^{r_0}(P_{el} + P_{inel})n_e 4\pi^2 r dr \tag{3.46}$$

$$\int_0^{r_0}(n_4 + n_5 + n_6)4\pi r^2 dr = N \tag{3.47}$$

The boundary conditions are

$$\left.\frac{dn_1}{dr}\right|_{r=0} = -\left.\frac{dn_e}{dr}\right|_{r=0} = 0$$
$$n_1(r = r_0) = n_w$$
$$n_e(r = r_0) = 0$$
$$n_5(r = r_0) = 0 \tag{3.48}$$

where $K(i, j)$ are rate constants for the process of excitation from state i to state j; D_a, D_1, and D_4 are factors of the ambipolar diffusion as well as the diffusion of metal atoms and inert gas atoms respectively; indices 1–6 relate to metal atoms in ground, excited, and ionized states as well as to atoms of the inert gas in ground, excited, and ionized states; τ_{eff2} and τ_{eff5} are effective lifetimes of excited states of metal and inert gas atoms; T_e and T_{et} are average temperature and "tail" temperature of the electron velocity distribution function; N is the total number of inert gas atoms in the volume of the lamp.

The following assumptions were made for the boundary problem (3.41)–(3.47):

(1) Ambipolar diffusion mode.

(2) The metal and inert gas atoms are considered as three-level systems.

(3) The velocity distribution function of the electrons is approximated by a two-temperature function [159]. The temperature for electrons with an energy lower than the first excitation potential is T_e, and T_{et} for more energetic electrons.

(4) The temperature of electrons throughout the volume of the lamp is constant.

The boundary problem (3.41)–(3.47) was solved for a mixture of rubidium and krypton vapor, located in a spherical bulb of diameter 13 mm. The solution was searched by the method of variation of values of the initial concentration of atoms and electrons, temperatures T_e and T_{et} and effective lifetimes τ_{eff2} and τ_{eff5}. The scheme of the solution algorithm is given in Figure 3.6.

The result of calculations for one of the modes is shown in Figure 3.7. From the plots it can be seen that rubidium atoms in the ground state are concentrated mainly at the walls of the lamp, and in the excited states at some distance away from it. The krypton atoms in the ground state are distributed practically homogeneously with insignificant increase in concentration at the walls of the lamp. Krypton atoms in excited states are also distributed inhomogeneously with a steep decrease in concentration near the walls. The character of the distribution of rubidium atoms throughout the volume is in accordance with [69, 113] but differs from [74], where the concentration of excited atoms at the walls of the lamp was presumed to be nonzero. Figure 3.8 shows the distributions of excited rubidium atoms, with parameter equal to the power of the discharge.

When increasing the power of the discharge the maximum of the distribution is quickly shifted to the wall of the lamp. Subsequently the distributions become narrower, leading to a decrease in the intensity of radiation I, noted in [85]. This was a result of calculations at a lamp temperature of 94 °C and electron concentration of 10^{13} cm^{-3} (Table 3.2).

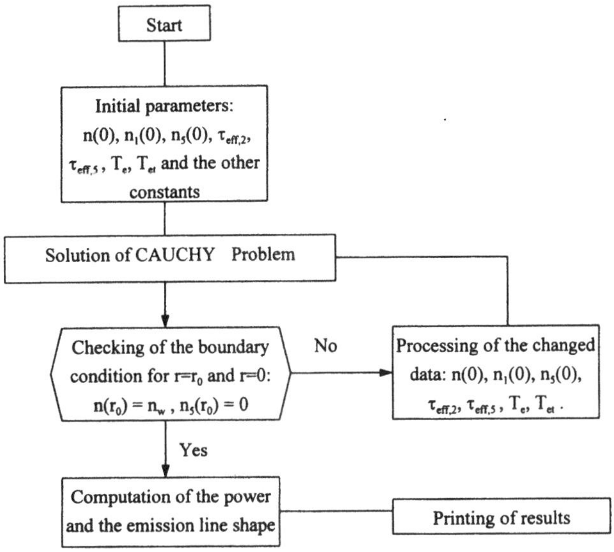

FIGURE 3.6. Flow chart of the solution algorithm for the boundary problem (3.45)–(3.51).

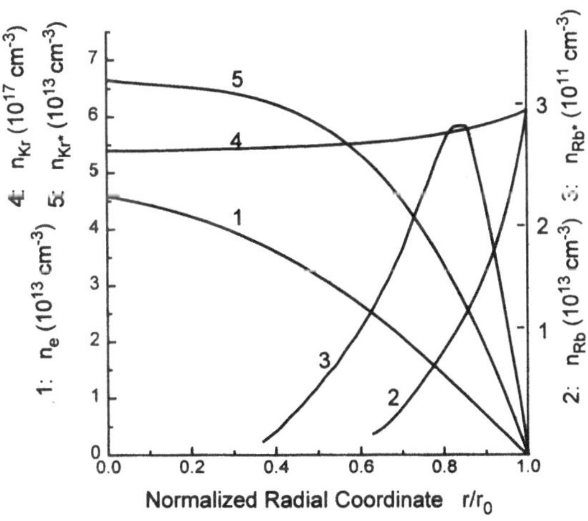

FIGURE 3.7. Calculated radial distribution in the lamp of: 1, electrons; 2, rubidium atoms in the ground; 3, excited states; 4, krypton atoms in the ground state; 5, in metastable states.

TABLE 3.2. Radiation Intensity I (Rb Resonance Radiation) for Different Values of the Discharge Power W^a

W [W]	0.627	0.752	0.987	1.231
I [mW/cm^2]	123	96.9	95.4	89.97

^{a}Lamp temperature, 94 °C; electron concentration, 10^{13} cm^{-3}.

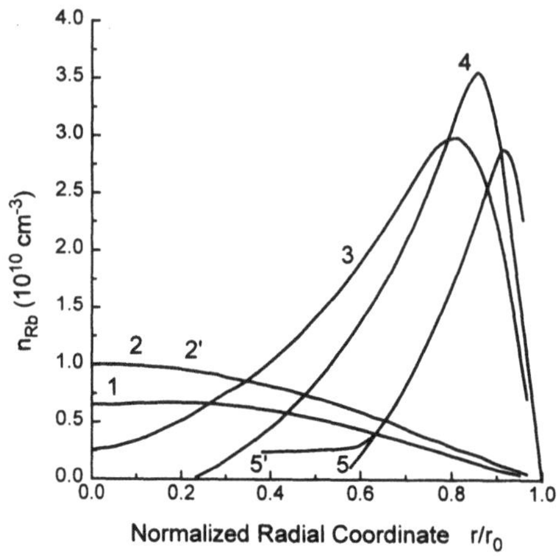

FIGURE 3.8. Calculated (1–5) and experimental ($2'$, $5'$) radial distributions of rubidium atoms in excited states: 1, power of the discharge $W = 0.08$ W; 2, 0.1 W; 3, 0.3 W, 4, 0.6 W; 5, 1.2 W; $2'$, (0.10 ± 0.05) W, E-discharge; $5'$, (1.5 ± 0.5) W, H-discharge.

The outcome of the theory is in good agreement with measurements of various distributions of discharge power and radiation intensity.

3.3. Shape of the Spectral Lines

3.3.1. Initial Profile of the Line

General questions concerning the formation of the shape of the lines were considered in Section 3.1 and it was shown that the initial profile is a Voigt function. The width of the lines is determined by collisions of the emitting atoms with foreign gas atoms and electrons, and therefore by the filling of the lamp and the concentration of electrons. Generally speaking, the electrical field existing in the lamp causes an additional shift of the spectral lines. However, under the discharge conditions in hf light sources this shift can usually be neglected.

In the plasma of an ED the following factors influence the radiating atoms: (1) the exciting electrical field; (2) the intraplasma field; (3) charged particles; (4) magnetic fields; (5) neutral atoms. The Stark constant for resonance lines of alkali atoms equals 30–300 kHz/(kV/cm)^2, so the shifts caused by the electric field as well as broadening due to this field are about 10^4 Hz (for exciting field strengths in the E-discharge up to 400 V/cm and 2–4 V/cm in the H-discharge, respectively). These values are much smaller than the observable values. However, the plasma field should be taken into account in a layer of the order of the Debye thickness, where it reaches an intensity of ~ 1 kV/cm. In this case the resulting shift is about 0.1 MHz.

The broadening caused by the electrons (the contribution of ions to the shift

TABLE 3.3. Calculated Shifts Δf_e and Broadenings γ_e for the Rubidium D_1 Line ($\lambda = 794.7$ nm) at Different Electron Concentrations n_e

n_e [cm^{-3}]	10^{12}	10^{13}	10^{14}
Δf_e [MHz]	0.081	0.81	8.1
γ_e [MHz]	0.093	0.93	9.3

and to the broadening of emission lines is much smaller) is calculated from the Weisskopf–Lindholm formula (see Section 3.1). The results of computing the shift and broadening for the long-wave components of the doublet of the first resonance lines of rubidium are given in Table 3.3.

The shift and broadening of spectral lines occurs also as a result of collisions with other atoms. This effect has been thoroughly investigated. Measurements of the shifts and broadening of the Rb D_1 line ($\lambda = 794.7$ nm) are shown in Table 3.4 (cf. Chapter 10).

The magnetic field causes a frequency shift for electric dipole transitions $\Delta\nu = \alpha B$, where $\alpha = 1$–10 MHz/(mT). The shift of the spectral lines is therefore no more than 1 MHz for fields up to 0.1 mT. Additionally, the magnetic field can directly influence the intensity of the plasma radiation of the discharge [128], but this effect will be considered in the chapter on intensity fluctuations.

3.3.2. The Influence of Anisotropy of the Atomic Distribution

As shown before, the processes in the ED are characterized by strong anisotropy in the distribution of atoms (3.2), resulting in a flow Γ_a of particles in the plasma. It is known that the flow caused by a gradient of the density of particles in the ground state equals

$$\Gamma_a = -D\frac{\partial n_a}{\partial r} \tag{3.49}$$

where D is the diffusion coefficient. If we set $\Gamma_a = nv$, it is possible to evaluate the velocity of emitting atoms:

$$v = -\frac{D}{n_a}\frac{\partial n_a}{\partial r} \tag{3.50}$$

TABLE 3.4. Measured Shifts Δf and Broadenings γ for the Rubidium D_1 Line (λ=794.7 nm) for Different Buffer Gases

	He	Ne	Ar	Kr	Xe
Δf [MHz/mbar]	1.65 ± 0.4	-2.8 ± 0.4	-5.3 ± 0.6	-6.3 ± 0.8	-5.3 ± 0.9
γ [MHz/mbar]	10.4 ± 0.2	5.3 ± 0.75	10.2 ± 1	10.2 ± 1	11.7 ± 1.3

TABLE 3.5. Frequency Shift Δf [MHz] of the Rubidium D_1 Line ($\lambda = 794.7$ nm) Due to Anisotropy of the Atomic Velocities as a Function of the Discharge Current and Krypton Pressure[a]

i	p [mbar]			
[mA]	1.3	2.1	4.3	12
130	−10.1	−6.3	−3.2	−1.1
160	−17.7	−11	−5.5	−1.8
180	−17.7	−11	−5.5	−1.8
230	−26.2	−17.8	−9.2	−3
230	−32*	−28*	−28*	−52*

[a] Values of the total shift caused by anisotropy and the pressure of the buffer gas are marked by an asterisk.

By using the calculated distribution of atoms throughout the volume of the lamp, one can obtain the frequency shift of the emitted line due to a Doppler shift (Table 3.5).

Thus, the shift caused by the movement of atoms due to the density gradient is comparable to the shift caused by the buffer gas, and the latter should be taken into account for exact model calculations. The value of the shift caused by anisotropy varies for different parts of the lamp, from the highest value at the center down to zero at the edge of the lamp. Hence, the effective value of the observed shift will also depend on the parameters of the optical system, i.e., the focus length of the lens, position of the lamp in relation to it, aperture of the optical system, etc. If this light source is employed in a quantum frequency standard, the so-called "light shift" will depend on the mode of the lamp in view of the given effect.

The values of the line shift depend additionally on the sort of buffer gas. In the case of argon and xenon the shift of lines due to perturbation by atoms of the buffer gas is the same as for krypton, but the diffusion factor in argon is 1.5 times larger and in xenon 1.5 times smaller (see Section 3.3.2). This effect can increase for lighter atoms like Li, Na, K, Cd, Mg, etc. Their coefficient of diffusion is much larger than that for rubidium atoms (see Table 3.1).

3.3.3. *Emission Line Profile*

The final shape of the emitted line was calculated under the one-beam approximation (assuming that the solid angle in which the studied radiation is

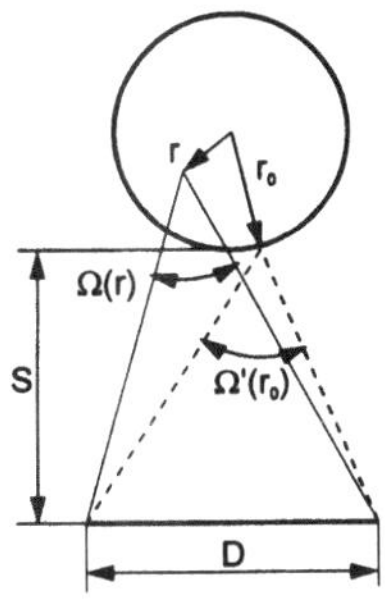

FIGURE 3.9. Illustration of the set of variables used in the calculation of the shape of the radiation line. S, distance of the collecting lens with diameter D.

collected by the optical system is small) from Eq. (3.51) (see Figure 3.9):

$$I(\nu) = I_0 \int_0^{r_0} 4\pi r^2 dr \int_0^{l_0} f_e[l(r)] P_e(\nu) \exp\left\{ -S \int_0^{l} P_a(\nu) f_a\left[l'(r)\right] dl' \right\} dl \qquad (3.51)$$

where f_e and f_a are normalized distributions of emitting and absorbing atoms; l_0 and l are chords, along which the beam of light is spread, where l_0 is the length of a pillar of the absorbing atoms and l is the length of a pillar of emitting atoms; P_e and P_a are line shapes of the emitting and absorbing atoms, respectively; S is a constant.

In accordance with Section 3.3.1, the initial shape for the calculations, was taken as a Voigt profile with a Lorentzian component which comprises natural and collision broadenings due to collisions with atoms and electrons. The Gaussian component was defined by the thermal movement of atoms. Additionally, collisional shifts were taken into account. At a temperature of $T = 130$ °C, the intensity of the radiation grows rapidly and reaches the highest value at $p = 2.4$ mbar with decreasing krypton pressure (Figure 3.10). For a further decrease in pressure, steady solutions only exist when the temperature of the lamp is raised.

The dependency of the emission line intensity on the temperature of the lamp is shown in Figure 3.11, from which it is clear that the highest intensity is realized at $T = 90$–100 °C, the linewidth (of a single hyperfine component) being 600–900 MHz. Regimes with narrower spectral lines could be observed under certain conditions in an H-discharge. In this case, the efficiency for the transformation of power of the high-frequency oscillator into emitted power attains 50–90% while for more powerful discharges this efficiency usually amounts only to 1–10%.

The shape and contrast of the radio-optical resonance is of interest for frequency standards. Figure 3.12 shows the dependence of the radio-optical resonance signal, which characterizes the density of the optical radiation of the pumping light, on the temperature of the lamp for an E-discharge. The maximum

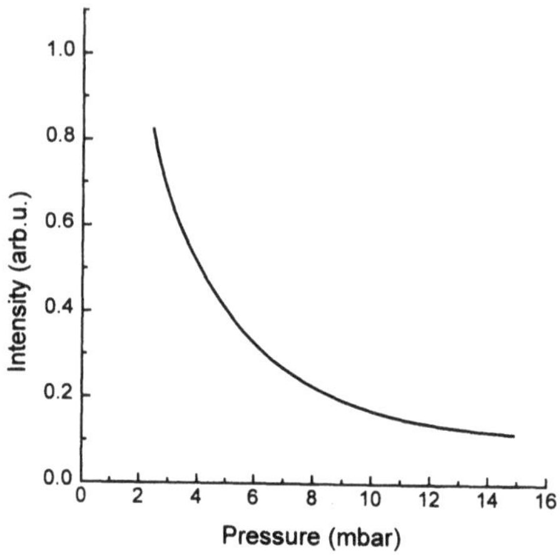

FIGURE 3.10. Dependence of the radiation intensity of rubidium on the krypton pressure: $r_0 = 6.5$ mm, $n_e = 10^{13}$ cm^{-3}, $T = 130$ °C.

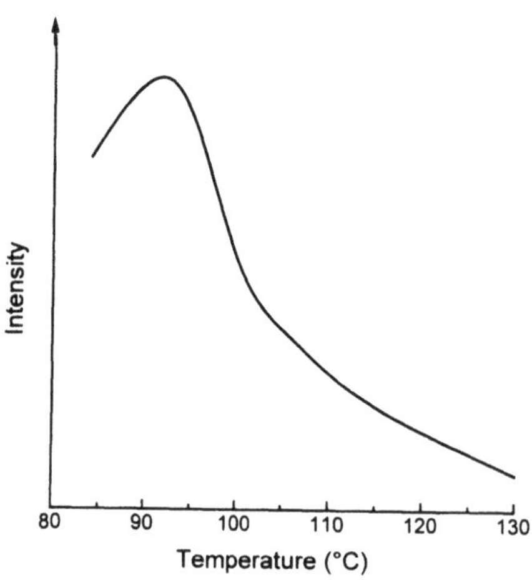

FIGURE 3.11. Dependence of the intensity of the rubidium D_2 line on the lamp temperature: $r_0 = 6.5$ mm, $n_e = 5 \times 10^{16}$ cm^{-3}, $p_{Kr} = 2.4$ mbar.

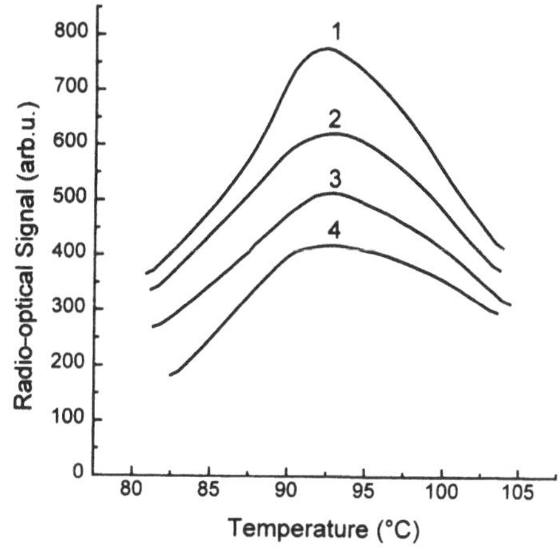

FIGURE 3.12. Dependence of the radio-optical resonance signal on the temperature of the lamp in the E-discharge mode at different power of the self-excited oscillator: 1, 1 W; 2, 0.8 W; 3, 0.75 W; 4, 0.63 W.

signal of resonance is observed in the same temperature range that was obtained by calculations for the highest intensity of the emitted spectral lines. The agreement of the calculated and observed values of the thermal and optical features confirms the applicability of the model for the description of the physical processes in the lamps.

Thus, the model gives the correct distribution of atoms throughout the volume, and even quantitatively the concurrence of the highest density of radiation and highest efficiency of optical pumping.

Figures 3.13 and 3.14 show the dependency of the radiation intensity I and ionic flow to the wall of the lamp on the lamp radius and gas pressure. The flow of ions, being the main reason for spectral lamp aging, depends strongly on the operating mode of the lamp. For example, an increase of the lamp radius from 10 to 20 mm decreases the flow by 2–2.5. The durability of the lamp is increased by the same factor. Increase in the pressure of the buffer gas from 1.3 to 4 mbar decreases the flow of ions by 6–6.5 times.

Some line shapes observed at peripheral and central parts at different lamp powers are presented in Figure 3.15.

When calculating the line shape with regard to the excitation of atoms of the inert gas based on the solution of Eq. (3.27), it turns out that a higher radiation intensity than in low-power discharges can be reached. At the same time, reabsorption begins at higher temperatures and concentrations of alkali atoms. From Figure 3.15 it can be seen that the line shapes in different parts of the lamp vary. Therefore self-absorption is much more present in contrast to low-power discharges. The width of the line as a function of the mode of the lamp varies from 600 to 900 MHz and more, where self-reversal begins.

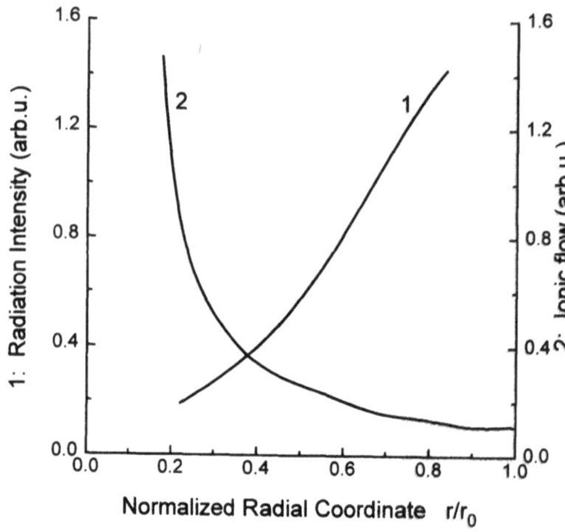

FIGURE 3.13. Dependence of the radiation intensity (1) and ionic flow (2) on the relative distance from the lamp's center at $n_e = 5 \times 10^{13}$ cm^{-3}.

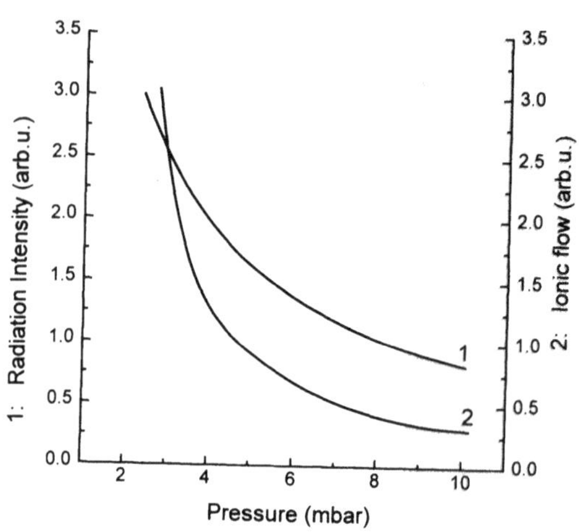

FIGURE 3.14. Dependence of the radiation intensity (1) and ionic flow (2) on the gas pressure.

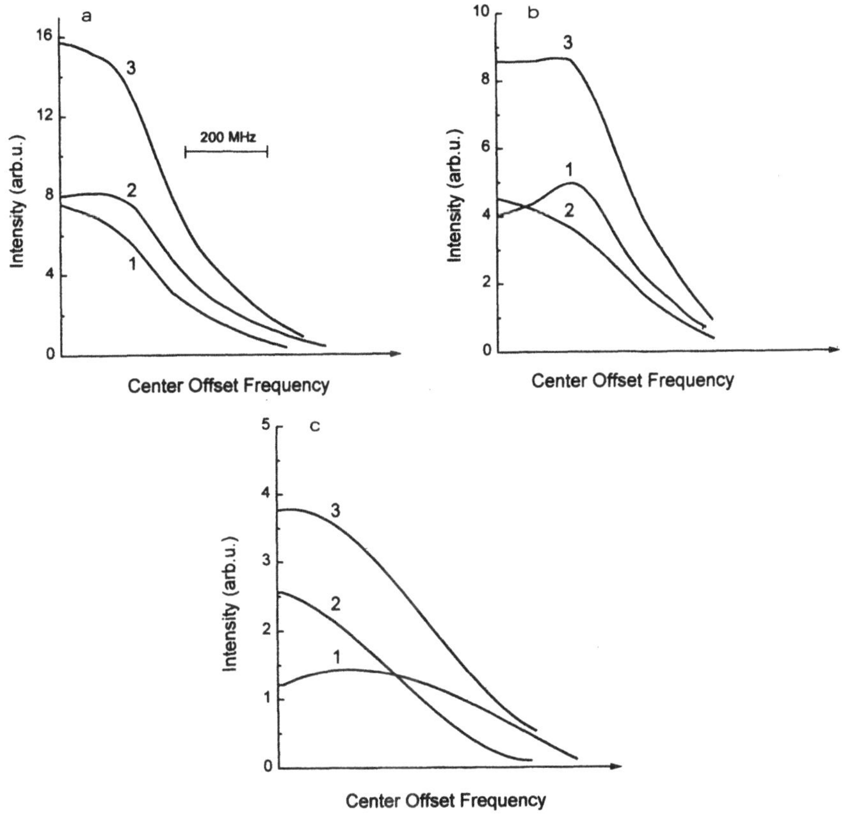

FIGURE 3.15. Shapes of the emission line from the peripheral (1) and central (2) parts of the lamp and the overall shape of the total light output (3): $R = 6.5$ mm, $T = 90$ °C, $n_e = 10^{11}$ cm^{-3}; (a) $W = 0.24$ W; (b) $W = 1.14$ W; (c) $W = 1.69$ W.

On increasing the concentration of electrons, the intensity of radiation decreases, and its maximum is shifted to the wall of the lamp. On increasing the temperature, the intensity grows and the area of light emission expands.

4

Spectral Characteristics of the Optical Radiation

4.1. Radiation Intensity

4.1.1. Radiation Intensity of High-Frequency Discharges in Inert Gases

The intensity of a spectral line is determined by the concentration of atoms in the excited state i, n_i and the probability of spontaneous emission to the lower level k, A_{ik} (see Chapter 3). On the other hand, the concentration of atoms in the excited state is determined by the concentration of atoms in the ground state, as well as by the concentration and temperature of the electrons, exciting these atoms by collisions.

In Table 4.1 the measured intensities I of krypton spectral lines (krypton is the most used gas in the hf EL in the modes of E- and H-discharges) are listed in comparison with some reference data [165].

The intensity was measured with the aid of a diffraction spectrometer and a photomultiplier. The error in the intensity measurement was no higher than 5%, while the measuring error of λ was no higher than 0.1 nm. The most intense lines in the investigated spectral region correspond to the excitation of and therefore to transitions from metastable states. As a whole the spectrum is not as rich in lines as is usually observed in gaseous discharges [166] and is mainly characterized by the excitation of low-lying levels. A rather weak dependence of the line intensity on the current through the generator, i.e., less than 5% was observed (except for the 587, 749, 758, 760, and 791 nm lines, showing intensity variations from 5 up to 20% when varying the current). A comparative analysis of the physical processes in the hf discharge and in the positive column of a low-pressure arc discharge shows that the physical conditions for the formation of spectral lines of atoms are similar. The differences are caused by the features in the distribution of plasma potentials and the higher efficiency of introducing energy into the discharge with the help of the hf fields.

To measure the integrated radiation intensity of Ne, Ar, and Xe lines the

TABLE 4.1. Intensities of Kr Lines for Different Regimes of E- and H-Discharges

	I from [165]	I [mW/cm^2]			
		E-discharge	H-discharge		H-discharge
		without rubidium vapor			with rubidium vapor
λ [nm]	[arb.u.]	100 mA	140 mA	200 mA	113 °C, 160 mA
427.4	1200	—	1.2	1.2	< 1
431.8	2400	—	1.6	1.6	1
557.0	2500	—	2.45	2.4	2
587.1	3000	—	3.8	4.3	< 1
749.4	50	—	0.28	0.33	< 1
758.7	1000	3.6	25	27	23
760.1	5000	13	90	96	99
768.5	1000	1.5	21	20	7
769.4	1000	2.1	11	11	10
774.6	150	1.5	0.38	0.35	1
785.4	800	2.25	26	25.5	14
791.3	200	< 0.75	0.78	0.52	1
805.9	1000	1.95	15	15	12
810.4	500	7.5	50	50	49
811.2	5000	19.5	204	211	140
819.0	3000	9	33	33	27
826.3	2000	4.5	42	45	39
829.8	5000	7.5	16	17	9.5

same setup as for the measurement of the ignition voltage of the discharge was used (see Section 2.2.1), but this time the radiation was chopped (80 Hz) and focused on a silicon photodetector (Figure 4.1). The photodiode was connected to the input of a phase-sensitive voltmeter. The power of the self-excited oscillator was sustained at a level of 7.0 ± 0.5 W by varying the current through the generator: the input resistance of the generator increased when the pressure was reduced. This rather high power was chosen in order to conduct measurements over a greater range of pressure without any change in the discharge mode. The efficiency of the self-excited oscillator (excitation of the H-discharge) was estimated to be $30 \pm 10\%$.

The experimental setup permits one to exclude the nonreproducibility of the sizes of lamps and their installation in the self-excited oscillator. Otherwise, with

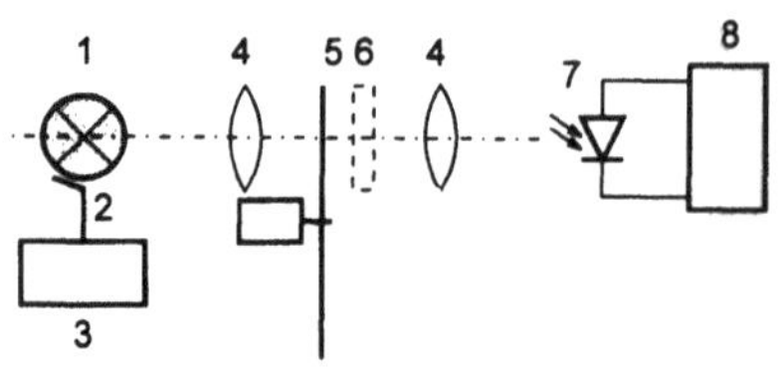

FIGURE 4.1. Experimental arrangement for recording the radiation intensity of the spectral lamps: 1, investigated lamp; 2, coil for excitation of the discharge; 3, generator; 4, lens; 5, chopper; 6, interference filter; 7, photodiode; 8, sensitive voltmeter.

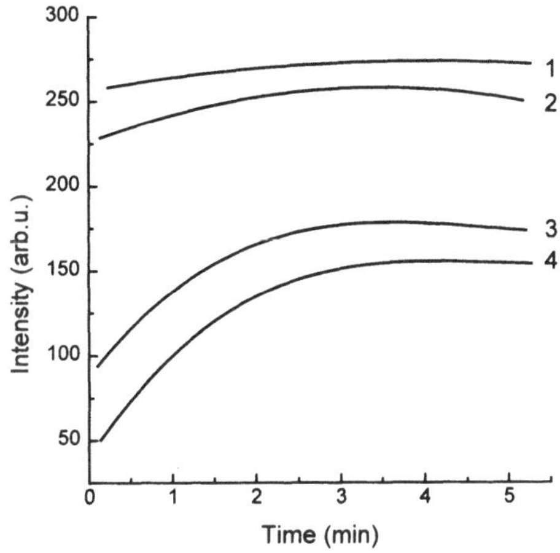

FIGURE 4.2. Dependence of the radiation intensity of the spectral lamps on the time for different pressures in the lamps: 1, $p =$ 0.4 mbar; 2, 0.5 mbar; 3, 1.7 mbar; 4, 3.1 mbar.

this arrangement a change in the concentration of atoms in the volume of the lamp could not be avoided because, due to warming-up of the gas during measurement, part of it leaves the volume. Depending on the pressure, this process takes several seconds to several minutes (Figure 4.2). The measured values of the radiation intensity of the inert-gas spectral lines as a function of pressure is given in Figure 4.3a–c, curves 1. The measurements were carried out after stabilization of the intensity and therefore for a constant gas pressure. The intensity decreases with decreasing pressure and a maximum can be observed between 0.1 and 0.7 mbar. The maximum intensity shows a slight decrease in the row of Ar, Xe, and Kr.

In addition to the experimental results, Figure 4.3 shows the pressure dependences of the integrated intensity, obtained by calculations [102] based on an earlier developed model (Chapter 3) of a discharge in a hf EL [100, 101] (curves 2). The model is based on the assumption of a Maxwellian velocity distribution of electrons, ambipolar diffusion in the discharge, and a self-consistent approach to the electromagnetic field and the discharge plasma parameters. Also the intensity distributions, electron temperatures, concentration profiles, population of levels, spatial distribution of temperature, and general power losses in the discharge have been calculated for different pressures and gases. It should be noted that actually the agreement between calculation and experiment is much better if we take into account that a constant pressure of the gas in the experiment and a constant concentration of atoms in the calculation were sustained [162]. The difference between the calculated dependence and the experimental one is no more than 3–5% (Figure 4.3), confirming the reliability of the model developed in [100]. In Figure 4.3d the calculated and measured dependences of the intensity of the continuous spectrum [102], emitted by hf ELs filled with krypton at a pressure of 0.7 mbar in the H-discharge mode are given. The calculation was based on the model pre-

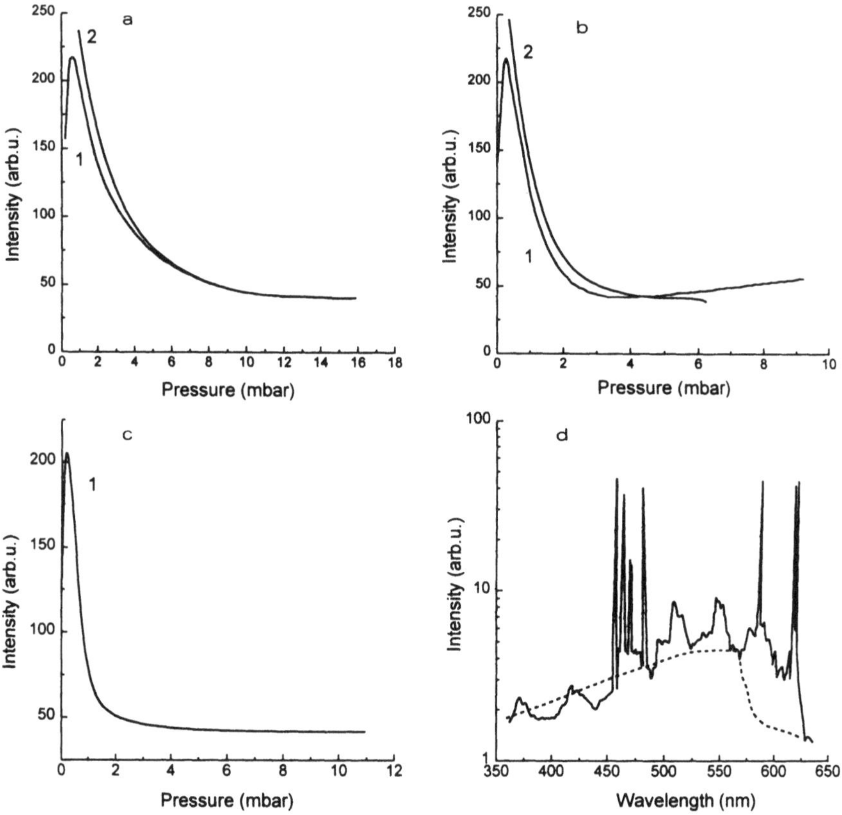

FIGURE 4.3. Variation in the relative radiation intensity for lines of inert gases with the pressure for different gases: 1, experiment; 2, calculation (not performed for krypton). (a) argon; (b) xenon; (c) krypton; (d) continuous spectrum with some Kr lines. Dotted line: calculated continuous spectrum.

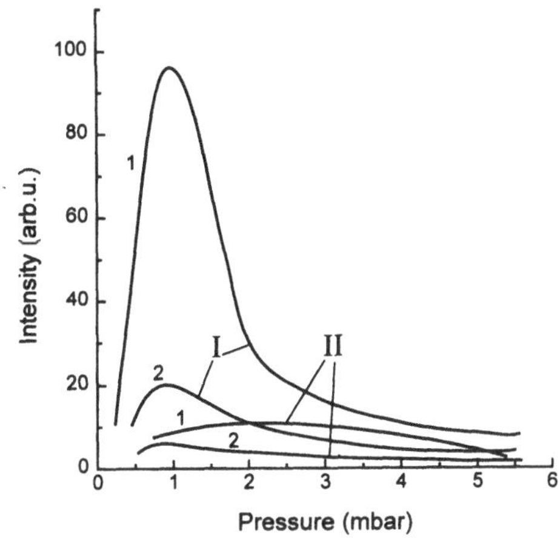

FIGURE 4.4. Dependence of the relative intensity of helium lines on pressure at a current through the self-excited oscillator of 210 mA (I) and 100 mA (II): 1, $\lambda = 438.8$ nm; 2, $\lambda = 667.8$ nm.

sented elsewhere [100], assuming that the continuous spectrum is produced by the recombination and bremsstrahlung of electrons, and shows satisfactory agreement with experiment.

Figure 4.4 shows the variation in the radiation intensity profiles of some helium lines due to a change of current through the generator. The dependence of current on the radiation is much stronger in helium than in krypton (cf. Table 4.1). This difference is connected, in our opinion, to a different efficiency of kinetic processes of excitation at different values of the specific power of the discharge. At high power of the hf oscillator, a restriction of the energy input into the plasma is possible, owing to the skin effect, and a growth of power does not necessarily cause a proportional growth of radiation intensity.

4.1.2. Intensity of Discharges in Metal Vapors

4.1.2.1. Hyperfine Structure of the First Rubidium Resonance Doublet

The most important metal filling used in hf ELs is rubidium, so we present in Figure 4.5 the hyperfine structure of the Rb D_1-($5\,^2S_{1/2} \leftrightarrow 5\,^2P_{1/2}$) and D_2-($5\,^2S_{1/2} \leftrightarrow 5\,^2P_{3/2}$) lines. The fine structure splitting of Rb is large, leading to well separated lines: $\lambda_1 = 794.76$ nm for the D_1-line and $\lambda_2 = 780.02$ nm for the D_2-line. Their theoretical intensity ratio is $I_2/I_1 = 2$. Natural Rb consists of two isotopes, having different nuclear spin quantum numbers: ^{87}Rb (27.8%), $I = 3/2$ and ^{85}Rb (72.2%), $I = 5/2$. Some data concerning these isotopes, including the hyperfine constants A and B of the levels involved, are given in Table 3.1.

The ground state, $5\,^2S_{1/2}$, as well as the upper state of the D_1-line, $5\,^2P_{1/2}$, splits into two hyperfine states, the upper level of the D_2-line, $5\,^2P_{3/2}$, splits into four hyperfine states. The splitting of the upper states is small compared with the

TABLE 4.2. Properties of Rb Radiation of a Discharge on Different Temperature[a]

T [°C]	$\frac{n_a(T)}{n_a(75°C)}$	I [arb.u.]	$k_0 l$	$\left(\frac{I_2}{I_1}\right)_{fs}$	$(k_0 l)_{fs}$	$\left(\frac{I_b}{I_a}\right)_{hfs}$	$(k_0 l)_{hfs}$	$k_0 l$ $n_e = 3 \times 10^{18}$ cm^{-3}
75	1	1						0.12
90	2.2	2.2		1.98	0.05			
100	4.5	4.5		1.85	0.4			0.6
105	7.2	7.0	0.1	1.75	0.8	1.2		1.2
110				1.56	2.2	1.2		2.1
115	13.5	10.5	0.75	1.50	2.4	1.2	4.0	3.0
120	19.0	13.0	1.2	1.40	3.2	1.16	5.0	3.6
125	30	16	5.3	1.28	5.8	1.15	5.3	6.6

[a] n_a is the number of Rb atoms, I the intensity, $k_0 l$ the optical density obtained from absorption measurements, $((I_2/I)_1)_{fs}$ the intensity ratio of the Rb D_2 and D_1 the lines, $(k_0 l)_{fs}$ the optical density obtained from $((I_2/I_1))_{fs}$, $((I_b/I_a))_{hfs}$ the intensity ratio of the Rb D_1 hyperfine component groups (cf. Figure 4.22), $(k_{0l})_{hfs}$ the optical density obtained from $((I_b/I_a))_{hfs}$. Additionally, $k_0 l$ is given for an electron density of $n_e = 3 \times 10^{18}$ cm^{-3}.

splitting of the ground state. Only the splitting of ^{87}Rb 5 $^2P_{1/2}$ is comparable with the Doppler width of the observed spectral lines and may be resolved under certain conditions using conventional, Doppler-limited spectroscopy (for example, high-resolution interference spectroscopy). Therefore the hyperfine structure of each electronic transition shows two groups of hyperfine components whose separation reflects, to a good approximation, the splitting of the ground state. The complete structures, taking into account the natural abundance, are shown in Figure 4.5a,b, together with a Doppler profile assuming $\Delta\nu_D = 800$ MHz. We label the high-frequency (short-wave) groups "a" and the low-frequency (long-wave) groups "b."

4.1.2.2. Dependence of the Radiation Intensity of Lamps with Rubidium Vapor on the Discharge Mode

As opposed to lamps with inert gases, the radiation intensity of lamps with metal vapors depends strongly on the temperature of the lamp. Table 4.2 contains results obtained for lamps of 13 mm diameter with a short glass seal, used to fasten the heat conductive holder to the thermostatically controlled surface, so the temperature of the vessel with Rb was about the same as the thermostat temperature. The power of the discharge was chosen to be 1.8 W when measuring the intensity of the lines, and 1.2 W when determining the intensity ratios between components of the fine and hyperfine structure.

In Table 4.2, the radiation intensity of the resonance line of Rb, $I(T)$, initially varies in proportion to the concentration of Rb atoms in the ground state, $n(T)$, when increasing the temperature of the lamp. Then, the growth of intensity is less than the growth of the number density. The reason for this is the re-absorption of radiation, which could be estimated by means of the intensity

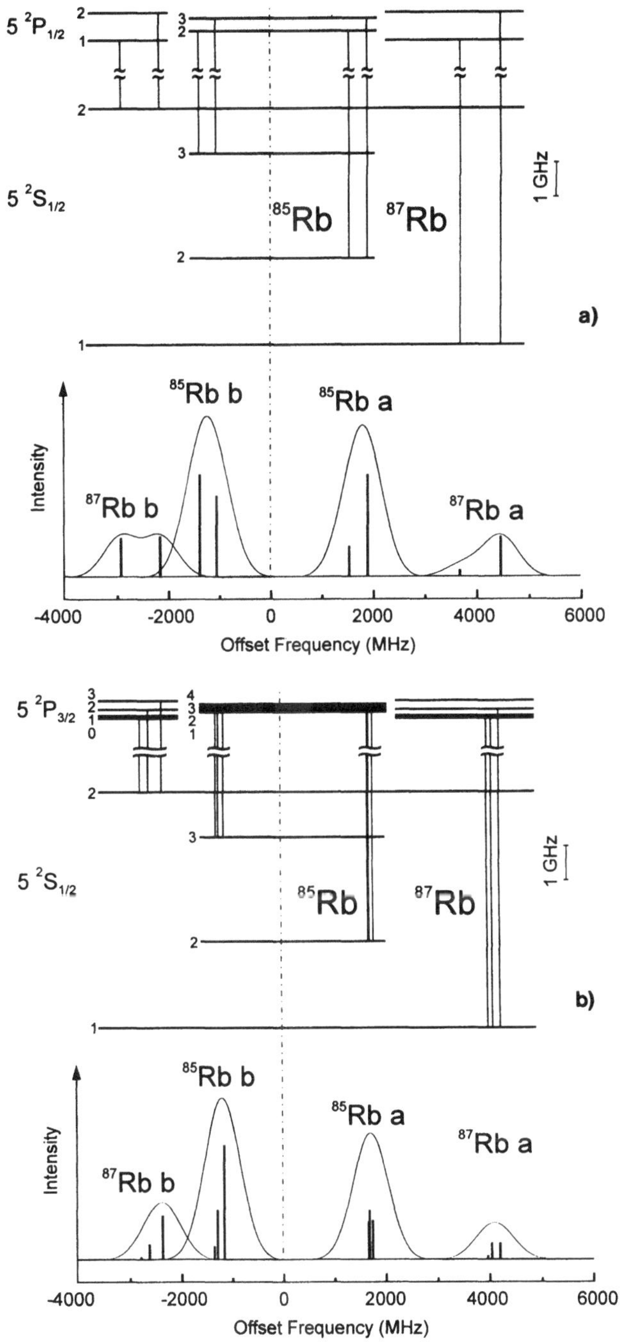

FIGURE 4.5. Hyperfine structure of (a) the Rb D_1-line ($\lambda = 794.76$ nm) and (b) the Rb D_2-line ($\lambda = 780.02$ nm) for the isotopes ^{85}Rb and ^{87}Rb. For the Doppler profile a FWHM of 800 MHz was assumed. Overlapping of neighboring hyperfine components has been shown for clarity only separately for each isotope.

the fine structure components $((I_2/I_1))_{fs}$ (5th column) and the hyperfine structure components $((I_b/I_a))_{hfs}$ (7th column), but can also be calculated using a model presented elsewhere [85]. Here, I_1 and I_2 are the intensities of the Rb D_1 and D_2 lines, while I_a and I_b are the intensities of the hyperfine structure groups, respectively. For the highest optical depth $k_0 l$ the results are close, but for the smallest optical depth the difference is significant.

The distribution of intensities of spectral lines of lamps filled with rubidium and krypton vapors, recorded by means of a spectrograph and a photomultiplier, within the range 400–830 nm [165] is given in Table 4.3, where I_Σ is the total intensity of all the spectral lines being recorded. The analysis of the data given in Table 4.3 shows an increase in the intensity of krypton and rubidium lines when increasing the power of the discharge. On increasing the temperature, the intensity of the second resonance doublet (420 and 421.5 nm; for these lines re-absorption is insignificant) grows in proportion to the concentration of Rb atoms. The intensity growth of the first resonant doublet is slower due to re-absorption at temperatures higher than 105 °C. The intensity of the krypton lines varies slightly up to this temperature, after which it decreases rapidly. At a power of the H-discharge of 4.5 W and a temperature of 105 °C about 30% of the radiation falls to the Rb resonance lines and, at 130 °C it is about 70%. When decreasing the power in the discharge down to 3 W, at temperature $T = 94$ °C, approximately 40% of the power within the wavelength range 400–830 nm falls to the resonance lines, at 120 °C 65%, and at temperatures higher than 137 °C more than 98% of the power is transformed into optical radiation of the resonance lines. In the E-discharge, despite a smaller power, the fraction of power transformed into the Rb resonance lines is lower: at $T = 83$ °C it is about 50%, at $T = 103$ °C about 70%, and at $T = 149$ °C about 92%. This indicates the difference between the processes of excitation in E- and H-discharges. The intensity ratio between the components of the fine structure of the head doublet, I_2/I_1, is 1.82 in the E-discharge at a temperature of 149 °C, and therefore close to the theoretical value ($I_2/I_1 = 2$) without re-absorption. This value considerably exceeds the corresponding ratios in the H-discharge (1.5 at a temperature of 105 °C and a current through the generator of 160 mA; 1.15 at 130 °C and 240 mA) and probably indicates the higher energy of electrons in the E-discharge, which predetermines the observable relation between the components of the fine structure of the head resonance doublet. The latter is also confirmed by the relatively higher intensity of lines with higher excitation energy (Table 4.3).

4.1.2.3. Dependence of the Radiation Intensity of Rubidium Lines on the Pressure and Type of the Buffer Gas

The variation of the intensities of the Rb spectral lines with the temperature of the lamp, and the power of the discharge, the pressure and type of gas was measured with a spectrograph and a photodetector connected to a microvoltmeter

TABLE 4.3. Intensities of Rb and Kr Lines for E- and H-Discharges at Different Temperatures[a]

	I [mW/cm^2]												
	H-discharge							E-discharge					
	temperature [°C]							temperature [°C]					
λ [nm]	94	110	118	128	137	150	158	83	98	103	109	149
[165]	$i = 140$ mA							$i = 100$ mA			$i = 180$ mA	
Rb 420.1	2	5	11	18	—	—	—	—	—	—	—	—
Rb 421.5	0.9	3	6	9	—	—	—	—	—	—	—	—
Kr 431.8	1.2	1.1	0.8	0.7	0.5	—	—	—	—	—	—	—
Kr 557	2.2	1.9	—	—	—	—	—	—	—	—	—	—
Rb 564.8	0.8	0.9	1	1.8	2	2	—	—	—	—	—	—
Rb 620.6	2.7	2.3	0.8	—	—	—	—	—	—	—	—	—
Kr 758	18	18	21	2	< 1	< 1	< 1	5.4	5.1	4.2	19	4.5
Kr 760	75	66	70	72	5	< 1	< 1	12.1	13	12.6	42	6
Rb 761.9	0.8	0.6	1	1.5	3	4	4	0.6	0.7	0.6	1	1
Kr 768.5	5	5	4.5	< 1	< 1	< 1	< 1	1.7	1.6	1.2	4.5	1
Kr 769.4	9	8	8	< 1	< 1	< 1	< 1	2.2	2.4	1.9	6.5	1
Rb 775.7	0.6	1.2	1	2	—	7	7.5	0.9	1.3	1.7	6	2
Rb 780	108	255	295	495	700	663	475	28	63	78	160	275
Kr 785.4	10	9	9.5	10	—	1.8	1.2	2.2	2.0	2.1	0.4	1
Rb 794.7	54	152	195	390	600	520	4200	18	35	42.6	93	150
Kr 805.9	7	7	7.5	6.6	2	1.5	1.5	1.8	1.8	0.7	5.7	1
Kr 810.4	30	27	27	25	7	< 1	< 1	7.5	8.4	6.6	21	0.6
Kr 811.2	123	111	110	105	21	< 1	< 1	19.2	20.4	19	48	10
Kr 819.0	26	27	23	22	2	< 1	< 1	8.1	7.8	7.5	21	4
Kr 826.3	40	40	39	38	3	< 1	< 1	4	5	4.2	15	3
Kr 829.8	9	10	9	7.5	< 1	< 1	< 1	6.9	7.8	7.2	33	5
I_2/I_1	2	1.7	1.4	1.3	1.2	1.3	1.4	1.5	1.8	1.8	1.7	1.8
$(I_1+I_2)/I_\Sigma$	0.4	0.6	0.6	0.8	0.98	0.99	0.99	0.5	0.65	0.7	0.6	0.92

[a] I_2/I_1 is the intensity ratio of the Rb D-lines, $(I_1+I_2)/I_\Sigma$ the intensity ratio of the resonance radiation and the overall emitted intensity.

input. The thermostat temperature was changed from 100 to 160 °C and the power of the self-excited oscillator from 1.5 up to 7 W, corresponding to a discharge power from 0.7 to 3.5 W. The construction of the light source is similar to that shown in Figure 1.1. The concentration of atoms was defined by the thermostat temperature, which was sustained with an accuracy of ±0.1 °C. The power consumed by the generator was stabilized with an accuracy of 1%. In order to determine the pressure dependency, lamps with different content of inert gas were placed in the self-excited oscillator. The daily reproducibility of the radiation intensity of the lamps was better than 5%. Systematic investigations were performed with lamps of 13 mm diameter filled with Rb vapor and an inert gas, i.e., argon, krypton, or xenon. Lighter gases, such as helium and neon, are inexpedient when used in ELs with metal vapors, as it is impossible to sustain a stable intense radiation due to diffusion through the glass and even due to reduction of pressure in the case of helium. For neon, there exists a difficulty in coupling the generator with the lamp. The range of pressure was chosen experimentally to insure a steady discharge mode. The frequency of the exciting field for all lamps was within the range 90–96 MHz.

The intensity variation of the first Rb resonance lines (780 and 794.76 nm) with the pressure of argon, krypton, and xenon at different temperatures is given in Figures 4.6–4.8. At rather low temperatures (lower than 110 °C) maximum intensity of Rb is realized at a gas pressure less than 2 mbar [25, 26, 167]. In this case the observed variations are close to those described in Chapter 2 and are approximated by the expression $I \sim A/p$, where A is a constant. However, on increasing the temperature up to 120–130 °C, a minimum intensity is observed at a pressure characteristic for each gas (about 4 mbar for argon, about 7 mbar for krypton, and about 15 mbar for xenon). On further increase in temperature, the intensity of these Rb lines grows and the character of the dependences changes. The intensity becomes almost constant within a wide range of the gas pressure. For xenon the maximum intensity is reached at 150 °C, for argon between 120 and 150 °C, and for krypton between 140 and 150 °C. The unexpectedly weak dependence of the highest intensity of the Rb lines on the gas pressure is explained by the possibility to achieve an optimum concentration of metal atoms and electrons by changing the temperature at each pressure. The maximum radiation intensities of Rb, when filling the lamps with different heavy noble gases, appear to be approximately the same with a slight decrease within the row Ar, Kr, and Xe.

Figure 4.9 shows the temperature dependency of the intensities of the Rb lines at different pressures as an example for Kr. It can be seen that a certain intensity can be reached at different pressures and for different temperature values. The temperatures which correspond to the maximum intensity, depending on the pressure and the sort of gas, are given in Table 4.4.

The observed dependency can be understood if one takes into consideration that the radiation intensity depends on the discharge parameters such as the con-

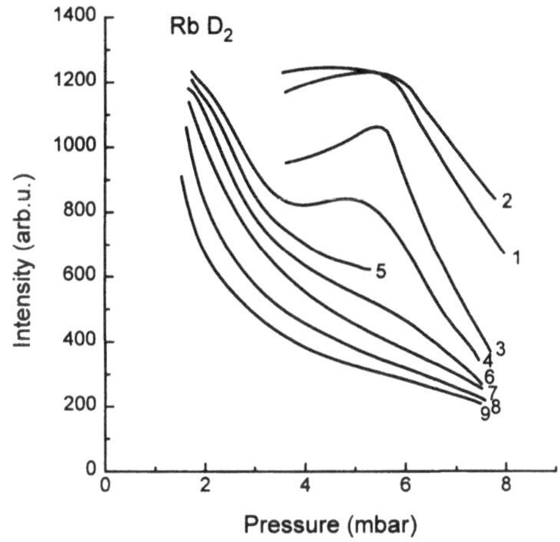

FIGURE 4.6. Dependence of the intensity of the Rb D_2-line, $\lambda = 780$ nm on argon pressure. 1, 150 °C; 2, 148 °C; 3, 143 °C; 4, 141 °C; 5, 131 °C; 6, 125 °C; 7, 116 °C; 9, 110 °C. $i = 250$ mA. (The D_1 line behaves similarly.)

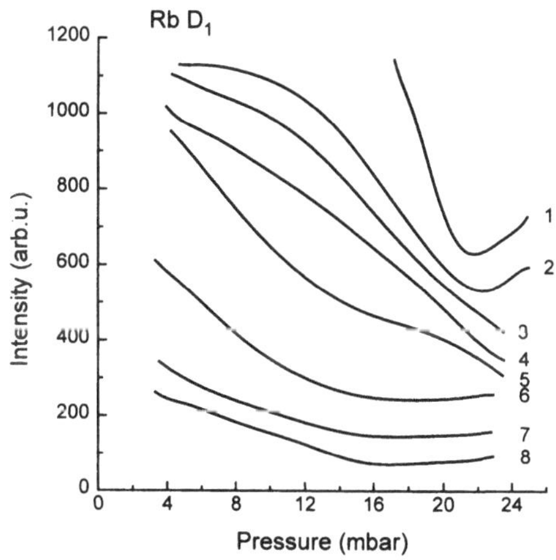

FIGURE 4.7. Dependence of the intensity of the Rb D_1-line $\lambda = 794.7$ nm on krypton pressure: 1, 154 °C; 2, 153 °C; 3, 152 °C; 4, 147 °C; 5, 143 °C; 6, 141 °C; 7, 131 °C; 8, 121 °C.

TABLE 4.4. Temperatures Corresponding to the Maximum Rb Intensity for Different Pressures and Different Gases

Ar		Kr		Xe	
p[mbar]	T [°C]	p [mbar]	T [°C]	p [mbar]	T [°C]
2.5	125	1.3	< 110	4	132
3.8	132	3.3	120	—	—
12	140	5.3	125	13	137
20	152	8	140	17	140

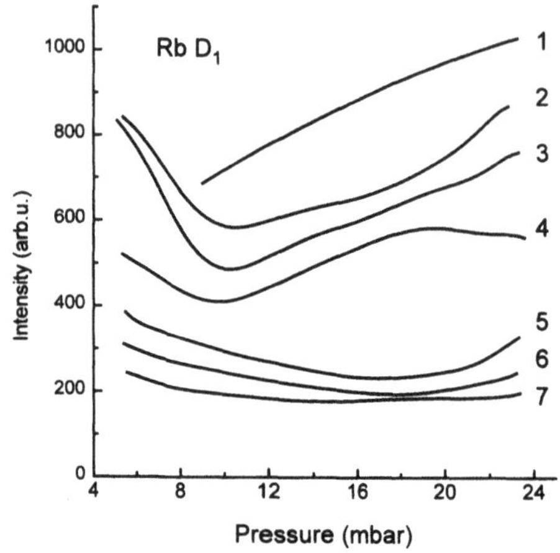

FIGURE 4.8. Dependence of the intensity of the Rb D_1-line $\lambda = 794.7$ nm on xenon pressure: 1, 147 °C; 2, 143 °C; 3, 131 °C; 4, 125 °C; 5, 121 °C; 6, 116 °C 7, 110 °C.

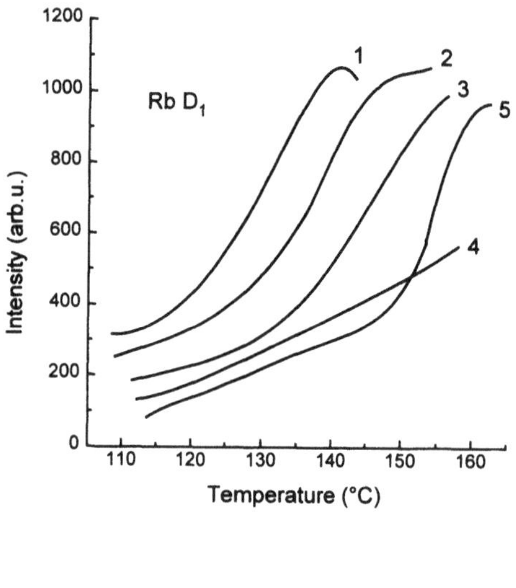

FIGURE 4.9. Dependence of the intensity of the Rb D_1-line $\lambda = 794.7$ nm on lamp temperature. Kr pressure: 1, 2.5 mbar; 2, 4 mbar; 3, 12 mbar; 4, 15 mbar; 5, 20 mbar.

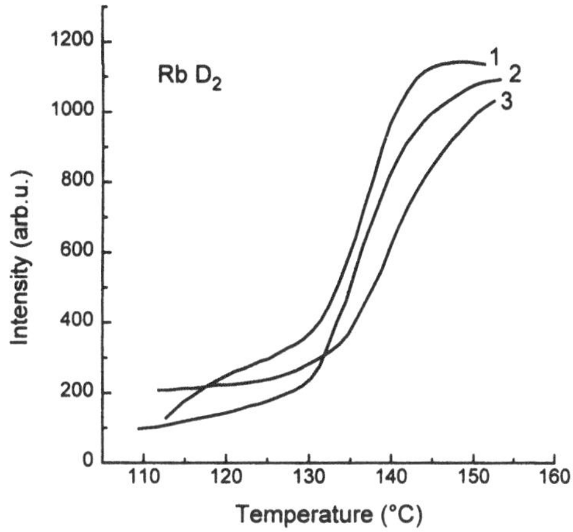

FIGURE 4.10. Dependence of the intensity of the Rb D_2-line $\lambda = 780$ nm on lamp temperature. Argon pressure 3 mbar. Current through the generator: 1, 160 mA; 2, 200 mA; 3, 250 mA.

centration of metal atoms and electrons in the volume. At small optical thickness the radiation intensity is proportional to the concentration of metal atoms in the ground state and to the concentration of electrons. The concentration of atoms is a result of a complex dynamic process, i.e., atoms are ionized in the volume and, as ions, diffuse to the walls of the lamp and recombine there. The inner surface of the lamp vessel with an excess of condensed metal is supposed to compensate the atoms which have already come in contact with the glass by means of evaporation. Since the speed of diffusion decreases with the increase in pressure, a higher temperature is required to sustain constant intensity.

In Figures 4.10 and 4.11, the dependencies of the Rb resonance line intensities are given as functions of the temperature at different currents through the generator (proportional to the power of the discharge), for Ar and Xe at different pressures. It is seen that with an increase in current, the dependency of the intensity on the temperature becomes similar to that for the pressure of saturated vapors (Xe, at 7 and at 13 mbar, cf. Figures 4.8 and 4.9). As a whole, the differences of the dependencies on the pressure and temperature, when changing the current, are not very pronounced at low temperatures, but when increasing the temperature, the highest intensity, specific for a given filling and regime, could be attained under a lower discharge power. For example, at $p = 2.5$ mbar and $i = 250$ mA, $T_{max} = 140$ °C; $i = 180$ mA, $T_{max} = 135$ °C; $i = 100$ mA, $T_{max} = 125$ °C.

The variation in the intensity of inert gas lines as a function of the temperature and power of the discharge is different qualitatively, i.e., the radiation intensity first slowly decreases with increasing temperature, but is quickly reduced when maximum Rb intensity is achieved (Figure 4.12).

As the discharge power increases the intensity of the inert gas lines grows. The reason for this effect is the decrease in the average energy of electrons with

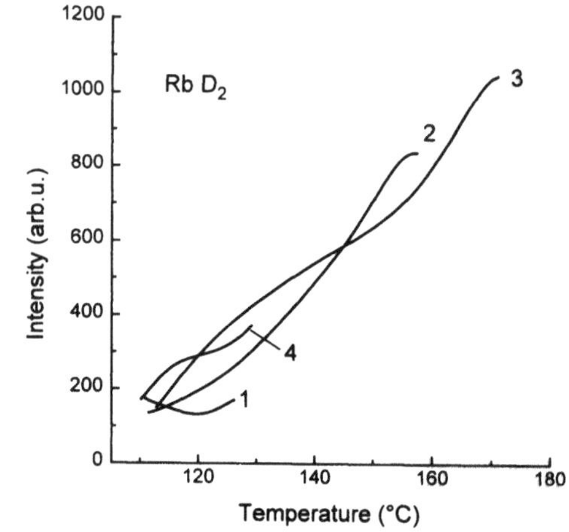

FIGURE 4.11. Dependence of the intensity of the Rb D_2-line $\lambda = 780$ nm on lamp temperature. Xenon pressure 13 mbar. Current through the generator: 1, 100 mA; 2, 150 mA; 3, 200 mA; 4, 250 mA.

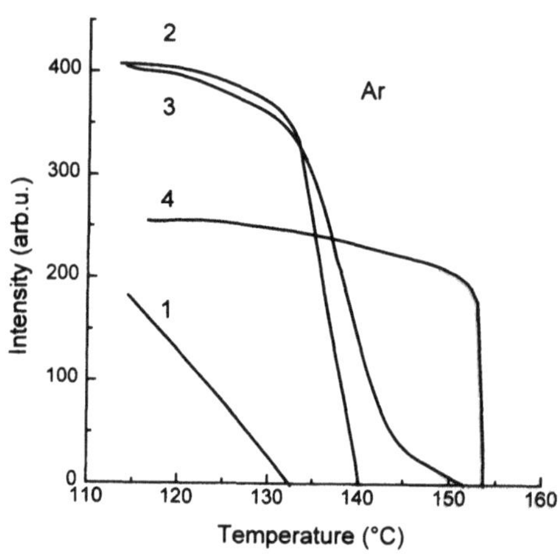

FIGURE 4.12. Dependence of the intensity of the argon line $\lambda = 811.2$ nm on lamp temperature. Argon pressure: 1, 1.3 mbar; 2, 4.4 mbar; 3, 7 mbar; 4, 10.5 mbar.

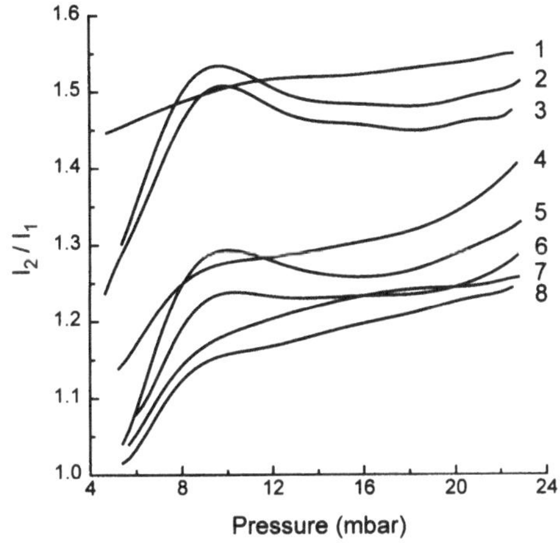

FIGURE 4.13. Dependence of the intensity ratio I_2/I_1 of the lines of the first Rb resonance doublet on Xe pressure. Temperature: 1, 116 °C; 2, 121 °C; 3, 125 °C; 4, 131 °C; 5, 141 °C; 6, 143 °C; 7, 147 °C; 8, 157 °C.

an increase in the concentration of Rb atoms.

Figure 4.13 shows the dependence of the intensity ratio of the first resonance doublet lines of Rb ($\lambda = 780$ and 794.76 nm) on the pressure of Xe for different temperatures. The ratio grows with increasing pressure in the H-discharge. This can be explained by the decrease in concentration of metal atoms, due to the reduction in the speed of their diffusion. When increasing the temperature, the ratio of the components of the fine structure doublet approaches 1, a consequence of the increase in the optical thickness of the absorbing layer near the wall of the lamp.

In contrast, in the E-discharge the intensity ratio of the hyperfine components remains approximately constant and is close to the theoretical value in all modes, but it slightly decreases at minimum temperatures. This different behavior of the intensity ratio illustrates the basic differences between the line formation processes for the two types of hf discharge. In the E-discharge it is possible to expect a manifestation of the potential of the plasma, which is much larger than in the H-discharge, and which increases the efficiency of excitation of the *P*-states (in comparison with destruction of these states when colliding with electrons) and re-absorption of photons.

Figure 4.14 shows the dependence of the intensity ratio of the Rb resonance lines and inert gas lines on the temperature. The ratio grows with increasing temperature, because in the H-discharge the intensity of the inert gas lines, after achieving the maximum intensity of the resonant lines, decreases quickly. At the maximum intensity more than 90% of the power, emitted within the wavelength range 400–900 nm, is represented by the resonance lines. For a further increase in temperature the intensity of the inert gas lines drops rapidly and a significant re-absorption of spectral lines appears. These processes result from the decrease

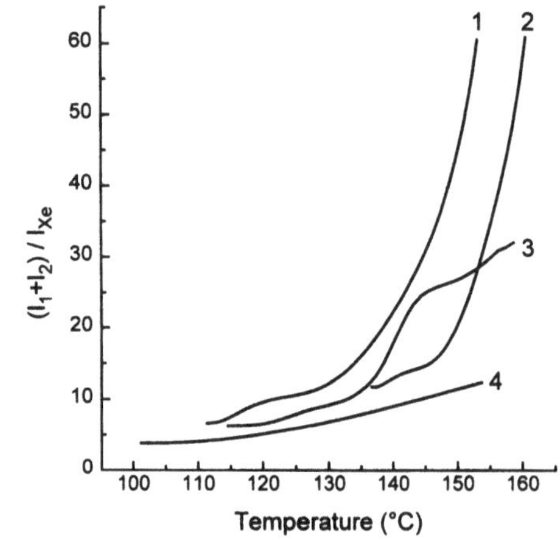

FIGURE 4.14. Dependence of the intensity ratio $(I_2 + I_1)/I_{Kr}$ of the rubidium and inert gas lines on temperature. Xe pressure: 1, 4 mbar; 2, 7 mbar; 3, 13 mbar; 4, 17 mbar.

in the average energy of the electrons and from volume recombination. The fact of rather fast reduction of the line intensity in the H-discharge can be used to stabilize the mode of the lamp [191]. In the E-discharge at maximal intensity, 60–70% of the whole power belongs to the resonance lines. The intensity of the inert gas lines decreases much slower than in the H-discharge. This also confirms the higher average energy of electrons for this type of discharge.

Figure 4.15 shows the dependence of the intensity of rubidium and krypton lines on the temperature for 100 mA current in an E-discharge. In this case maximum intensity is reached at $T = 95$ °C; this is significantly lower than in the H-discharge. Having analyzed the dependencies of the maximum intensity on the power W in the discharge and the corresponding vapor pressure of atoms (p_m), it

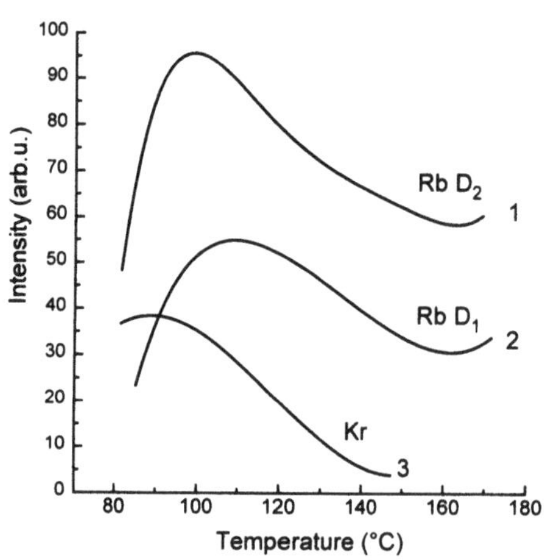

FIGURE 4.15. Dependence of the intensity of rubidium and krypton lines on temperature for the current through the generator of 100 mA. 1, $\lambda = 780$ nm; 2, $\lambda = 794.7$ nm; 3, Kr lines.

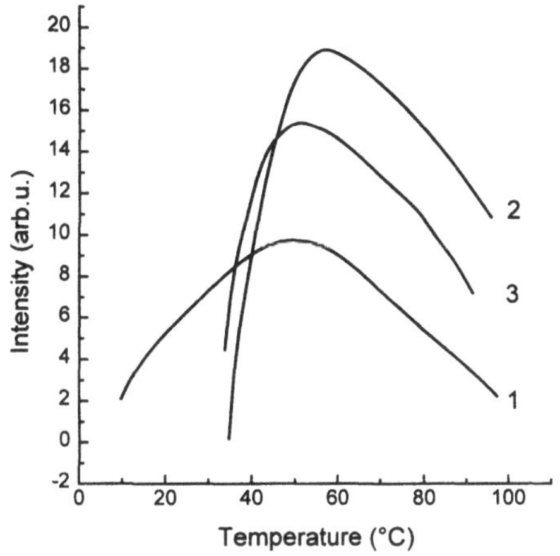

FIGURE 4.16. Dependence of the intensity of the mercury atomic line $\lambda = 253.7$ nm and the ionic line $\lambda = 194.2$ nm on temperature of the mercury reservoir. Current through the generator: 1, 100 mA (253.7 nm); 2, 120 mA (194.2 nm); 3, 160 mA (194.2 nm).

is possible to give an empirical relation between these parameters: $\log(p_m/1.33 \text{ mbar}) = 10^{-3} \times \log(W/1 \text{ W})$. This approximation is valid for lamps (1 cm^3 in volume) filled with rubidium vapor and krypton. In the case of other elements one should expect similar dependencies, but with slightly different parameters.

Figure 4.16 [5] presents the intensity of mercury radiation as a function of the thermostat temperature. It can be seen that the highest intensity of the resonance line (253.7 nm) is achieved between 50 and 70 °C [corresponding to vapor pressures of $1\text{–}4 \times 10^{-2}$ mbar]. At these temperatures a strong reduction in the intensities of the inert gas lines is observed (Figure 4.17). In the case of mercury vapor, rather intensive ionic lines have also been observed [5]. The highest intensity of these lines was detected at temperatures corresponding to a vapor pressure of mercury between 8×10^{-3} and 10^{-1} mbar with a maximum at a pressure of 3.5×10^{-2} mbar ($T = 60$ °C). Subsequently. as for the resonance lines of Rb, when increasing the current through the generator, the maximum intensity of ionic lines is reached at a larger concentration of atoms. The intensity variation of resonant and ionic lines of mercury with the current through the generator is given in Figure 4.18.

4.1.3. Integral Radiation Intensity of Lamps

The radiation intensity of a lamp is defined by the brightness of its surface:

$$dI = B(r, \vartheta, \varphi, r_0) dS_1 \tag{4.1}$$

where B is the brightness of a surface element dS_1, depending on the coordinates of the surface element (r, ϑ, φ) and the radius of the lamp r_0.

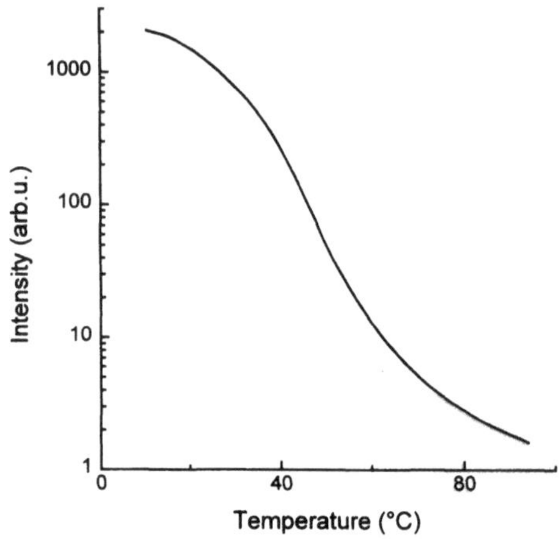

FIGURE 4.17. Dependence of the intensity of the krypton lines on the temperature of the mercury reservoir.

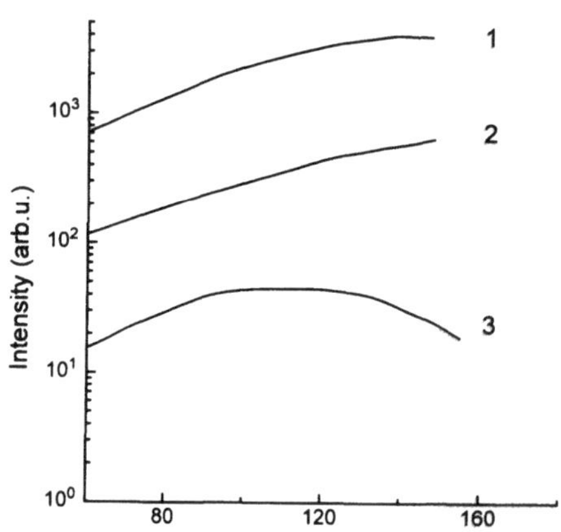

FIGURE 4.18. Dependence of the intensity of the Hg line $\lambda = 253.7$ nm (1) and of the mercury ionic lines, 185 nm (2) and 194.2 nm (3) on the current through the generator.

TABLE 4.5. Radiation Power W_{rad} and Brightness for Lamps of Different Sizes, Filled with Rubidium and Krypton Vapors[a]

$2R$ [mm]	W [W]	W/V [W/cm^3]	W_{rad} [W]	$\frac{W_{rad}}{W}$	N/t [photons/s]	$N/(tS)$ [photons/(s m^2)]
26[b]	40	4.4	8	0.2	4.2×10^{19}	2.0×10^{18}
13	7	6.1	0.51	0.073	2.6×10^{18}	6.4×10^{17}
13	4.2	3.7	0.20	0.048	1.4×10^{18}	3.2×10^{17}
13	2.8	2.4	0.14	0.05	0.7×10^{18}	1.7×10^{17}
10	2.4	4.6	0.1	0.042	0.7×10^{18}	2.2×10^{17}
6	3.8	34.5	0.2	0.052	1.4×10^{18}	1.2×10^{18}

[a] W is the generator power, W/V the generator power per volume unit, N/t the total photon flux per second, $N/(tS)$ the photon flux emitted per second and per surface unit.
[b] From [23].

Experimentally determined values for the radiation power and the brightness for lamps of different sizes filled with rubidium and krypton vapors, and for modes with maximum radiation intensity at different powers W are given in Table 4.5.

Absolute intensities in the characteristic modes for spherical and cigar-shaped lamps, which allow one to evaluate the highest power of radiation with regard to a known dependence of the radiation power on the operating mode and filling of a lamp, were measured. For lamps with Rb the highest total photon flux appeared to be 7×10^{17} photons/s. The appropriate values of radiation power are given in Table 4.5 and also elsewhere [90].

There exists less information about lamps with other fillings. In [24] some data for a lamp (diameter 30 mm) filled with cesium are given, where a total photon flux of 10^{19} photons/s was obtained (which was not yet the maximal level of the radiation output). For cesium lamps of diameter 10 mm the integrated intensity reached 9×10^{17} photons/s, corresponding to a power of the resonance lines of 0.12 W. The absolute radiation intensities of mercury lines from the unit surface measured elsewhere [5, 42] are given in Table 4.6.

TABLE 4.6. Absolute Radiation Intensity of Mercury Lines from the Unit Surface [5, 42] of a hf EL

λ	$N/(tS)$ [photons/(s cm^2)]		
[nm]	$i = 100$ mA	$i = 160$ mA	$i = 210$ mA
253.7	6.26×10^{16}	1.21×10^{17}	1.86×10^{17}
404.7	0.56×10^{16}	1.21×10^{16}	2.30×10^{16}
435.8	1.31×10^{16}	2.86×10^{16}	5.02×10^{16}
546.1	1.54×10^{16}	3.48×10^{16}	5.33×10^{16}
194.2[a]	10^{14}	10^{14}	10^{14}

[a] Ionic line.

The intensity of a lamp shaped like a pressed-in sphere is given in [81]. In such a lamp very high intensities (up to 30 mW) of the second resonance doublet of Rb were attained.

The experimental data show that the majority of lamps has an efficiency (ratio of the power of radiation to the power of the self-excited oscillator) between 4 and 7%. In [23], the highest efficiency, 25% for a lamp of diameter 26 mm and operated with Rb vapor, was reported. The correctness of this result was confirmed by Franz [24], where efficiencies exceeding 15% in a lamp of the same diameter, filled with cesium, were obtained and where the possibility of a further increase in efficiency was pointed out. With regard to the efficiency of the self-excited oscillator, which is usually about 40–60% of the efficiency of the power transformation, the efficiency of lamps with diameter 6–13 mm is 6–28%, and more than 50% for lamps of diameter 26 mm.

The intensity of the mercury ionic line $\lambda = 194.2$ nm was 2×10^{14} photons $cm^{-2}\ s^{-1}\ sr^{-1}$ for a lamp of diameter 15 mm and excitation power 20 W [5]. Thus, only 0.006% of the power supplied to the lamp is emitted in this ionic line.

When reducing the radius of the lamps, at first the brightness decreases and then grows. However, when reducing the size of the lamp the H-mode requires higher effective power, and the achievement of an appropriate thermal mode could become a problem.

Only few data exist for light sources operated in the mode of an E-discharge. The integral radiation power of evacuated cylindrical lamps in the mode of an E-discharge is 5 mW for a consumed power of 0.6 W, and for lamps of diameter 5 mm it is equivalent to 0.3 W. A study of spherical lamps in the E-discharge mode has shown that the maximal intensity is about a factor of 5 less than in the H-discharge mode for the same temperature of the lamp. The main advantage of light sources based on E-discharges is the opportunity to receive an intense radiation at a lower temperature and power of the discharge. This makes it possible to develop technical solutions for compact and economical devices with low power of the discharge and rather low working temperature. Light sources in the E-discharge mode possess a high efficiency in the transformation of power delivered to the discharge into radiation power, which theoretically attains values of 50–90%. However, the real efficiency appears to be significantly lower (7–20%) with regard to a generator efficiency between 20 and 50%.

When reducing the lamp diameter from 13 mm to 6 mm at the same pressure of 2.5 mbar, the radiation power is decreases, while the brightness of the lamp grows (Table 4.7).

The maximum emitted power depends slightly on the pressure of the gas (see Section 4.2.2). As follows from Table 4.7, the highest brightness is reached when reducing the size of the lamp. Unlike in the H-discharge, the lamps in the E-discharge mode work stably over a wide range (including those which are substantially less stable in the H-discharge).

TABLE 4.7. Radiation Power and Brightness for Lamps of Different Sizes, Filled with Mercury and Krypton Vapors

$2R$ [mm]	W [W]	W/V [W/cm^3]	W_{rad} [W]	$\frac{W_{rad}}{W}$	W_{rad}/S [W/cm^2]
13	0.8	0.8	0.08	0.1	0.025
10	0.8	1.6	0.06	0.08	0.038
6	0.8	4.0	0.05	0.06	0.060

In conclusion, the shape of a hf EL influences weakly the brightness of its surface. Therefore, the shape of the lamp can be chosen to meet technological and constructive requirements. A decrease in the distance between the walls of the cylinder of the lamp requires higher concentration and temperature of electrons to sustain the discharge, and causes a growth in brightness of the spectral lines.

4.1.4. Experimental Study of the Spatial Distribution of Atoms in Ground and Excited States in the Plasma

For typical concentrations of electrons in the ED the theory predicts a nonuniform spatial distribution of atoms of easily ionizable elements (e.g., alkalis) throughout the volume of the lamp (see Section 3.2).

In experiments, the observed radiation originates from different areas within the lamp bulb. From the observed intensity, the distribution of atoms throughout the volume of the lamp was calculated with the aid of Abel's transformation. For a reliable determination of the distribution of atoms, the line under study should not be re-absorbed. Hence, the conversion from the measured transversal intensity distribution $b(x)$ to the radial intensity distribution $\varepsilon(r)$ is reduced to solving an Abel's integral equation, which mathematically is a Fredholm equation of second order:

$$b(x) = 2\int_x^{r_0} \frac{\varepsilon(r)rdr}{\sqrt{r^2 - x^2}} \tag{4.2}$$

The solution has the form

$$\varepsilon(r) = -\frac{1}{\pi}\int_x^{r_0} \frac{\frac{db(x)}{dx}dx}{\sqrt{r^2 - x^2}} \tag{4.3}$$

where r_0 is the radius of the boundary of the discharge.

This problem relates to the class of incorrect problems. Since the real experimental curve is always known with limited accuracy, a set of theoretical curves $\varepsilon(r)$ exists, that fits the experimental profile of brightness $b'(x)$. The problem consists of finding a selection algorithm for a set of solutions of Eq. (4.2).

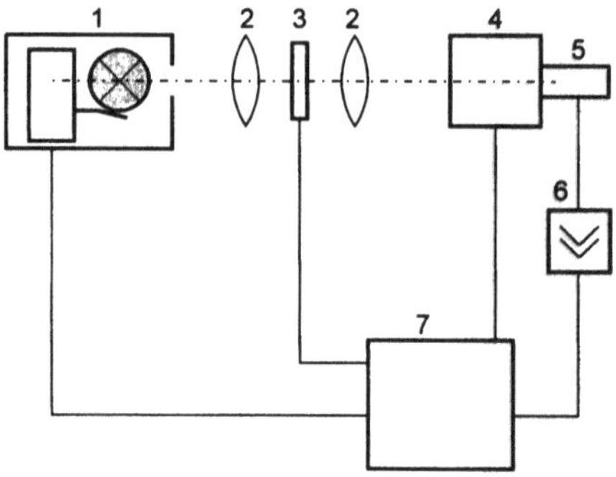

FIGURE 4.19. Block diagram of the experimental setup: 1, light source under research being moved transversely to the optical axis by means of a step-motor device; 2, lens; 3, closing device; 4, diffraction spectrograph; 5, photomultiplier tube; 6, amplifier; 7, microcomputer.

It can be obtained by using a regulatory method, which is reduced to smoothing six neighboring points of the experimental curve $b(x)$ with the aid of an approximating polynomial followed by calculation of the integral (4.3).

Since specific algorithms for the profiles are absent, the statistically indiscernible groups of profiles have been specified and the radial transformation for every profile from the group has been undertaken. The result was averaged and a root-mean-square deviation from the average was evaluated. This deviation $s(r_k)$ characterizes the "band" of errors of the result and is evaluated for the formula

$$s(r_k) = \sqrt{\frac{\sum_{i=1}^{n} [\overline{\varepsilon}(r_k) - \varepsilon_i(r_k)]}{n(n-1)}} \tag{4.4}$$

where $\overline{\varepsilon}(r_k) = \frac{1}{n}\varepsilon_i(r_k)$, r_k is the point of the radius, i the number of an individual structure, n the number of averaged structures, and $\varepsilon_i(r_k)$ the value of the ith radial profile of brightness at the point r_k.

The experimental setup on which the measurements were carried out was an automatically recording spectrometer arrangement (Figure 4.19). It included a diffraction spectrometer, a control computer, a display module and interface circuits. This device not only permits one to measure the radiation intensity, to record the spectrum in the range 300–830 nm, and to measure the radiation stability, but also to record the transversal intensity distribution of the radiation and to determine the distribution of atoms at a certain level throughout the volume of the lamp.

Figures 4.20–4.22 show the cross structures of the brightness and the result of their radial transformation for nonre-absorbed lines of Rb and Kr combined in appropriate groups.

Despite a large scatter of results, in the function $b(x)$ close to the center of the lamp a nonmonotonic part of $b(x)$ can be found in almost all plots. This phenomenon cannot be explained by statistical errors, but results from the deviation in lamp symmetry from axial symmetry, caused by the tank with condensed metal.

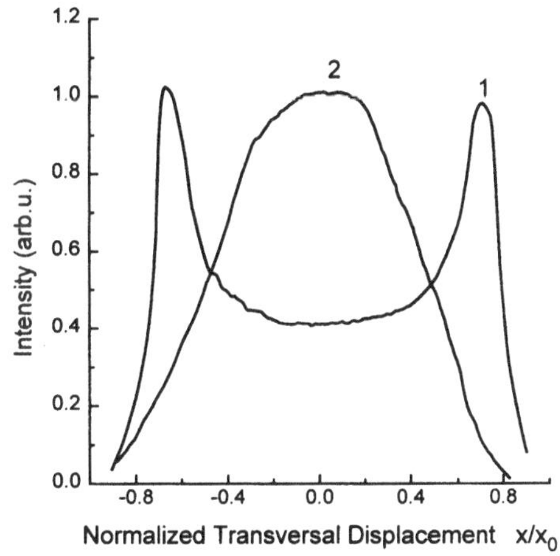

FIGURE 4.20. Transverse distribution profiles of the brightness of lamp emission. Filling of the lamp Rb + 4 mbar Kr: 1, distribution profile of rubidium lines; 2, distribution profile of krypton lines.

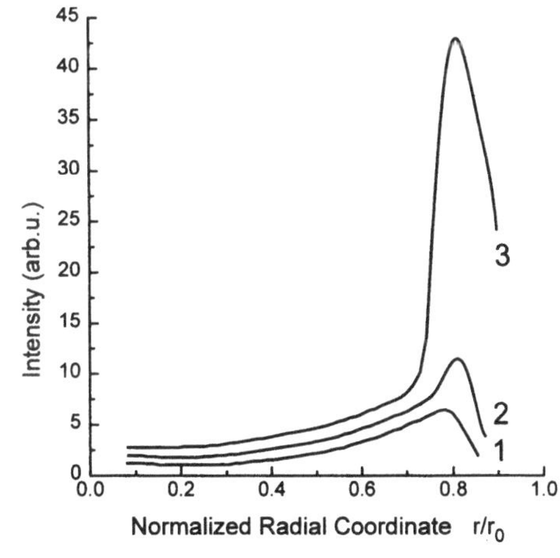

FIGURE 4.21. Distribution of excited Rb atoms throughout the volume of the lamp as a function of the mode of the discharge: 1, $T = 88\,°C$; 2, $T = 101\,°C$; 3, $T = 124\,°C$. The power of the self-excited oscillator is 2.9 ± 0.2 W.

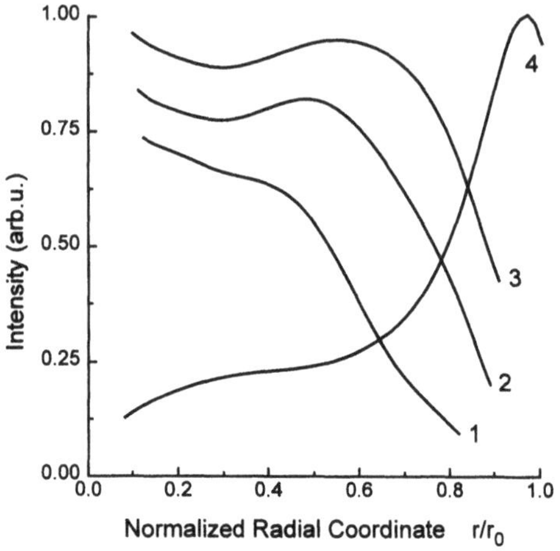

FIGURE 4.22. Distribution of krypton atoms in a lamp filled with Rb and Kr throughout the volume of the lamp as a function of the current through the generator: 1, 100 mA; 2, 140 mA; 3, 200 mA; 4, 240 mA.

At higher pressure the maximum intensity was observed in the center of the lamp (Figure 4.22). This can be explained by volume recombination, which increases when the pressure of the buffer gas is increased.

Thus, a study of the radial distribution confirmed the very inhomogeneous distribution of atoms of easily ionizable elements throughout the volume of the lamp. A comparison with the calculated distribution in the absence of spatial recombination (i.e., at a pressure of the inert gas lower than 5 mbar), with allowance for significant uncertainty of some parameters (mainly the electron concentration and temperature), shows the adequacy of the discharge model for real processes in the mixture of vapors of the metal and the inert gas (see Section 3.2).

4.2. Shape of Radiation Lines of Electrodeless Spectral Lamps

4.2.1. Experimental Studies of the Profiles of the Emitted Spectral Lines by Means of a Fabry–Perot Interferometer

The study of the profile of an emission spectral line requires the use of a device with high spectral resolution. One of the best optical devices with a high resolution power is the plane Fabry–Perot interferometer, which was used to obtain most of the results mentioned here. The Fabry–Perot interferometer consists of two plane-parallel plates coated with mirror layers possessing the required high reflectivity. When an axially-symmetric light beam falls on these plates, an interference pattern of concentric fringes appears in the focal plane of an objective lens. The radiation of any gas-discharge light source contains many emission lines. To avoid superposition of different fringe systems, the line under research is isolated with the help of a monochromator. When changing the optical thickness τ between the interferometer plates in one or another way with time, where $\tau = nd$ (n is the refraction index, d is the plate spacing), the fringes swell or shrink and, after passing the radiation through a narrow diaphragm positioned in the center of the interference fringe system, it is possible to record the time dependence of the amplitude of the signal with the help of a photomultiplier. The time scale can be easily calibrated into a scale of wavelengths.

The resolution of this interferometer exceeds substantially the resolution of other high resolution devices. The free spectral range is given by

$$\Delta\nu = \frac{c}{2nd} \tag{4.5}$$

The resolvable frequency difference $\delta\nu$ is determined by the finesse F (which principally is defined by the ratio $F = \Delta\nu/\delta\nu$):

$$F = \frac{\pi\sqrt{R}}{1-R} \tag{4.6}$$

where R is the reflexivity of the interferometer plates, and the resolvable frequency difference is given by

$$\delta\nu = \frac{\Delta\nu}{F} = \frac{c}{2nd}\frac{(1-R)}{\pi\sqrt{R}} \tag{4.7}$$

It is not difficult to obtain that for $n = 1$, $d = 10$ mm, and $R = 0.96$ the theoretical resolution is $\delta\nu = 200$ MHz, and for $R = 0.90$ the resolution is $\delta\nu = 500$ MHz.

Usually these values are slightly larger due to imperfections of the plate surfaces and range practically from 300 to 800 MHz in various experiments, as can be checked with the help of a laser of appropriate wavelength. This resolution is comparable with the hyperfine splitting of the first excited state $5\,^2P_{1/2}$ of ^{87}Rb atoms (Figure 4.5) and does not really allow the width of the hyperfine components of the emitted spectral lines to be measured. However, knowledge of the true line shape is not only necessary to study the light sources, but also for solving applied problems, primarily, for modeling the process of optical pumping. Therefore, in order to obtain reliable spectral information, a reduction in the instrumental width of the interferometer is necessary, either by increasing the distance between the mirrors, by magnetic scanning techniques or by computer processing of the interferogram.

In [168] an interferometer with a free spectral range smaller than the hyperfine splitting of the ground level (6.8 GHz) was used to investigate Rb lines. The interferometer had a mirror spacing of 40 mm. Therefore the feasible resolution reached 250 MHz, but the free spectral range was 4.3 GHz. Evaluating the overlapping interference orders was rather time-consuming work, but it allowed one to determine that the width of the lines of the hf EL radiation within the investigated mode was not larger than the Doppler width.

In order to measure the width of the lines it is also possible to use the method of magnetic scanning (see Chapter 10). The instrumental function in this case is defined by the width of the hyperfine components. It was about 400 MHz in an experiment described elsewhere [69, 169]. An advanced technique of magnetic scanning with computer processing of the results, as shown in [169], permits one to process the line shape and to determine its parameters with high reliability.

A method suggested and applied in [55], in which a conventional Fabry–Perot interferometer was used with subsequent computer processing, appeared to be very promising. Its correct use requires reliable knowledge of additional information about the instrumental line shape and parameters of the interferometer.

The best results were obtained by using a complex Fabry–Perot interferometer (multiplex), based on the combination of two interferometers of different thickness [55]. In this case the resolution of the interference pattern is defined by the "thick" interferometer and the free spectral range by the "thin" one. The spacings should have a ratio m/n, where m/n is a rational fraction. A disadvantage of this instrument is the necessity to strictly adjust the device and its low light transmission. In [55] the ratio of the thicknesses of the multiplex was 1 : 8. The width of the instrumental transmission function was determined by the 64-mm interferometer (about 140 ± 20 MHz), measured with the aid of a He–Ne laser. The free spectral range was determined by the 8-mm interferometer, 18,750 MHz, and therefore high enough to avoid overlapping of interference orders when recording the hyperfine structure. For deconvolution of the Doppler part of the broadening, the interferograms were processed by a computer that allowed the resolution to be increased and the instrumental function to be overcome.

4.2.2. Experimental Techniques

For our experimental studies a hf EL of spherical shape with diameter 13 ± 0.5 mm, filled with krypton at a pressure of 4.3 ± 0.3 mbar and Rb isotope vapors, was chosen. The lamp had a lengthened branch, placed in a thermostat to collect the surplus metallic Rb. This branch was the coldest part of the lamp during excitation of the discharge. The temperature in the thermostat was sustained with an accuracy of 0.5 °C.

A hf H-discharge was excited in the lamp with the help of a self-excited oscillator. The frequency of the exciting field was 95 ± 5 MHz, the current in the generator 120–180 mA, corresponding to a discharge power of 1–3 W, and the range of thermostat temperatures 80–140 °C. At hf capacitive excitation the current through the generator was 70–120 mA and the range of thermostat temperatures was 80–120 °C.

The D_1-line of Rb, excited in the source, has a hyperfine structure given in Figure 4.5a. This structure consists of four hyperfine components for each of the two stable isotopes arranged in two subgroups: a short-wave one (group a) with an intensity ratio of the hyperfine components for ^{87}Rb (^{85}Rb) of 50 : 10 (35 : 10) and a long-wave one (group b) with an intensity ratio of the hyperfine components 10 : 10 (8 : 10). As mentioned above, the emission intensity shape has a Voigt profile with a small Lorentzian constituent. The profile of the components is broadened for different reasons, but mainly by the Doppler effect. The average value of the temperature of the atoms along the axis of the lamp under the given discharge conditions is about 450–500 K. The Doppler broadening can be determined by Eq. (3.12). At the temperatures mentioned above this width is up to $\sim$ 500 MHz.

The second reason for the line broadening is the pressure broadening by collisions with the buffer gas in the lamp. Under the conditions mentioned above the buffer gas causes a contribution of the Lorentzian constituent having a width not exceeding 80 MHz [97].

The broadening due to the pressure of the Rb vapor itself can be neglected under the given conditions, as the vapor pressure of Rb is several orders smaller than the pressure of the buffer gas. The natural width of the hyperfine structure components is much smaller than the pressure broadening. The last two kinds of broadening have a small effect on the central part of the shape of the hyperfine structure component.

These evaluations of the width of the Doppler and Lorentzian constituents of the Voigt profile show that almost full overlapping of the hyperfine structure components (having for the ^{87}Rb frequency separations of 818 MHz due to hyperfine splitting of the upper state $5\,^2P_{1/2}$) occurs in the light source. The total hyperfine structure splitting of the D_1-line of this isotope amounts to 7647 MHz, which surpasses the hyperfine splitting of the ground state $5\,^2S_{1/2}$ by only 11.9%.

The line shape of the Rb D_1-line, depending on the temperature of the lamp and the power of the discharge, was studied in particular [69, 170]. A block diagram of a typical experimental installation is presented in Figure 4.23. The radiation of the spectral lamp was focused on the entrance slit of a monochromator and then into a Fabry–Perot interferometer. The distance between the mirrors of the Fabry–Perot interferometer was either 8 or 10 mm, and the reflectance of the dielectric mirrors was 92%. The transmitted radiation was detected by a photomultiplier with subsequent amplification and recording. The spectrum was scanned with the help of a special system, varying the air pressure in the interferometer; the reproducibility and linearity of the records were no worse than 2%.

When modeling processes in hf ELs with metal vapors it was shown that radiation from different parts has various spectral profiles, due to the nonuniform spatial distribution of the metal atoms at certain generator modes. For discharges in a mixture of Kr and Rb the rubidium atoms emit their radiation from the peripheral

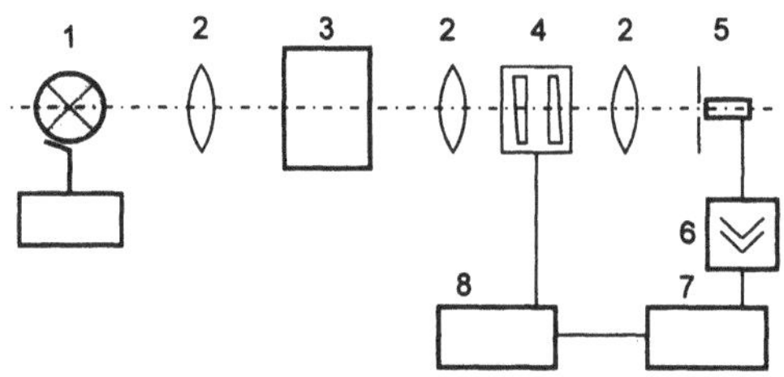

FIGURE 4.23. Block diagram of the experimental setup for the study of spectral characteristics: 1, light source under research; 2, lens; 3, monochromator; 4, Fabry–Perot interferometer (pressure scanned); 5, pinhole for selecting the fringe center with a photomultiplier placed behind; 6, direct current amplifier; 7, two-channel recorder; 8, synchronizing device for pressure scan.

areas of the lamp while the krypton atoms emit from the center. Therefore, for each type of discharge the dependence of the spectrum on the thermostat temperature was investigated for different currents through the generator and for different sections of the lamp. When selecting the required sections of the lamp special diaphragms were used:

(1) H-discharge: Two round diaphragms (2.5 mm in diameter) cutting out two areas of the source—the center or the periphery (the center of the diaphragm was situated at a distance of 5.5 mm from the center of the lamp).

(2) E-discharge: Three diaphragms of diameter 1.5 mm located along the radius of the lamp, the first placed at the lamp center.

The lamps show a strongly nonuniform distribution of Rb atoms along the radius of the lamp while the processes of self-absorption and self-reversal, which differ in different sections of the bulb, strongly deform the shape of a single line. Under these conditions the description of the line shape in analytical form is not simple, and the main evident and measurable value is the linewidth of an isolated hyperfine structure component of the line under study. This full width at half maximum (FWHM) was measured only in those cases where the line profile was not appreciably re-absorbed. The error in determining the width of the hyperfine component was limited by the nonlinearity of the air flow through a capillary into the chamber of the interferometer, by errors in the evaluation of the records, as well as by the instrumental function and the free spectral range of the interferometer. The total error was no more than 3.5%.

The observable shape of the emission line, $f(\nu)$, is a convolution of the instrumental function $a(\nu)$ and the true line shape $v(\nu)$: $f(\nu) = a(\nu) \otimes v(\nu)$. Thus, to obtain the true width of the line shape it is necessary to invert this equation, what cannot always be done analytically. An empirical evaluation with the help of a single-frequency He–Ne-laser gave the instrumental width $\Delta\nu_{ap} = 900 \pm 100$ MHz. Buger and Van-Sittert [171] suggest a full linewidth (of the Voigt profile) $W = 0.64\Delta\nu + 0.98\Delta\nu_{ap}$ when $1 < \Delta\nu/\Delta\nu_{ap} < 3$. As the latter condition is fulfilled in our measurements, this relation was employed subsequently.

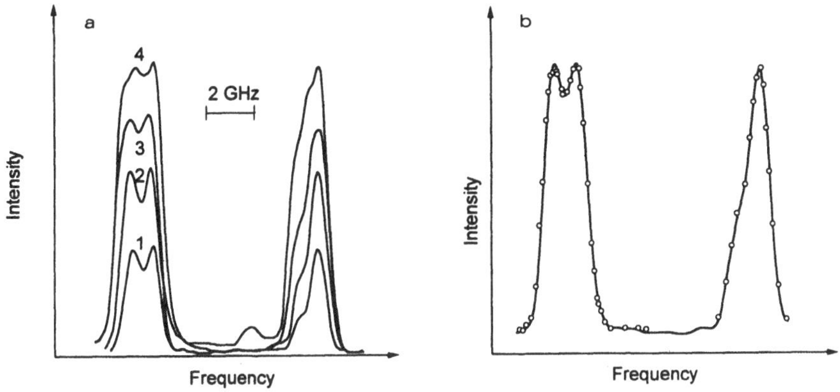

FIGURE 4.24. Interferograms of the Rb D_1-line obtained from an H-discharge obtained with the help of the multiplex interferometer. The lamp was filled with a highly enriched ^{87}Rb isotope probe. (a) the power of the self-excited oscillator $W = 2.8$ W, temperature: 1, 103 °C; 2, 110 °C; 3, 122 °C; 4, 135 °C. (b) $W = 2$ W, $T = 110$ °C; continuous line, calculation; dots, experiment.

4.2.3. Initial Profile of the Radiation Line

The following experiments have been performed using a lamp which had a filling with a highly enriched probe of ^{87}Rb metal. With the help of the complex interferometer the shape of lines emitted by hf ELs was investigated, depending on the working mode (Figure 4.24a).

At the lowest power and temperature the hyperfine structure of the excited state of ^{87}Rb can be resolved. On increasing the temperature, broadening of the line shape takes place and, at temperatures higher than 120 °C, self-reversal begins. An inequality between the intensities of the hyperfine components being emitted at the transition from the $5\,^2P_{1/2}$ state to the $5\,^2S_{1/2}$ state (Figure 4.24b) should be considered. In this experimental situation the self-pumping effect, which appears during self-absorption within the discharge plasma, is apparent [131, 172]. Extraction of the Gaussian component in analyzing the line shape allowed the temperature of the atoms in the discharge to be determined. For this purpose an integral equation was solved:

$$\int_A^B K(x,S)y(S)dS = f(x) \tag{4.8}$$

where $K(x,S) = \Delta\nu/[\Delta\nu^2 + 4(x-f)^2]$ is the kernel, describing the instrumental function of the multiplex, $\Delta\nu = 140$ MHz; $f(x)$ is the interferogram and $y(S)$ the line shape required.

The equation was solved by means of a specially developed iteration method:

$$y_{i+1}(x) = \lambda_i(x)f(x) \tag{4.9}$$

where

$$\lambda_i(x) = \frac{y_i(x)}{\int_A^B K(x,S)y_i(S)dS} \tag{4.10}$$

while A and B are numbers defined by "zero" of the integrand function. The regulating parameter is the number n of iterations. This iteration method converges rapidly for $n = 15$ and the even solution is obtained within a limit:

$$\varepsilon = \left\{ \frac{\sum_{i=1}^{200} \left[\int_A^B K(x,S)y_i(S)dS - f(x_i) \right]^2}{200} \right\}^{1/2} < 0.01 \tag{4.11}$$

With the use of conventional methods we did not succeed in deriving the solution to an accuracy better than 0.1.

The initial function was approximated by a Voigt profile, taking into account the theoretical relation of the four hyperfine components of the D_1-line:

$$z(x,k_0l,C,y_0,\Delta\nu_D) = y_0\varphi(x,C_1,\Delta\nu_D)\exp[-k_0ly(x,C_1,\Delta\nu_D)] \tag{4.12}$$

with

$$\varphi(x,C_1,\Delta\nu_D) = \sum_{k=1}^{4} a_k I(a,x,C_k) \tag{4.13}$$

where

$$I(a,x,C_k) = \int_{-\infty}^{\infty} \frac{e^{-y^2}}{a^2 + (xt_k - y)^2} dy \tag{4.14}$$

I is the Voigt profile; $a_1 = a_3 = a_4 = 1, a_2 = a_1/5, C_3 = C_1 - 818, C_3 = C_1 + 6835$, and $C_4 = C_3 + 818$.

The functional

$$F(k_0l,C_1,y_0,\Delta\nu_D) = \sum_{i=1}^{128} [z(x_i,k_0l,C_1,y_0,\Delta\nu_D) - y_n(x_i)]^2 \tag{4.15}$$

was minimized by the Newton–Ruffin method. The Lorentzian part of the broadening takes natural and collisional broadening into account.

The width of the initial shape, depending on the discharge mode, was determined to be within the limits 740–880 MHz at a current through the generator of

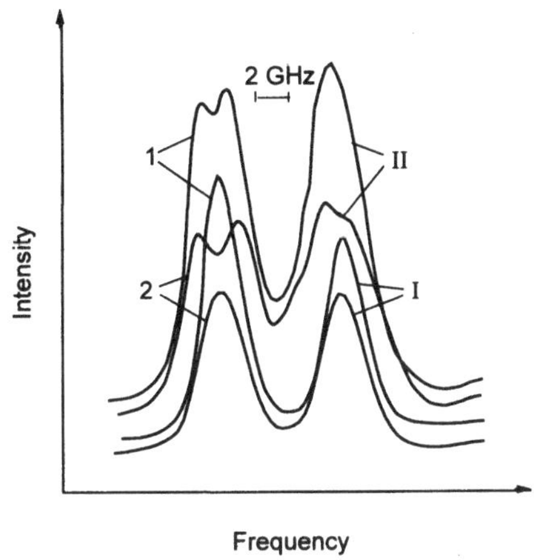

FIGURE 4.25. Interferograms of the Rb D_1-line from (1) central and (2) peripheral areas of lamps (current 140 mA). The lamp was filled with a highly enriched ^{87}Rb isotope probe: I, T = 120 °C; II, 140 °C.

140 mA and temperatures ranging from 110 to 130 °C. Separation of the Gaussian component allowed one also to determine Doppler broadening and optical thickness for these operating modes. The width of the Doppler shape varied from 574 ± 25 to 590 ± 30 MHz with a simultaneous change in optical thickness from 0.95 to 1.30 ± 0.6. These calculated changes of the linewidth correspond to a temperature change from 145 ± 40 to 170 ± 50 °C. The determined temperatures are in agreement within the error bars with the temperatures of the surface of the lamp measured with the aid of a thermocouple. It implies that the radiation of the hf ELs for the studied discharge modes originates mainly from surface layers of the lamp.

4.2.4. *Spatial Characteristics of the Radiation of Spectral Lamps*

4.2.4.1. *H-Discharge*

As in Section 4.2.3, here highly enriched ^{87}Rb has been used for the lamp fillings. Interferograms of the radiation of the Rb D_1-line (794.76 nm) from the central and peripheral zones of an H-discharge, at a current through the generator of 140 mA, are presented in Figure 4.25. In the peripheral area of the lamp the line had a lower integral intensity.

The results of processing the interferograms are given in Table 4.8 and Figure 4.26. For each location of the lamp the relative intensities and widths of hyperfine structure components have been measured within a thermostat temperature interval of of 80–140 °C. For all locations both components are strongly self-reversed at 140 °C, and at 80 °C the intensity ratio differed from the theoretical value: $I_a/I_b = 0.7$ (center) and $I_a/I_b = 0.8$ (periphery), confirming the appreciable self-absorption

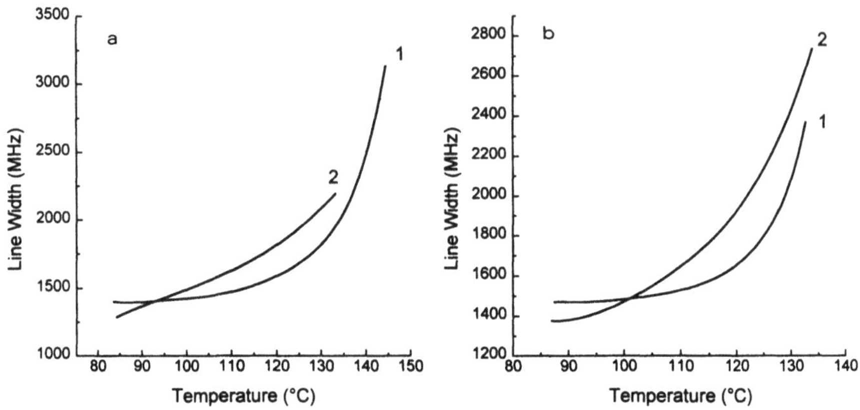

FIGURE 4.26. Dependence of the linewidth of D_{I}-radiation on temperature for radiation from (1) central and (2) peripheral areas of lamps (current of 140 mA): (a) long-wave component (group a in Figure 4.5); (b) short-wave component (group b in Figure 4.5).

even in the already given mode. For some records inversion of populations was observed at 130 °C. From Figure 4.26 it is seen that a strongly nonlinear growth of the width of each hyperfine structure component is observed with increasing temperature for the central and peripheral parts of the source.

The ratio of the widths of short-wave and long-wave components depends weakly on the temperature, and the observed weak reduction of this ratio with increasing temperature is in accordance with the growing re-absorption. For the peripheral part of the source the absorption has a stronger effect than for the central part at a given temperature. This transition from the central areas of the

TABLE 4.8. Characteristics of the Radiation of the Rb D_{I}-Line (794.76 nm) in the H-Discharge for the Central and Peripheral Areas of the Lamp at $i = 140$ mA[a]

T	Center				Periphery			
[°C]	$\frac{I_a}{I_b}$	$\frac{\Delta\nu_a}{\Delta\nu_b}$	$\Delta\nu_a$ [MHz]	$\Delta\nu_b$ [MHz]	$\frac{I_a}{I_b}$	$\frac{\Delta\nu_a}{\Delta\nu_b}$	$\Delta\nu_a$ [MHz]	$\Delta\nu_b$ [MHz]
80	0.70	0.87	1390	1600	0.79	0.87	1340	1540
100	0.71	0.88	1440	1640	0.83	0.86	1520	1760
110	0.76	0.90	1520	1690	0.90	0.86	1660	1920
120	0.78	0.90	1580	1750	0.99	0.85	1850	2190
130	1.03	0.83	2160	2600	1.10	0.80	2180	2730
140	Re-absorption		3130		Re-absorption			

[a] Subscripts a and b stand for the short- and long-wave hyperfine component groups (cf. Figure 4.5).

TABLE 4.9. Characteristics of the Radiation of the Rb D_1-Line (794.76 nm) in the H-Discharge for Central and Peripheral Areas of the Lamp at $i = 160$ mA[a]

	Center				Periphery			
T [°C]	$\frac{I_a}{I_b}$	$\frac{\Delta\nu_a}{\Delta\nu_b}$	$\Delta\nu_a$ [MHz]	$\Delta\nu_b$ [MHz]	$\frac{I_a}{I_b}$	$\frac{\Delta\nu_a}{\Delta\nu_b}$	$\Delta\nu_a$ [MHz]	$\Delta\nu_b$ [MHz]
80	0.70	0.87	1320	1525	0.79	0.78	1310	1675
100	0.71	0.90	1420	1590	0.84	0.85	1470	1730
110	0.74	0.90	1490	1660	0.89	0.86	1630	1900
120	0.80	0.88	1520	1730	0.95	0.90	1920	2140
130	0.85	0.89	1730	1950	1.05	0.80	2100	2610
140	Re-absorption		2470		Re-absorption			

[a] Subscripts a and b stand for the short- and long-wave hyperfine component groups (cf. Figure 4.5).

lamp to the peripheral areas under the same conditions corresponds to a broadening of the hyperfine structure components, especially appreciable in the temperature interval 90–120 °C. Similar changes in the emission profiles are observed when varying the temperature by 5–15 °C for a fixed location of the lamp. Processing of experimental values showed a dependence of the reproducibility of the absolute linewidth values on the thermostat temperature: for temperatures within 80–100 °C the error was 3–4%, for 110–120 °C about 10%, and for strong absorption 18–20%. The behavior of the relative intensity for currents of 160 and 140 mA is identical (see Table 4.9).

Examination of dependence of the linewidth of the components on the mode and the location (Tables 4.8 and 4.9) shows that the change in temperature has a stronger effect on the line shape than the change in the current mode. As a general rule the width of a line grows with increasing temperature. With increasing current, the re-absorption of a line decreases and the spectral line becomes narrower at the same temperature.

4.2.4.2. E-Discharge

Examples of records and results of the processed shape of the radiation over the whole surface of the lamp under various operating conditions of the generator are given in Figure 4.27 and Table 4.10. The current was varied over an interval of 70–120 mA, where the thermostat temperature was 80–120 °C. When increasing the generator current appreciable self-absorption is not observed below 90 °C. At 110 °C, self-reversal of hyperfine structure components is observable. The increase in the current leads to an insignificant growth of the integral intensity, which is practically identical for both hyperfine structure component groups. The variation in the generator current over the whole interval yields only a 100 MHz increase in the width of each hyperfine component, but that is within the limits of

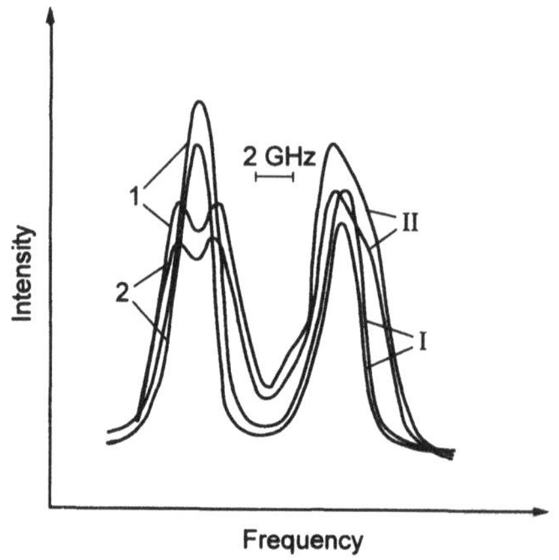

FIGURE 4.27. Interferograms of the Rb D_1-line obtained from the whole surface of the lamp at different currents through the generator (I, $T = 90$ °C; II, $T = 110$ °C) in an E-discharge. Current through the generator: 1, 120 mA; 2, 100 mA.

the absolute error of the measurement. A change in the thermostat temperature causes self-reversal, and a change in the generator current leads to a quantitative change in the radiation intensity.

The results of the data processing, corresponding to temperatures of 80, 90, and 100 °C, are given in Table 4.11. The values obtained indicate the growth in radiation intensity with increasing thermostat temperature, while the intensity of the short-wave component increases growing faster than the long-wave component. That is a point in favor of the absorption processes. The distinctive characteristic of the E-discharge is the weak growth in the linewidth of each component with growth in temperature.

The interferograms of the radiation from different areas of the lamp and the results of their processing at a current of 80 mA are given in Figure 4.28 and Table 4.12. As the thermostat temperature increases the radiation intensity and linewidth of each component grows alike for all parts of the lamp. It is noteworthy

TABLE 4.10. Dependence of Intensity and Width $\Delta\nu$ (FWHM) of the Component Groups a and b of the Rb D_1-Line (794.76 nm) on the Current through the Generator ($T = 90$ °C)

I [mA]	I_a [arb.u.]	I_b [arb.u.]	$\Delta\nu_a$ [MHz]	$\Delta\nu_b$ [MHz]
70	1.0	1.0	1520	1630
80	1.06	1.08	1535	1725
90	1.18	1.20	1565	1755
100	1.34	1.36	1635	1760
110	1.44	1.46	1685	1770
120	1.53	1.55	1620	1735

TABLE 4.11. Dependence of Intensity and Width $\Delta\nu$ (FWHM) of the Component Groups a and b of the Rb D_1-Line (794.76 nm) on the Thermostat Temperature ($i = 100$ mA)

T [°C]	I_a [arb.u.]	I_b [arb.u.]	$\Delta\nu_a$ [MHz]	$\Delta\nu_b$ [MHz]
80	1.0	1.0	1620	1600
90	1.67	1.59	1650	1730
100	2.67	2.35	1700	1850
110		Re-absorption		
120				

that the linewidth of each component at a given temperature is the same for all locations of the lamp, within the limits of the measurement error. The results of processing the interferograms of the radiation from different parts and zones of the lamp at a current of 100 mA are given in Table 4.13.

Comparison of the results for the E- and H-discharges, yields the following conclusions. The integral intensity of the first resonant doublet in the E-discharge is about 4–5 times less than in the H-discharge. Self-reversal of the investigated spectral lines in the E-discharge appears at lower thermostat temperatures (110 °C) than in the H-discharge (> 130 °C). An essential difference between the spectral characteristics of the H-discharge and E-discharge is connected with the dependency of radiation from various areas of the source located at different points along the radius from the center of the lamp. For the E-discharge ($T < 110°C$) the relative intensities and the widths of the hyperfine structure components from

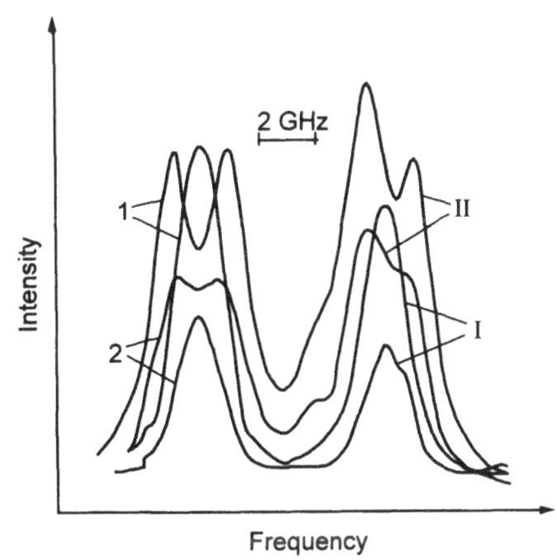

FIGURE 4.28. Interferograms of the Rb D_1-line obtained from (1) central and (2) peripheral parts of the lamp for a current of 80 mA in an E-discharge; I, 90 °C; II, 110 °C.

TABLE 4.12. Radiation Intensity and Linewidth of the Component Groups a and b of the Rb D_1-Line ($\lambda = 794.76$ nm) for Different Locations on the Lamp Bulb ($i = 80$ mA)[a]

Location	T [°C]	I_a [arb.u.]	I_b [arb.u.]	$\Delta\nu_a$ [MHz]	$\Delta\nu_b$ [MHz]
Center	80	23.8	29.8	1160	1482
	90	43.5	53.2	1321	1544
	100	72.8	85.0	1482	1719
	110		Re-absorption		
	120				
2nd diaphragm	80	22.0	29.5	1217	1326
	90	43.5	52.8	1488	1604
	100	69.0	78.5	1681	1810
	110		Re-absorption		
	120				
3rd diaphragm	80	14.7	18.5	1185	1496
	90	29.5	36.0	1321	1625
	100	52.8	59.0	1496	1820
	110		Re-absorption		
	120				

[a] For positions of diaphragms, see text.

TABLE 4.13. Radiation Intensity and Linewidth of the Component Groups a and b of the Rb D_1-Line ($\lambda = 794.76$ nm) For Different Locations on the Lamp Bulb ($i = 100$ mA)[a]

Location	T [°C]	I_a [arb.u.]	I_b [arb.u.]	$\Delta\nu_a$ [MHz]	$\Delta\nu_b$ [MHz]
Center	80	1.0	1.0	1080	1335
	90	1.91	1.85	1262	1455
	100	3.13	2.76	1430	1640
	110		Re-absorption		
	120				
2nd diaphragm	80	1.0	1.0	1175	1435
	90	1.63	1.69	1365	1575
	100	2.65	2.40	1480	1670
	110		Re-absorption		
	120				
3rd diaphragm	80	1.0	1.0	1185	1375
	90	2.04	1.92	1360	1615
	100	3.45	3.02	1535	1660
	110		Re-absorption		
	120				

[a] For positions of diaphragms, see text.

different radial zones of the lamp do not differ from one another. The radiation efficiency of the peripheral zones in the H-discharge is lower than in the central zones while broadening of the hyperfine structure components in the peripheral zones is higher than at the center. For some operating modes of discharge a nonlinear radial intensity distribution is observed.

Thus, the conditions governing excitation of resonance Rb lines in the E-discharge are practically identical for various zones of the lamp ($T < 110$ °C), while these conditions are noticeably variable along the radius for the H-discharge. These features could be explained by temperature differences between various locations of the lamp. In the H-discharge the temperature in the peripheral parts of the lamp is 10–15 °C higher than in the central part. This fact is important when evaluating the parameters of the emission line profile of a lamp knowing only the thermostat temperature.

5
Intensity Fluctuations of Emitted Spectral Lines

5.1. General Problems of Intensity Fluctuation Studies

The main features of light sources for optical pumping on the basis of high-frequency electrodeless spectral lamps are the high intensity and stability of radiation of the emitted spectral lines. However, there are only a few works devoted to the study of fluctuation characteristics of the radiation of such light sources. Fluctuations in the frequency domain lower than 1 Hz and in the time domain have not yet been sufficiently investigated. Nevertheless, in a number of cases (for example, in gas cells for quantum frequency standards) the parameters of the device are determined by the entire spectrum of fluctuations. The sensitivity of radio-optical resonance for a measurement period of 1 s is conditioned by the spectral density of radiation fluctuations at modulation frequencies which are usually within the range from 30 to 300 Hz. The sensitivity for a measurement period longer than 10^3 s depends upon the intensity fluctuations for the corresponding time. As in spectroscopic experiments and optical devices, various types of hf discharge light sources operating in various modes are used, the problem of optimization of the light sources becomes very significant and requires preliminary studies of the fluctuation characteristics.

In the present section the results of an experimental study of radiation intensity fluctuations of hf ELs with rubidium vapor within the range of frequencies from 300 down to 0.1 Hz at measurement periods from 1 to 10^6 s are presented. Additionally, we give results for mercury vapor lamps for a measurement period from 1 to 10^3 s in the modes of the E- and H-discharges. The spectral density of fluctuations was measured in the frequency domain, while in the time domain everywhere the mean square deviation (called an Allan variation) during the measurement time τ_m was studied.

Let us recall the connection between the fluctuation characteristics in the time and frequency domains. Usually, photodiodes are used to record the emitted radiation. Their photocurrent i is proportional to the rate n_p at which photons strike the photodiode during a period of 1 s: $i = \eta n_p e$, where e is the charge of the electron, and η the quantum efficiency of the photodiode. For a description of random processes, the spectral density $g(\omega)$ in the frequency domain and the dispersion of fluctuations σ_D^2 in the time domain are used. The dispersion can be determined by the known spectral density as follows:

$$\sigma_D^2 = \int_{-\infty}^{+\infty} g(\omega) d\omega \tag{5.1}$$

Strictly speaking, the dispersion does not depend on time, but actually an averaging of the random process during the measurement time τ_m always occurs. When evaluating experimental results the mean-square relative deviation $\sigma_D^2(\tau_m)$, averaged over the measurement time, is used:

$$\sigma_D^2(\tau_m) = \frac{1}{\bar{i}} \sum_{k=1}^{N} \frac{(i_k - \bar{i})^2}{N} \tag{5.2}$$

where i_k is the average value of the photocurrent during the measurement time τ_m at the kth sampling instant, and $\bar{i}$ is the arithmetic-mean value of the photocurrent during the observation. If the radiation is detected by two independent photodiodes, the two-sample arithmetic-mean relative deviation $\sigma_{va}^2(\tau_m)$ can be calculated:

$$\sigma_{va}^2(\tau_m) = \frac{1}{2\bar{i}} \sum_{k=1}^{N} \frac{(i_k - i_{k-1})^2}{N-1} \tag{5.3}$$

It is shown in [173] that the mean-square deviation can be calculated in terms of the spectral density of fluctuations, if they are represented by rectangular pulses with a duration τ_m as follows:

$$\sigma_D^2(\tau_m) = \int_{-\infty}^{+\infty} g(\omega) \frac{\sin^2(\omega\tau_m/2)}{(\omega\tau_m/2)^2} d\omega \tag{5.4}$$

where $g(\omega)$ is the relative spectral density of fluctuations. For the two-sample mean square deviation $\sigma_{va}(\tau_m)$ the connection with the spectral density is somewhat different:

$$\sigma_{va}^2(\tau_m) = \int_{-\infty}^{+\infty} g(\omega) \frac{\sin^4(\omega\tau_m/2)}{(\omega\tau_m/2)^2} d\omega \tag{5.5}$$

As is known, the spectral density can often be represented in the form of a power law as a reasonable and accurate model for random fluctuations in an oscillator:

$$g(\omega) = \sum_{k=-\infty}^{\infty} h_k \omega^{-k} \tag{5.6}$$

We can limit ourself to the following terms of the expansion to describe the real processes:

$$g(\omega) = h_{-2}\omega^2 + h_{-1}\omega + h_0 \tag{5.7}$$

Having substituted (5.7) in (5.5) and integrating, we obtain a representation for the random process in the time domain expressed through its characteristics in the frequency domain:

$$\sigma_{va}^2(\tau) = (2\pi)^2 h_{-2}\tau/6 + 2\ln 2 h_{-1} + h_0/(2\tau) \tag{5.8}$$

The first terms in (5.7) and (5.8) are usually called the spectral density of the random walk process, while the second and third flicker noise and white noise, respectively. The white noise is a consequence of the discrete nature of the photocurrent and its stochastic character. Therefore the fluctuations conditioned by it are practically the limit of noise. This type of noise predominates at frequencies above 10–40 Hz. The random walk process is the representation of the relaxation of an investigated object to its equilibrium state. Deviation from balance arises as a consequence of the manufacturing process of a spectral lamp. In the lamps there exist nonequilibrium states in glass, in the balance gas-monolayer on the internal surface of the bulb due to gas dissolved in the material. Any effect, such as a change of temperature, voltage supply, atmospheric gas pressure, magnetic field intensities, etc., results in larger or smaller deviations from the balance, followed by relaxation to the balance. Thus, a random walk process should be expected with large probability when investigating light sources in the time domain. As a whole, observable fluctuations in the photocurrent are the result of a number of stochastic processes such as fluctuations in the radiation intensity caused by different effects (such as instability of the oscillator, external electromagnetic fields, etc., apart from processes in the lamp), fluctuation in the parameters of the photoreceptor itself, and the system of recording. In the frequency domain, at rather high frequency of the analysis, it is possible to correlate fluctuations of the photocurrent to a given mode of the lamp with a certain confidence. In the time domain the situation is far more complex and it is more reasonable to speak about "observable" fluctuations, at least until the analysis of the nature of the investigated stochastic processes is carried out.

By means of the connection between fluctuations in the time and frequency domains it is possible to evaluate the limiting instability in the time domain, which is in accordance with the shot noise of a photocurrent. It is easy to see that $\sigma_{va}^2(\tau_m) = h_0/(2\tau_m) = e/(i\tau_m)$, and therefore at a photocurrent $i = 100\mu\text{A}$ the two-sample relative root-mean-square deviation will be $1.6 \times 10^{-15}\tau_m^{-1}$, and at $i = 1\mu\text{A}$ the value is $1.6 \times 10^{-13}\tau_m^{-1}$. We will see further that in the time domain the terms proportional to τ_m are predominant, corresponding to the random walk process. Expressions (5.7) and (5.8) allow one to approximate and analyze the

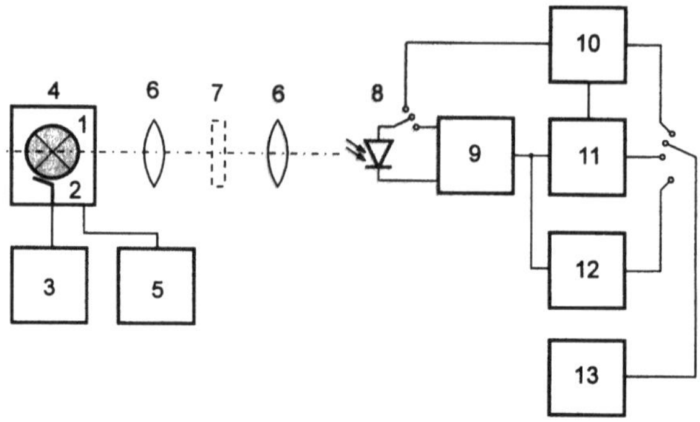

FIGURE 5.1. Block diagram of the equipment for studying the fluctuation spectra: 1, studied lamp; 2, inductor; 3, hf oscillator; 4, thermostat; 5, thermoregulator; 6, lens; 7, neutral density filter; 8, photodiode; 9, head amplifier; 10, dc voltmeter; 11, selective voltmeter; 12, spectrum analyzer; 13, recorder.

experimental dependences and compare the results obtained in the frequency and time domains.

The block diagram of a typical setup for analyzing the spectrum of fluctuations in the source is given in Figure 5.1.

The radiation of the lamp (1) excited by the inductor (2) and the hf power supply (3) is investigated. The lamp is placed in the thermostat (4), where the temperature is varied by changing the current through the heating element of the thermostat. A constant temperature is sustained by the thermoregulator (5). The radiation is imaged by means of lenses (6) to photodiode (8) and can be attenuated by neutral density filters (7). Its photocurrent is fed to a matching amplifier (9). The constant component of the amplifier output voltage is measured either by a spectrum analyzer (12) and plotted by a two-coordinate recorder, when measuring the spectral density, or by a differential voltmeter (10) connected to the recorder, when measuring the variance of the process. Furthermore, when analyzing the fluctuations in the light source with the thermostat switched off, i.e., with a predominance of radiation of krypton atoms, we used a double-channel measurement technique. In this case the radiation of the lamp was divided into equal parts by a beam splitter and entered two photoreceptors, each of them connected to a voltmeter and an electronic recorder. This allowed us to investigate the relative contributions of the photoreceptors and the lamp separately.

The spectral density of fluctuations in the light flux may be represented as

$$g_{sp}(\omega) = |k_l(i\omega)|^2 \left|k_{mf}(i\omega)\right|^2 |k_p(i\omega)|^2 \times |k_{os}(i\omega)|^2 \left|k_{ph}(i\omega)\right|^2 [g_n(\omega) + k_a g_{ps}(\omega)]^2 \qquad (5.9)$$

where $g_n(\omega)$ and $g_{ps}(\omega)$ are the spectral density of natural fluctuations of generator and power supplies; $k_l(i\omega)$, $k_{mf}(i\omega)$, $k_p(i\omega)$, and $k_{os}(i\omega)$ are the transfer functions of lenses and glasses, systems of thermostatic control and external fields, plasma of discharge, and the self-excited oscillator in relation to the fluctuations of the power supply, while $k_{ph}(i\omega)$ and $k_a(i\omega)$ are the transfer functions of the photodiode and the matched amplifier.

All functions are characterized by various time constants and are displayed in different frequency ranges. The lenses and glasses change their transmittance rate slowly, leading to a gradual reduction in the light intensity. The influence of the thermostatic control system, having a time constant of about several seconds in the pulse mode of the thermoregulator, is conditioned by the flow of current heating the thermostat. The voltage or current stabilizers, which feed the oscillator, can predetermine the spectral density of fluctuations over all frequency ranges in which measurements are being carried out. In the present section, when investigating the fluctuation characteristics of the photocurrent, we will primarily be interested in the fluctuation characteristics of the lamp $g(\omega)$. When performing studies of the spectral density of fluctuations, it is important to account for the transmission function of the interconnecting amplifier as well as to estimate the contribution of the power supply. The technical fluctuations conditioned by radioelectronics, etc., which are inherent in light sources, will be discussed in Section 5.4.

5.2. Spectral Density of Fluctuations at Frequencies between 330 and 0.1 Hz

5.2.1. Frequency Range from 300 down to 30 Hz

Already in the first papers devoted to the research of light sources based on hf EL, it was stated that the lamps are characterized by a shot noise [23, 138, 174]. However, the experimental technique was not described and systematic studies of lamps in various modes were not carried out. At the same time it has been shown [10] that the introduction of an easily ionizable component into the plasma of the discharge can increase the spectral density of the noise.

Measurements of the spectral density of noise in lamps with vapors of rubidium and krypton in the E- and H-discharge modes are available [25]. The spectral lamp in the light sources based on the H-discharge had a diameter of 13 mm and was filled with krypton at pressure 4.1 ± 0.3 mbar and saturated rubidium vapor. The excess of rubidium was placed in a lengthened reservoir the temperature of which corresponded to the temperature of the thermostat of the lamp. Its temperature was defined within an error of 0.5 °C and sustained with an accuracy of 0.2 °C. The measurements were carried out over temperature ranges of the thermostat from 50 up to 140 °C for the H-discharge and from 90 up to 120 °C for the E-discharge. Three stages of the H-discharge corresponded to these ranges: from 50 to 80 °C, the discharge in krypton; from 90 to 130 °C, the discharge in a mixture of krypton and rubidium vapor; and more than 130 °C, the discharge in rubidium (see Figure 4.9).

The current of the self-excited oscillator was varied from 80 to 200 mA. In the light source based on the E-discharge the lamp had the same size, but was filled with krypton at a pressure of 2.4 mbar. The self-excited oscillators of the light source in the H-discharge worked at a frequency of 92 ± 2 MHz, but in the E-discharge at 160 ± 5 MHz. Their voltage–current characteristics coincided with an error of not more than 1% in the modes being investigated.

The oscillator was powered from a stabilized power supply. The intensity of radiation was monitored with the aid of a silicon photodiode in front of which neutral filters were placed. The latter were permitted to attain constant light intensity, when changing the mode of the lamp, with an accuracy better than 0.1%. The signal from the photodiode was fed to the amplifier, whose output was connected either with the spectrum analyzer or with the microvoltmeter. In order to determine the spectral density of fluctuations the microvoltmeter was calibrated with the help of a noise generator. External mechanical interferences were excluded by placing the light source on a massive duraluminum plate, the latter being located on a porolone cushion.

Investigations showed that in the E- and H-discharges the spectral density of the noise does not depend on the current of the generator and the temperature. However the spectral density appeared to be different: in the H-discharge $g(\omega) = 29 \pm 3\ \mu\text{V/Hz}^{1/2}$, in the E-discharge $g(\omega) = 38 \pm 3\ \mu\text{V/Hz}^{1/2}$, i.e., in the E-discharge the spectral density appeared to be 1.3 times higher than in the H-discharge.

The installation of a tungsten band lamp instead of a spectral lamp results in a spectral density of $29 \pm 3\ \mu\text{V/Hz}^{1/2}$, which corresponds to the value of the shot current calculated by the known photocurrent transfer function of the optical system and the matching amplifier. Thus, the value of the spectral density did not depend on the frequency with which the analysis was carried out. Different sensitivity of the light sources to the fluctuations of the feeding voltage oscillator was obtained: $\partial I/I\partial V = 0.016V^{-1}$ in the H-discharge, and $\partial I/I\partial V = 0.13V^{-1}$ in the E-discharge. The spectral density of fluctuations of the stabilizer was estimated experimentally by replacing it with a noise generator and it was found to be $2.5\ \mu\text{V/Hz}^{1/2}$. The voltage–current characteristic allows one to see that the noise caused by the stabilizer is lower than the shot noise in the H-discharge, but higher than in the E-discharge [124].

5.2.2. Frequency Range from 30 down to 0.1 Hz

When reducing the frequency below 30 Hz the spectral density of fluctuations of the light intensity usually grows quickly [124]. However, there exist only

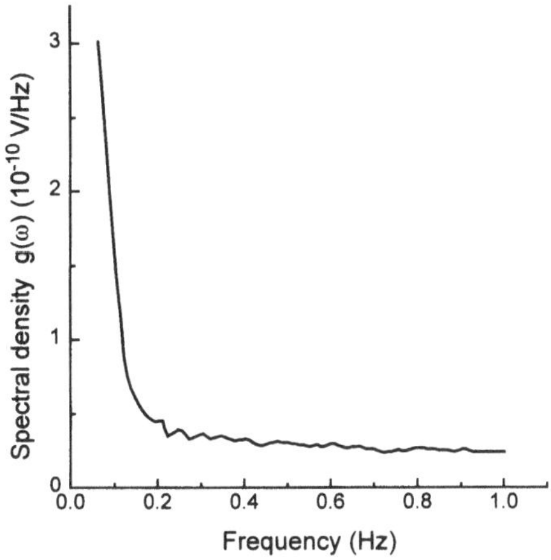

FIGURE 5.2. The spectral density of fluctuations of spectral lines in the H-discharge measured by a spectrum analyzer.

few papers devoted to research on fluctuations in this frequency range, despite its importance both for the better understanding of the physical processes in the lamp, that cause intensity instabilities, and for predictions in the time domain.

We have carried out studies on the spectral density of fluctuations in the frequency domain on H-discharges. The spectral lamp (again 13 mm in diameter) was placed in an inductor which comprised a two-section coil with its potential turns reversed to the lamp.

The lamp together with an inductor was placed in a thermostat with an aperture of 13 mm closed by three glasses. The lamp was fastened to the wall of the thermostat (temperature 120 °C) with the aid of a thermal-conducting holder. Between the thermostat and the board of the generator a (12 mm thick) layer of foam plastic was placed. The generator was assembled according to a Clapp circuit and consumed 2.8 W, where 1.5 W were transformed into thermal and optical power. A more detailed description of the design is given elsewhere [85, 122]. The intensity of radiation was recorded with a silicon photodiode connected to an amplifier, having a constant factor of amplification down to 0.001 Hz. The spectral density of fluctuations was measured by a spectrum analyzer and recorded by a double-channel electronic recorder.

The measurements showed a constant level of the spectral density of fluctuations (for the H-category it equals to the shot noise) down to frequencies of 5–10 Hz, and then a slow increase. At frequencies below 0.2 Hz a steep increase takes place (Figure 5.2). Thus, apart from the flicker noise, a random walk is observed, the latter surpassing the flicker noise for the analyzed frequencies below 10 Hz and the white noise at frequencies below 3 Hz.

For an E-discharge the results of measuring the spectral density of fluctuations, performed with the help of a spectrum analyzer (observation scope of 20

TABLE 5.1. Spectral Density of the Fluctuations in Light Intensity (arbitrary units)

Type of voltage supply	Frequency[Hz] 0.6	0.9	1.2	1.6	1.9	2.2	2.8	3.2	4.1	5.5	7.5	20
Stabilizer	1.74	1.09	0.79	0.62	0.53	0.43	0.36	0.34	0.28	0.26	0.21	—
Accumulator	1.07	0.66	0.58	0.52	0.40	0.36	—	0.30	0.25	0.28	0.22	0.2

Hz) on the output of the matching amplifier while feeding the oscillator from the voltage stabilizer and from a storage battery, respectively, are shown in Table 5.1.

It is seen that the level of fluctuations at low frequencies is much higher when feeding from the stabilizer than from the accumulator, and hence it is defined by the fluctuation characteristics of the power supply. By referring to the corresponding equation, these data enable one to define the factor in the expression for the spectrum of fluctuations of the power source voltage $A/|\omega|^\gamma$, which appears to be $2.9 \times 10^{-12}/\omega^2$ for the stabilizer and $10^{-12}/\omega^{1.4}$ for the accumulator. Comparison with published data [289] shows that the power factors are similar for the accumulator and stabilizer but the constants A are different.

Figure 5.3 shows the spectral densities of radiation intensity fluctuations depending on the mode of the E-discharge. These investigations were carried out using a light source [10, 58, 69] with a lamp 13 mm in diameter filled with the rubidium vapors and krypton at pressure 2.4 mbar (light source number 3 in Table 9.1), placed in a Dewar vessel filled with heat-conducting vapor (cf. Chapter 9). The thermoregulator worked in a smooth mode.

The curves in Figure 5.3 show that the spectral density of fluctuations grows for increasing temperature and current through the generator.

The study of the spectral density at frequencies lower than 1 Hz lies outside the capabilities of our spectrum analyzer, consequently more careful measurements were carried out by another technique (measurement of the dispersion over various periods of observation). For this purpose the intensity of radiation of the light source was recorded with the aid of an electronic recorder (Figure 5.4a), followed by standard processing with a computer program, after which the correlation function, the trend, and the spectral density of fluctuations, were computed. The results of the measurements and calculations are given in Figures 5.4b and 5.4c. A growth in the spectral density of fluctuations appears when the power of the discharge is reduced. Further, a dependence of the spectral density of fluctuations on the setup of the oscillator was detected. This method is promising and was also used to study the fluctuation processes for a period $t > 10$ s. According to Eq. (5.6) the approximation of the spectral density of the fluctuations, as well as the case of a light source operated in an H-discharge mode, has the form $g(\omega) = A\omega^{-2} + B\omega^{-1} + C$. The values of the factors depend strongly on the operating mode of the lamp and are not well investigated in this area of frequencies.

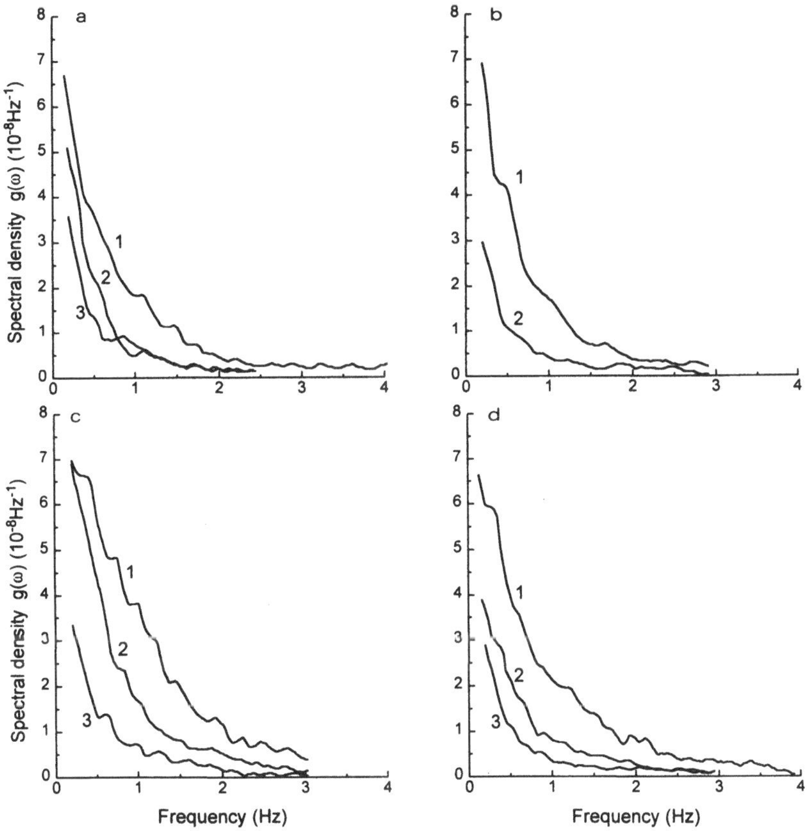

FIGURE 5.3. The spectral density of fluctuations of the spectral lines in the E-discharge. Curves a–d refer to different light-source temperatures; the parameter is the photocurrent: (a) $T = 84$ °C, 1, 185 μA; 2, 160 μA; 3, 141 μA; (b) $T = 92$ °C, 1, 204 μA; 2, 159 μA; (c) $T = 95$ °C, 1, 210 μA; 2, 185 μA; 3, 165 μA; (d) $T = 101$ °C, 1, 172 μA; 2, 147 μA; 3, 129 μA.

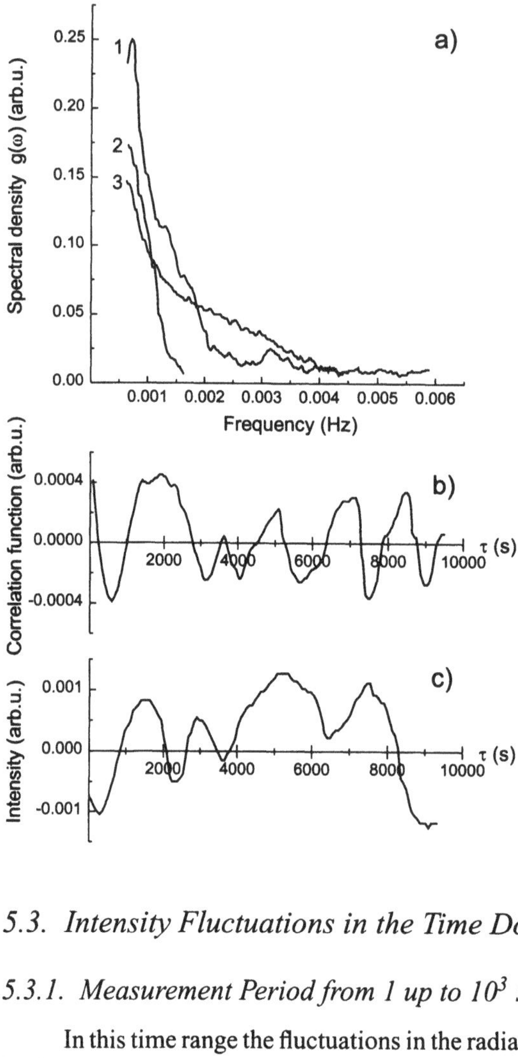

FIGURE 5.4. The spectral density of (a) fluctuation and (b) correlation function calculated using (c) the experimental dependence of the radiation intensity on the measurement period. Supply voltage of the generator: 1, 8 V; 2, 10 V; 3, 10 V (another adjustment of the generator). (b) and (c) are shown for curve 2 in (a).

5.3. Intensity Fluctuations in the Time Domain

5.3.1. Measurement Period from 1 up to 10^3 s

In this time range the fluctuations in the radiation intensity of the light source in the E- or H-discharge mode were monitored with the help of a differential voltmeter connected to the photodiode, terminated by a low-resistance load with minimum temperature factor. The recording was carried out with a recorder connected to the voltmeter output. The period of observation comprises ten periods of measurements. From the results of the measurements the root-mean-square two-sample relative deviation, as well as the root-mean-square relative deviation and the drift of intensity for different times of observation at various operating modes were calculated. To reduce the influence of a change in environmental temperature on the sensitivity of the photodetector the latter was placed in the thermostat,

TABLE 5.2. Dependence of the Allan Variance and the Relative Root-Mean-Square Two-Sample Deviation (*) of the Light Intensity Fluctuation $\sigma \times 10^5$ on the Measurement Time τ and on the Thermostat Temperature T

T	τ [s]			
[°C]	1	10	90	900
50	1.2 ± 1.0	5 ± 2	8 ± 2	20 ± 2
	0.6*	2.6*	5.5*	11*
103	5 ± 2	8 ± 2	4	60
	2.7*	3.8*	2*	45*
120	3 ± 2	19 ± 8	43 ± 9	—
	2.6*	7*	20*	—
130	4 ± 2	12 ± 4	20 ± 5	17 ± 5
	2.5*	7*	18*	20*

where the temperature was sustained with an accuracy better than 0.1 °C. The thermoregulator worked in a smooth regulation mode (to reduce induction) and the photoreceptor was placed at a significant distance from the light source to exclude its heating by the lamp. To exclude mechanical effects, the light source and the thermostat of the photoreceptor were connected by a titanium tube in which the optical system was placed and mounted on a massive duraluminum plate, the latter placed on a thick porolone cushion. The measurements for the H-discharge are shown in Table 5.2 and Figure 5.5. The dependences observed at different temperatures are characterized by an extremum of the level of fluctuations, which appears at lower temperature when the period of measurements is increased. The highest fluctuations are measured when the discharge is burning in the rubidium and krypton vapor. The discharge parameters are chosen to correspond roughly to the condition of maximum intensity of the rubidium line.

The measurement error is determined as the root-mean-square deviation from the arithmetic-average value, obtained upon tenfold repetition of the measurements under the given conditions.

Figure 5.5 shows that at first the variances of fluctuations grow with increasing temperature, unlike the spectral density of the noise, which is constant between frequencies of 50 and 300 Hz. After transition to the mode of primary radiation of the rubidium lines, they decrease. The dependence of fluctuations on the observation period can be approximated by $\sim A\tau^{0.50 \pm 0.1}$ only for $1 > \tau > 90$ s and $T = 50$ °C.

In the mode of the E-discharge in the inert gas (krypton at a pressure of 4.5 mbar) the fluctuations of the light source appeared to be independent of the

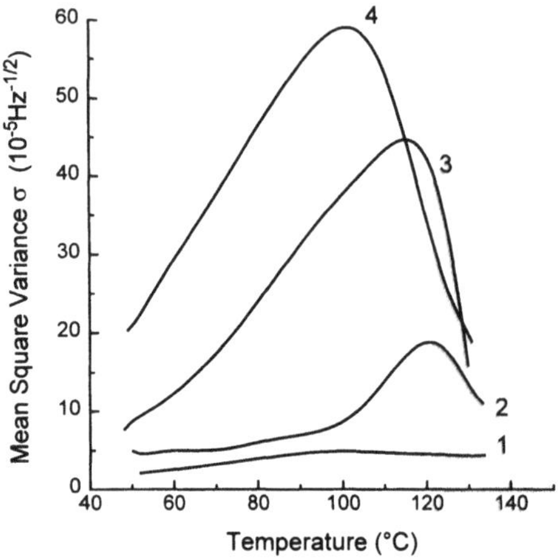

FIGURE 5.5. Dependence of the relative root-mean-square variance of the light intensity on the temperature for different measurement times. H-discharge, $W = 1.5$ W: 1, $\tau = 1$ s; 2, $\tau = 10$ s; 3, $\tau = 90$ s; 4, $\tau = 900$ s.

discharge power (within the error limits) when varying the consumption power of the generator from 0.5 to 1.2 W, and was approximated by a polynomial $\sigma_{va} = 7 \times 10^{-6}\tau^{1/2} + 3 \times 10^{-6}$.

The observed fluctuations depend, as expected, on the adjustment of the self-excited oscillator and appear to be by far greater for a badly adjusted setup. By readjusting of the generator much higher stability was attained (Table 5.3).

Though a reduction in the dispersion and two-sample variance is achieved by adjusting the generator, the relation between the fluctuation characteristics and

TABLE 5.3. Dependence of the Allan Variance and the Relative Root-Mean-Square Variance of the Light Intensity $\sigma \times 10^5$ of the Light Source in the H-Discharge and in the E-Discharge, Electric Filament Lamp, the Voltage Supply Unit and the Stabilizer, on the Measurement Period

Type of noise source	τ [s] 1	10	90	900	Additional parameters
H-discharge	$0.4 \pm 0.1^*$	$0.6 \pm 0.3^*$	$3.3 \pm 2^*$		1.7 W
H-discharge	$0.4 \pm 0.2^*$	$0.7 \pm 0.2^*$	$4 \pm 2^*$		1.9 W
Electric filament lamp	3.1 ± 0.7	7.1 ± 2	8.5 ± 4	22 ± 3	
Supply					
10 B	0.51 ± 0.1	3.0 ± 1.5	2.5 ± 1		
28 B	0.40 ± 0.2	0.8 ± 0.2	1.7 ± 0.8		—
Voltage stabilizer	0.18 ± 0.08	0.36 ± 0.17	0.59 ± 0.36		—

the spectrum of the high-frequency excitation field is not revealed. At the same time measurements show that the instabilities of the voltage at the wings of a particular harmonic of the hf field lead directly to an abrupt growth in intensity instabilities. Results of these measurements are given in Tables 5.2 and 5.3. These instabilities were eliminated by tuning the generator to a proper frequency.

Comparison with an incandescent wire lamp shows that the electrodeless discharge has essentially higher long-term stability and smaller dependence on environmental temperature changes.

The standard power supply units have higher stability for periods longer than 10 s in comparison to the stability observed for the electrodeless discharge. Stabilizers assembled on microcircuits show much higher stability. In Table 5.3, results are given of tests of the stabilizer connected to the generator. The light sources driven in the E-discharge mode were tested with the stabilizer.

The best results indicated in Table 5.3 are not limiting ones. During the simultaneous recording of the intensity fluctuations by two different photoreceptors (two voltmeters and recorders) it turned out that the greater part of the two observable fluctuation patterns does not correlate for a discharge power higher than 0.5 W.

The approximation of the best observed dependence has the form $\sigma_{va} = 1.5 \times 10^{-6}\tau^{1/2} + 10^{-6}$. This is rather close to the fluctuations in the voltage stabilizer, which have been approximated during experiment by the function $\sigma_{va} = 1.2 \times 10^{-6}\tau^{1/2}$. Therefore, these results cannot be considered as limiting fluctuations of a high-frequency electrodeless spectral lamp.

We used experience gained from the development and study of light sources with alkali metal vapors, when developing a light source with mercury vapor. A compact light source (lamp diameter 13 mm) operating in the E-discharge mode was taken as basic unit. Its design and parameters are given in more detail in Section 9.1.2 (light source 5, rubidium vapors; light source 7, mercury). For the study of fluctuations, silicon photoreceptors, which record the radiation from 700 to 1000 nm, and a solar-blind photoreceptor with a sensitivity in the range from 200 to 350 nm were used. Figure 5.6 presents the realization of a stationary mode of the light source with the mercury lamp, monitored by these photoreceptors. The intensity of the mercury resonance line is seen to grow when the lamp is warmed up. The intensity recorded by the silicon photodiode is reduced more than tenfold. This is caused mainly by fading of the krypton lines in accordance with an increase in the mercury vapor density (see Chapter 4). Thus, the dependence of the krypton line intensities on external conditions is much stronger than for resonance lines of the metal. In this connection, the recorded fluctuations appeared to be much stronger than in the light sources operating with rubidium vapors. Additionally, at measurement period $t = 1$ s the relative variance is $\sigma = 5 \times 10^{-5}$, at $t = 10$ s it is $\sigma = 2 \times 10^{-4}$, and at 50 s it is $\sigma = 5 \times 10^{-4}$.

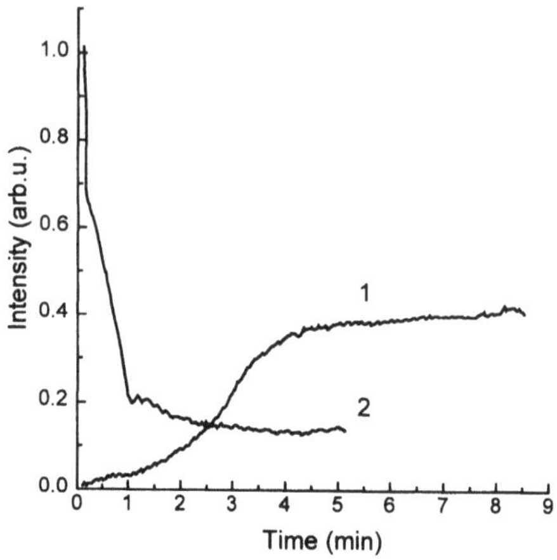

FIGURE 5.6. Realization of the stationary mode of the light source with a mercury vapor lamp: 1, photomultiplier recording the radiation lines from 200 up to 300 nm; 2, photodiode recording the radiation lines from 700 up to 1000 nm.

5.3.2. Measurement Period from 10^3 to 10^6 s

The variance of fluctuations of the light source intensity was measured with a photodiode terminated with a low-resistance load, connected to a digital voltmeter. It was established that the variance at a measurement period of 7.2×10^3 s and a observation period of 1.5×10^5 s varies from 10^{-3} up to 5×10^{-3} for different samples of similar light sources, when a discharge power of 1.4 ± 0.2 W and a thermostat temperature of 120 ± 3 °C are used. In [91], the Allan variance was found to vary from 3×10^{-5} at $t = 10^{-3}$ s up to 10^{-3} at $t = 10^6$ s.

In this time scale the measurements were also conducted with photodiodes terminated again with a low-resistance load, but connected to a strip-chart recorder with subsequent computer processing. The period of the individual readout was 150 s. The arrays with approximately 500 values were subject to standard statistical processing by the following algorithm: 1. Exception of a linear trend. 2. Construction of a histogram. 3. Calculation of the autocorrelation function. 4. Calculation of the spectral density. A characteristic result is shown in Figure 5.7.

In order to statistically process the data it is necessary to install a set of light sources at the same time and to use multichannel recorders, to cope with the long duration of the tests. However, the insufficient accuracy of the chart writer readout considerably reduces the practical value of this method. Nevertheless, it was possible to obtain a qualitative picture of fluctuations and an approximation of the dependence of the spectral density on the measurement period in the form A/ω^2.

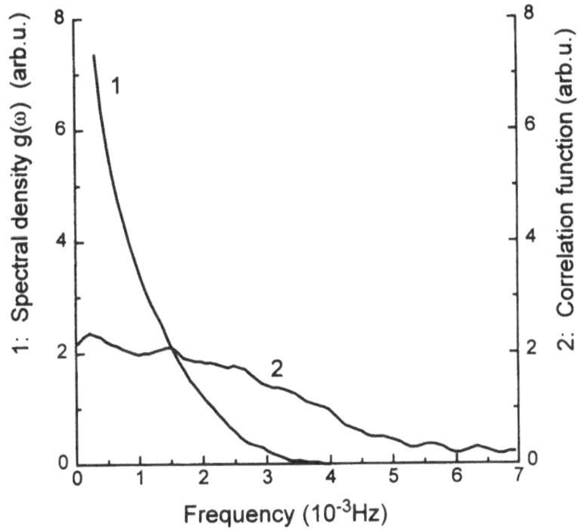

FIGURE 5.7. 1, spectral density of fluctuations; 2, correlation function of the intensity of the lamp in the H-discharge mode; $10^3 \ll \tau \ll 3 \times 10^6$ s.

5.3.3. Measurement Period from 10^5 to 10^7 s

The study of long-term intensity stability is extremely important for the application of high-frequency discharge spectral sources in autonomous satellite-borne systems for the frequency support.

In this time domain (up to 5000–75,000 h) some sets of spectral lamps were tested. In fact, the dependence of both the fluctuations and the drift of the radiation intensity on the technology of manufacturing, the filling, and the operating mode of the lamp as well as on the design of the light source was investigated. In that case the intensity was observed with a portable photoreceptor. The measuring error of the intensity was below 1%. It was found that the variance at the measurement period 10^5–10^7 s and the observation period 2×10^7 s was 0.02, while the intensity drift was 3–5% for 1000 h for lamps 10 mm in diameter. In experiments performed for this duration, lamps 3 mm in diameter showed a drift of 0.5–1.5% for 1000 h. When using lamps on a metal holder with a reservoir for the excess of metal, the drift appeared to be much smaller and was about 0 to 0.2% for 1 month (720 h). In certain modes (H-discharge at a rubidium tank temperature lower than 95 °C or at a temperature of 30 °C and a discharge power of 1.2 W) a change in neither the radiation intensity nor filling of the lamps was observed during 8000 h. As a whole, the tests which were carried out during 40,000–70,000 h showed for qualitatively high manufactured lamps retention of serviceability and insignificant change of parameters. Thus for low-power discharges the long-term stability is very good and stable work for 10 years and more can be achieved.

One can indirectly judge the long-term fluctuations in the light sources by measuring the light shift by means of a quantum frequency standard. Using an E-discharge for 15 months showed that the drift of the real value of the frequency is

6×10^{-12} 1/month. The root-mean-square deviation of frequency can be evaluated to be 3×10^{-12} 1/month. Thus, at a value of the light shift equal to 10^{-9}, the drift makes 6×10^{-12} for 1 month, and the root-mean-square two-sample deviation over the measurement period of 1 month was 3×10^{-3}. We should note that this value is lower than that received, when extrapolating the experimental dependences found in the range 1–10^5 s through direct intensity recording (according to the above-stated technique).

During the measurement, the drift of the light intensity in the H-discharge for times longer than 10^6 s, generally surpasses the fluctuations. However, for the correct choice of mode the systematic intensity drift can be reduced to less than 0.2% per month. In this case the ratio between fluctuations and drift will completely depend on the design of the light source, the parameters of the power supply, and the operating conditions. Basically, the fluctuations can be equal to or even more than the value of the intensity drift.

Due to the smaller power of the E-discharge systematic changes of intensity are substantially smaller and the intensity fluctuations prevail.

5.4. Sources of Technical Fluctuations of the Radiation Intensity

The nature of observable fluctuations can vary as the plasma is sensitive to every possible external perturbation. It is affected by external electric $\vec{E}$, magnetic $\vec{B}$, electromagnetic and thermal fields $T(r,t)$ and gravitational forces g, which are capable of producing a convective motion and mechanical and acoustic effects $F(W)$. When studying lamps with metal vapor it can be seen that the constancy of the vapor density, ensured by the system of thermostatic control, plays an important role, as alterations of the vapor density can be a source of strong perturbations.

Thus, the intensity of radiation is a function of a number of factors:

$$I = I\left(\vec{E}, \vec{B}, T(r,\tau), W, \tau, F(W), g, p, m_{\text{Rb}}, \xi\right) \tag{5.10}$$

where ξ is the factor of glass transparency, m_{Rb} the quantity of condensed rubidium, W the power of the discharge, p the pressure of the gas in the lamp, and τ the life duration.

The spectral density of fluctuations in the light intensity is equal to

$$\begin{aligned} g_I(\omega) = {} & k_E \frac{\partial I}{\partial E} g_E(\omega) + k_H \frac{\partial I}{\partial H} g_H(\omega) + \frac{\partial I}{\partial T} g_{T(\omega)}(\omega) + \frac{\partial I}{\partial F} g_{F(\omega)}(\omega) \\ & + \frac{\partial I}{\partial W} g_W(\omega) + \frac{\partial I}{\partial g} g(\omega) + \frac{\partial I}{\partial p} g(\omega) + \frac{\partial I}{\partial m_{\text{Rb}}} g(\omega) + I \frac{\partial \xi}{\partial t} \xi(\tau;\chi) \end{aligned} \tag{5.11}$$

where k_E and k_H are factors of shielding against external electric and magnetic fields, $g_j(\omega)$ is the spectral density of fluctuations of quantity j, and χ represents all possible effects on the material of the windows which cause their aging.

Hence, for a known spectral density of the corresponding processes and a known dependence of the intensity on the change of given parameters, it is possible to evaluate their influence on the spectral density of the process. Practically we recognize g_E as fluctuations of an exciting hf field that are usually in proportion to fluctuations of both the feeding voltage and other fields. For example, the perturbations induced by external circuits can be recorded. So, working television or radio stations create a field with strengths up to 100 μV/m and frequencies of some dozens of MHz at a distance of several kilometers. This field strength results in the following power release in a plasma of conductivity σ:

$$W = jEV = \sigma E^2 V \tag{5.12}$$

which is, as can be easily reconstructed to be convinced, of the order of 10^{-12} W. But the induction on a conductor of 0.5 m length will be about 0.05 mV and this can be the reason for an appreciable error. For example, when measuring the intensity of light in the gate mode of the photodiode, the signal is about 0.05 V and hence the relative error can reach 2.5×10^{-4}. This is too high for precise measurements.

The induction from hf generators at a low quality of the shielding can be 10 mV/m and more. Accordingly, the measuring errors are increased under the influence of external or internal generators. When testing and using the light sources it is necessary to take special measures to decrease the level of perturbations and to prevent inductions (suppressing of the receiving ability) in the signal circuits.

Another reason for fluctuations can be the thermal electromagnetic force which arises from contamination at the places of contacts between different metals, and from use of mechanical connections in the hf circuits instead of soldering.

The influence of external magnetic fields on the lamps was not investigated systematically but, as shown in [5], the relative change in the field strength can reach 10^{-4}. Therefore, good magnetic shielding is a necessary condition for high accuracy of intensity measurements and for stable work of the lamp. This is essential, because the thermostatic control system is an inevitable source of magnetic fields, the strength of which varies rapidly or smoothly. The magnetic inductions are reduced by the use of bifilar windings, which effectively permit one to reduce the level of perturbations and practically exclude the influence of these fields on the lamp.

Nowadays, transistors are often used for heating. Although in this case it is possible to attain very effective heating, the level of interference needs to be specially checked every time. The current in wires should necessarily be balanced. To exclude man-made or natural external fields the spectral lamp should

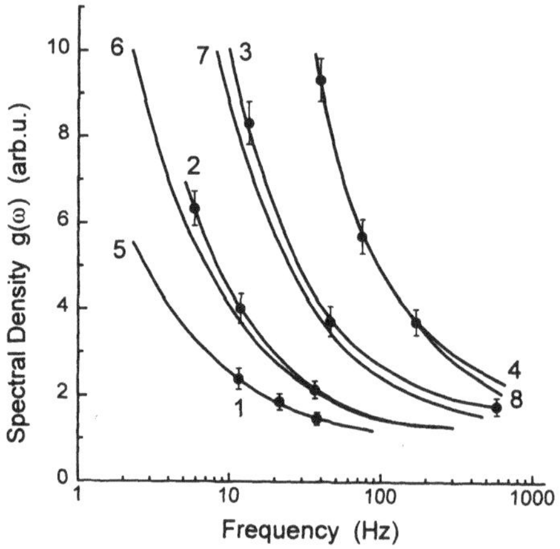

FIGURE 5.8. The dependence of the relative spectral density of the intensity fluctuations on the analysis frequency. Parameter is the power of the hf generator: 1–4 experimental dependencies; 5–8, calculated from the results of measuring the two-sample deviation: 1, 0.95 W; 2, 0.61 W; 3, 0.13 W: 4, 0.07 W; 5, 0.9 W, $\sigma(\tau) = 5 \times 10^{-6}\,\tau^{1/2}$; 6, 0.5 W, $\sigma(\tau) = 10^{-6}\,\tau^{1/2}$; 7, $3 \times 10^{-5}\,\tau^{1/2}$; 8, $\sigma(\tau) = 10^{-4}\,\tau^{1/2}$.

be shielded by a faultless permalloy screen. Otherwise, instabilities of the order of 0.1–0.3% are inevitable. Actually, for the permalloy screen to work effectively the field should be reduced to the optimum value for the given grade of permalloy, i.e., installation of an additional screen of soft iron is required. The calculation of the design of the magnetic screens will be examined in Section 7.4.4.

The temperature of the environment influences the operating conditions of the oscillator, lamp and photoreceptor [third term on the right-hand side of Eq. (5.11)]. Therefore, careful thermostatic control of the lamp and precise long-term measurements of the generator and photoreceptor are required. The thermostatic control system should work in a smooth mode to exclude the influence of changing currents on the radiation intensity. The temperature factor of the light source under thermostatic control usually amounts to about 2×10^3 °C^{-1}, and can be reduced by special preventive measures to 10^{-3} °C^{-1}. Therefore, a double thermostatic control of the light source is usually indispensable for precise devices.

The temperature factor of silicon photoreceptors reaches 5×10^{-4} °C^{-1} and, when long-run measurements are required, they should be thermostatically controlled with required accuracy. Besides, heating by the light source should be avoided.

The fourth term on the right-hand side of Eq. (5.11) takes into account the influence of the mechanical and acoustic effects on the light source. To exclude them it is necessary to fix all components carefully.

While operating a light source in the H-discharge it is inadmissible to use inductors of a type shown in Figure 2.2a, in order to achieve vibration and shock stability.

Acceleration can influence the operating mode of the lamp due to convective motion in the lamp. However, this question has not yet been be solved analytically and so the sixth term on the right-hand side of Eq. (5.11) is introduced for completeness only.

During operation, pressure fluctuations in the lamp are possible and they can be the reason for fluctuations in the radiation intensity. This process, which is basically known, has not been studied either. Obviously, this effect should depend on the quality of the glass bulb, absorption, adsorption, purification and type of buffer gas. Besides, a gradual reduction in gas pressure occurs and will cause intensity drifts. A reduction in the amount of metal and in the glass transparency also results in a gradual reduction of intensity.

When reducing the frequency employed below 30–10 Hz an increase is observed in the noise amplitude and the component proportional to ω^{-2}(random walk, relaxation) prevails. In this frequency range one should take into account the fluctuations in the photoreceptor, the measuring equipment, and changes in the external environment which all influence the measurement results. This is confirmed in [124], where the relative spectral density in fields of low frequencies appeared to be dependent on the intensity of irradiation.

This problem becomes more important when studies are carried out in the time domain. Direct recording of a split light beam with two different photoreceptors shows that the recorded fluctuations are weakly correlated and determined practically by the recording technique in the photoreceptors. Therefore, when analyzing such measurements, it is expedient to assume that the data presented at frequencies below 1 Hz and for times greater than 1 s are the upper estimate of fluctuations. This is why one should only speak about "observable fluctuations of the light source."

Figure 5.8 shows the typical dependence of the relative spectral density of the intensity fluctuations on the frequency at different power consumption of the oscillator. This is measured directly in the frequency domain and calculated from measurements of the two-sample intensity deviation in the time and frequency domains. We can see the surplus of noise which depends on the discharge power and the frequency.

6

Determination of the Quantity of the Working Element in Spectral Lamps; Methods of Dosage

6.1. General Remarks

The radiation intensity, its stability, and the reliability of all lamps working with alkali metal vapors depend on the quantity of the introduced metal, its chemical activity in relation to the filling gas, and the material of the lamp bulb. An interdependence between radiation parameters and the quantity of the condensed alkali metal has been known for a long time; for example, its influence on the radiation stability of the positive column of a dc discharge has been described [133]. It was noted that the condensed metal on the cylinder wall causes a nonuniformity in the radiation. The metal can move over the surface and be deposited in various areas of the lamp under the action of the discharge, resulting in an instability of the radiation output. In electrodeless discharges the metal shows a similar behavior but has a stronger influence on the discharge parameters. Therefore, already in the early papers, serious attention was paid to the migration of the metal over the volume of the lamp. To overcome these problems, the excess of metal was captured in a special reservoir away from the discharge itself. The reservoir must have a lower temperature than the lamp. This temperature determines the partial pressure of the alkali metal vapor. In this case the requirements on the dosage of metal are reduced. Nevertheless, for use in quantum devices lamp bulbs without excess of alkali metal are desired. In the beginning of the 1980s, when testing the quantum frequency standards, a number of failures of light sources was noticed, resulting in series of experiments on determination of the optimum amount of rubidium in the lamps. Some methods to control the dosage of a certain amount of metal in the lamp bulbs have been developed and are reviewed in this chapter.

6.2. Measurement of Alkali Metal Content in Spectral Lamps

Here we present a method developed to control the rubidium vapor content of spectral lamps [68, 69, 111], now used in technology and research. However, it can be used for other elements with rather high vapor pressure, such as mercury, cesium, potassium, indium, tellurium, selenium, and some other easily vaporizable elements. The principle of the method, first suggested in [111], implies the calculation of the amount of metal necessary to obtain a certain vapor density in a known volume after total evaporation, which happens at temperature T_{cr}. The vapor density is measured via the absorption of light passing through the lamp during its gradual heating. A block diagram of the experimental setup is shown in Figure 6.1.

When the light of a band lamp or an incandescent filament lamp (1) emitting a continuous spectrum is passing through the electrodeless lamp containing metal vapor (2), the photoreceptor (3) records the signal

$$A = \int_0^\infty \phi_l(\omega) v(\omega) d\omega \tag{6.1}$$

where $v(\omega)$ is the spectral sensitivity of the photoreceptor; $\phi_l(\omega)$ is the spectral distribution of the recorded light:

$$\phi_l(\omega) d\omega = \phi_0(\omega)[1 - \exp(-kl)] d\omega \tag{6.2}$$

where $\phi_0(\omega)$ is the flow of light falling on the investigated lamp, kl the optical density, $k = k'n, k'$ is the reduced factor of absorption by one atom, and n is the density of atoms.

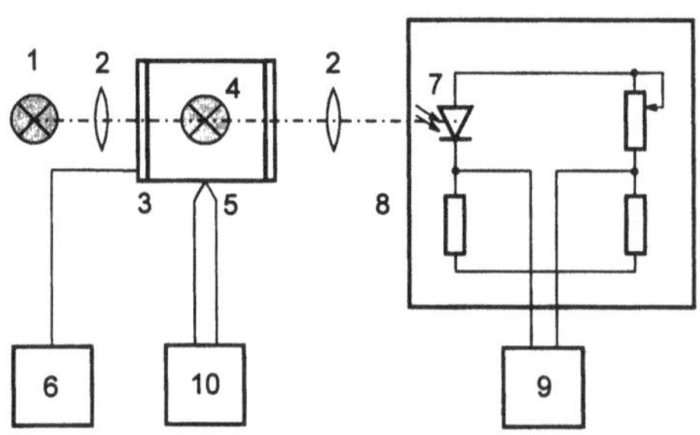

FIGURE 6.1. Block diagram of the experimental setup for measuring the amount of the metal. 1, incandescent filament lamp; 2, lenses; 3, thermostat; 4, lamp being researched; 5, thermocouple; 6, voltage regulator; 7, photodiode; 8, photoreceptor; 9, dc millivoltmeter.

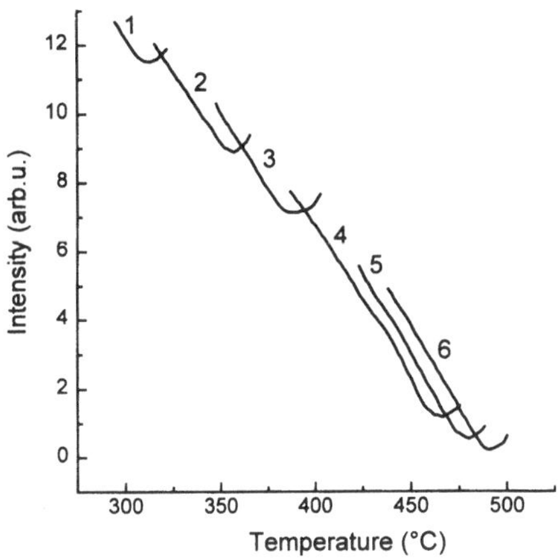

FIGURE 6.2. Intensity of light transmitted through the lamp depending on the temperature tested with different amounts of metal: 1–6, different quantities of metals providing different critical temperatures.

As already shown elsewhere [151], the value of complete absorption A at small optical density $k_0 l$ is proportional to the concentration of atoms n ($n^{1/2}$ respectively for high concentration), and therefore a similar dependence for A on the concentration of metal atoms can be expected. The concentration of rubidium atoms n in the saturated vapor in the closed volume depends on the temperature [155]:

$$n = \frac{1}{k_B T} 10^{(A+B/T+CT+D\lg(T/1\mathrm{K})} \tag{6.3}$$

where $A = 15.88$, $B = -4529.6$ K, $C = 0.00058$ K^{-1}, and $D = -2.991$. When increasing the temperature of the lamp the concentration of atoms of rubidium grows rapidly and the intensity of light passing through the lamp decreases. When increasing the temperature over T_{cr}, at which all metal is evaporated, no further change in the intensity of the passing light can be observed (Figure 6.2) Therefore, having recorded the temperature T_{cr} of complete evaporation, the amount of metal M in the lamp can be evaluated:

$$M = m_{\mathrm{Rb}} n(T_{cr}) V = \frac{m_{\mathrm{Rb}} V}{k_B T_{cr}} 10^{(A+B/T_{cr}+cT_{cr}+d\lg(T_{cr}/1\mathrm{K})} \tag{6.4}$$

where V is the volume of the lamp.

This method has its highest sensitivity at maximum slope of the dependence of the light intensity on the temperature at the moment of evaporation of all condensed metal. It can be increased by filters, which cut out the part of the radiation spectrum of the incandescent filament lamp surrounding the atomic resonance lines of the studied element. In particular cases it is also possible to use selective photoreceptors. The high temperature of the lamp was established by means of

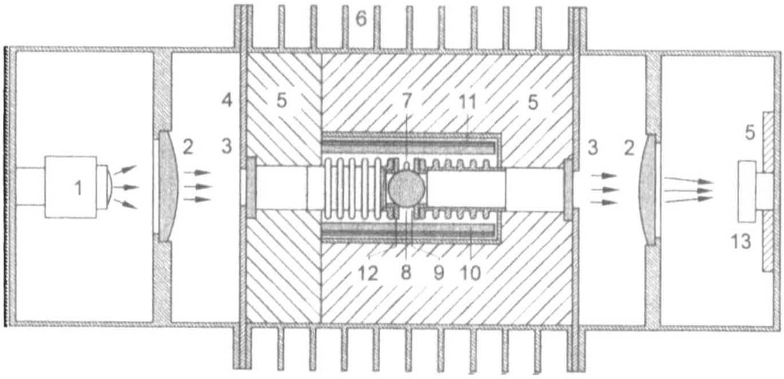

FIGURE 6.3. Design of the thermostat for measuring the amount of metal in the lamp: 1, incandescent lamp; 2, lens; 3, glass disc; 4, removable cover of thermostat; 5, asbestos cement thermo-insulator; 6, radiator; 7, siphon; 8, spectral lamp; 9, thermocouple; 10, quartz cylinder; 11, heater; 12, lamp holders; 13, photodiode.

a special thermostat (Figure 6.3), used for fast warming up to temperatures of 400–550 °C with simultaneous control of the transmitted light intensity. The use of a spectral lamp as background light source is not suitable, since the width of the absorption line at temperatures higher than 200 °C is much larger than the width of the radiation line. Consequently, a continuous light source was used. The dependence of the intensity of the transmitted light on the temperature is given in Figure 6.4. It can be seen, that the temperature at warming up and cooling differs slightly, due to both an error of the temperature measurement and absorption of rubidium by the glass at high temperatures. The amount of metal in various lamps, manufactured without special amount control, was found to be between 0.0003 and more than 0.2 mg, i.e., up to 600 times the smallest amount. This is the range of the amount of metal which is practically measurable for lamps with 1 cm^3 volume. When increasing the sizes of the lamp, the range of the measured quantity of metal naturally grows. For lamps of 10 cm^3 the range will be from 0.0001 mg up to more than 2 mg, and for a volume of 100 cm^3 from 0.00003 mg up to 20 mg.

The total measuring error in the amount of metal determined by this method is composed of: an error due to the tolerance in size when manufacturing the lamps and therefore in the volume V — 16%; an error due to the difference between the real temperature from the measured one — 3%; an error in the determination of the temperature by the thermocouple — 0.5%; an error in the determination of the temperature T_{cr} from the curves — 4%; an error determined by the burn-out of rubidium during the measurements — 4–8%. Hence, the total measuring error of the metal quantity is about 27–32%. Compared to the observable spread of the amount of metal this error is rather small. As the first two contributions are systematic errors, the total error can be reduced to 5% by repeated measurements

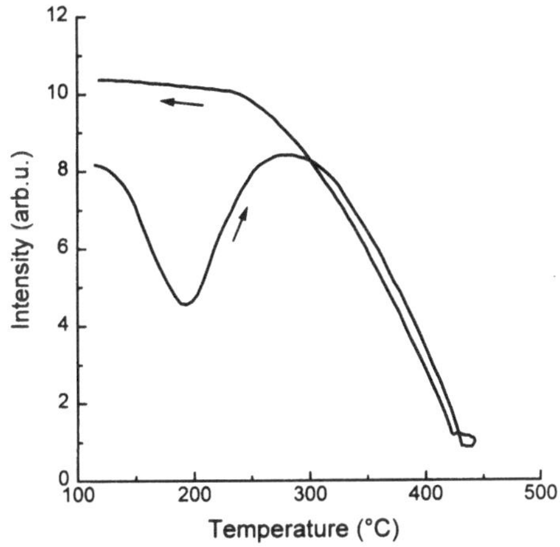

FIGURE 6.4. Dependence of the light intensity A, passing through the investigated lamp, on the temperature. Notice the different behavior when increasing/decreasing the temperature.

performed on one lamp without taking it out of the thermostat.

We will discuss in more detail the error connected to the burn-out of metal during the measurements. Under uniform heating of the lamp to a temperature T_{fin} the amount of metal in the volume of the lamp will change by

$$\Delta M = \text{const} \times \tau \int_{T_{in}}^{T_{fin}} f(T)n(T)dT \tag{6.5}$$

where τ is the duration of the measurement process, $f(T)$ that part of the atoms interacting with the glass per second at the temperature T, T_{in} and T_{fin} are the initial and final temperatures during measurements.

Since $n(T) \ll n(T_{cr})$ at 20 °C, the main error is caused by the process of measurement at $T > 20$ °C and especially at $T > T_{cr}$. Therefore it is necessary to optimize the time of measurement at $T > T_{cr}$. The experimental dependence of the error of measurement of the amount of metal M for the temperature T_{cr} is given in Figure 6.5.

The minimal amount of metal which can be determined is given by the design of the optical circuit and it is practically unlimited by the technology of lamps, when using filtered white light or discharge lamps as light source. The maximum amount of metal which can be measured, M_{max}, is determined by the volume of the lamp and the highest temperature T_{max} applicable to the lamp bulb containing the element under study:

$$M_{max} = m_{\text{Rb}} n(T_{max}) V \tag{6.6}$$

The temperature necessary to achieve a certain density of vapor is given in Table 6.1.

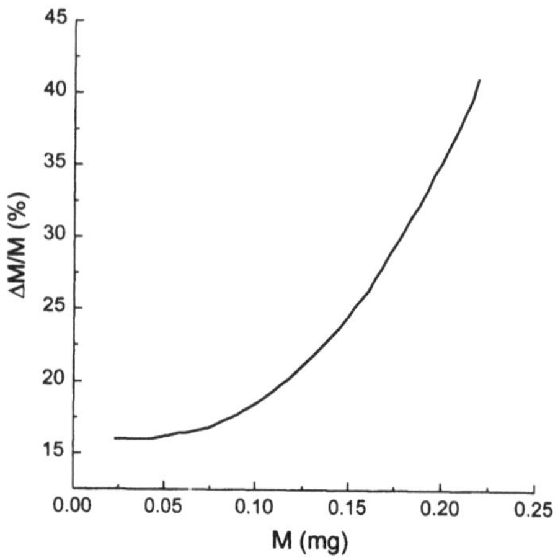

FIGURE 6.5. Experimental dependence of the measuring error on the amount of metal in the lamp.

When using this method for determining the amount of such elements as mercury, tellurium, or selenium it is necessary to use special light sources and photoreceptors, as the atomic resonance lines of these elements are located in the violet and ultraviolet parts of the spectrum.

6.3. Dosage of Metal in High-Frequency Lamps

The method described in Section 6.2 allows one to relate the properties of the lamps to the quantity of metal. However, to manufacture lamps with an exactly specified content of metal we require a reliable way to insert its dosage into the glass bulb. In [5] the dosage was carried out by sublimation of metal and redistribution of saturated vapor over the volume of the bulb. In this case the amount of metal dosed into the lamp appears to be proportional to $n(T)V$, as in the case of measuring the metal content described in Section 6.2.

The restrictions concerning the amount of substance dosed are similar to the restrictions in the above-considered control method. In practice the accuracy appears to be low, because of the nonuniform yield of metal when cooling the lamps, due to different heat capacities of various areas of the lamp and elements of the pump circuit in which the dosage is performed. Therefore in [5] an essentially advanced method was proposed, based on the introduction of vapor in the system with a buffer gas, that hinders diffusion from the lamp volume when cooling. Later on the technique was improved by using the sealing off of lamps with the help of an arc discharge at high temperature.This method appeared to be rather effective but complex. In case of dosage of rubidium into the lamps a more simple method, using equipment similar to that for measuring the amount of metal in the

TABLE 6.1. Temperatures [K] for Obtaining Different Vapor Pressures of Some Elements

	Vapor pressure [mbar]								
Element	1.3×10^{-5}	1.3×10^{-4}	1.3×10^{-3}	1.3×10^{-2}	1.3×10^{-1}	1.3×10^{0}	1.3×10^{1}	1.3×10^{2}	1.3×10^{3}
K	364	396	434	481	540	618	720	858	1070
Na	428	466	508	562	630	714	825	978	1175
Rb	336	367	402	446	500	568	665	802	1000
Cs	322	351	387	428	482	553	643	775	980
Fr	306	334	368	410	462	528	620	760	980
Hg	246	266	289	319	353	398	458	535	642
Se	406	437	472	516	570	636	719	826	972
Te	515	553	596	647	705	791	905	1065	1300
Zn	482	520	565	617	681	760	870	1010	1210
In	937	1015	1110	1220	1350	1520	1740	2030	2430

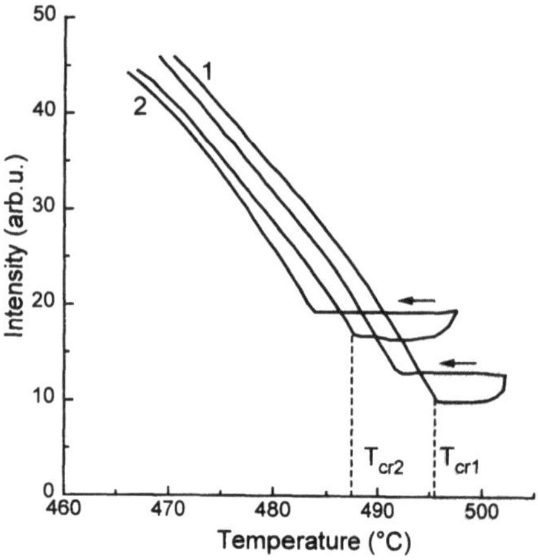

FIGURE 6.6. Change in the light intensity transmitted through the lamp during the dosage: T_{cr1}, T_{cr2}, temperatures at which all the metal is evaporated; 1, 2, different amounts of metal in the lamps.

lamp, was applied in production [111].

The idea of this method consists in the dosage of an obviously redundant amount of metal into the bulb before it is separated from the vacuum circuit and placed into the light source. The method is similar to the one described in Section 6.1. The lamp bulb is transilluminated by light of an incandescent lamp when simultaneously warming it to a temperature of 500–580 °C. Increase in temperature shifts the liquid–vapor equilibrium to the side of saturated rubidium vapor. At temperatures higher than 530 °C the penetration speed is great, and absorption of all condensed rubidium is possible within 5–20 minutes.

If the amount of metal is of the right order of magnitude, it is possible to control the burning out of metal by the transparency of the lamp. Until condensed rubidium is in the lamp, the absorption of metal increases rapidly with growing temperature and remains constant above a temperature T_{cr} (now all Rb is evaporated). While heating further and observing the transmitted light intensity (which increases due to penetration of the vapor into the glass), one can determine the amount of rubidium in the vapor phase. When achieving the desired amount of rubidium the warming up is ceased and the lamp is cooled. If necessary, a repeated check permits one to confirm the amount of metal in the lamp.

Figure 6.6 shows the experimental dependences which present the process of dosage of the metal. The accuracy of dosage appears to be of the same order as the accuracy of determining the amount of metal in the lamp. The lamps in which the metal is dosed in such a way are characterized by high stability of their parameters. A loss of metal after dozens of thousands of hours of operation is not observed. Moreover, when burning in a large amount of metal, during operation an increase in the metal concentration within the lamp can be observed, due to partial yielding out from the walls. The study of the radiation spectrum of the lamp did not show any contamination with other elements or impurities at a level of more than 0.1% of the total amount of atoms in the lamp.

7

Precise Measurement of Pressure in High-Frequency Electrodeless Spectral Lamps

7.1. Introduction

The pressure of a buffer gas present in high-frequency electrodeless lamps predetermines in many respects their main features: radiation intensity, radiation stability, longevity, ignition voltage, reproducibility of parameters, etc. On the other hand, the gas-pressure is a parameter that can be used for the diagnostic of spectral lamps, i.e., the proof of similarity of different lamps, for evaluating theoretical predictions compared to real processes, and for monitoring of the aging of lamps. Therefore a method, based on double radio-optical resonance techniques, was developed to determine the buffer gas pressure with high accuracy.

The principle of the radio-optical resonance can be easily explained in the following way. When an ensemble of atoms is optically pumped by absorption of light resonant to one of the hyperfine transitions from the ground level, a nonequilibrium population of the hyperfine levels of the ground state is formed. This leads to an increase in transparency for the pumping light beam. By applying a radio-frequency field resonant to the hyperfine splitting of the ground state, transitions between the hyperfine levels are induced, establishing again equilibrium population. In this way these resonant radio-frequency transitions can be clearly detected by a decrease in the transparency of the cell. This double radio-optical resonance is characterized by a very high accuracy, which is used in many precise optical devices [52].

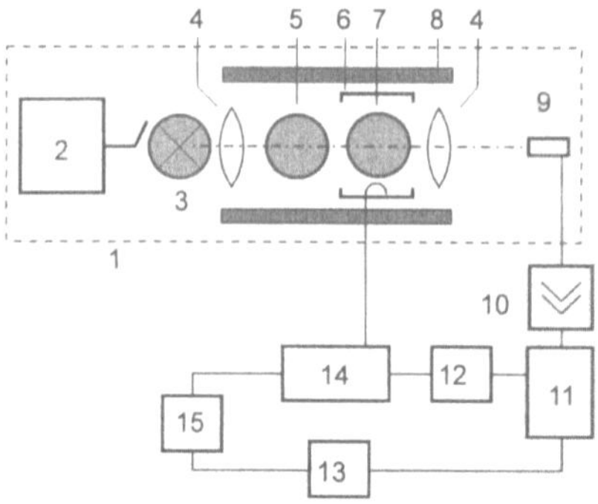

FIGURE 7.1. Scheme of the experimental setup for the observation of radio-optical resonance: 1, radiospectroscope; 2, hf oscillator; 3, lamp; 4, lens; 5, isotope filter cell; 6, resonator and heater; 7, cell with working element; 8, magnetic screen and magnetic system; 9, photodetector; 10, amplifier; 11, synchronous detector; 12, low-frequency oscillator; 13, quartz generator; 14, frequency multiplier; 15, frequency synthesizer.

7.2. Measurement of Gas Pressure by Means of Double Resonance Techniques

Double resonance techniques allow one to measure with high accuracy the frequency of atomic transitions, using the efficiency of optical pumping in a saturated vapor to create nonequilibrium populations in the sublevels of the ground state of the atoms. Knowing the shift of the transition frequency caused by the pressure of the buffer gas present, a precise determination of this pressure can be performed. In atoms, having a $^2S_{1/2}$ as ground state level, hyperfine transitions $F, m = 0 \longleftrightarrow F-1, m = 0$ show minimal dependence of the transition frequency on external fields and therefore are used as reference transitions in frequency stabilization techniques. Figure 7.1 shows a block diagram of the experimental set-up. The double resonance signal is formed in the radiospectroscope (1), consisting of light source (3), an isotope filter-cell (see below) (5), a gas cell (6) in which the working element is contained, which undergoes optical pumping, and a photodetector (9). The cell (7) is placed in a thermostat and irradiated by the resonant radio-frequency field. Usually this field is created by transformation of the frequency of a precise crystal oscillator (13). The resonance signal is recorded by a photoreceiver (9), detected by a synchronous detector (11) and serves for stabilization of the crystal oscillator by the frequency of the atomic transition. The output frequency of the crystal oscillator is measured by a frequency meter.

Vapor of the ^{87}Rb isotope is generally used as the working element in quantum frequency standards. In this case radiation of a ^{87}Rb lamp is filtered by a cell filled with ^{85}Rb vapor. The long-wave component of the hyperfine structure of the resonant doublet of ^{87}Rb coincides practically with the long-wave component group (a) of ^{85}Rb, while the short-wave group (b) differs in frequency. Owing to this situation the group (a) of ^{87}Rb is absorbed and pumping light, which enters the working cell, contains only one short-wave hyperfine group (b). The principle of the hyperfine filtering is shown in Figure 7.2. In this way only transitions from one of the hyperfine ground levels are excited by the pumping

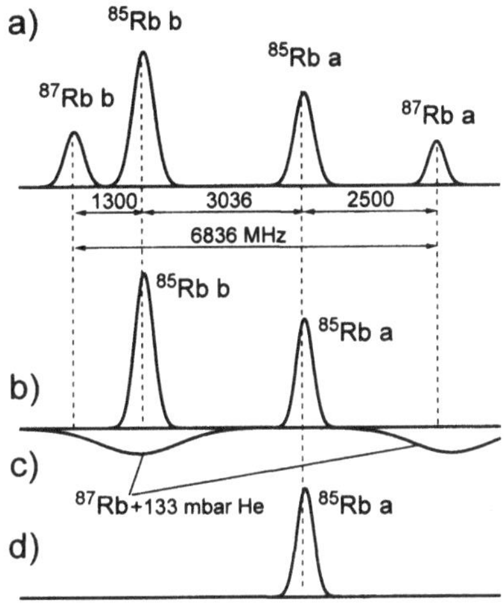

FIGURE 7.2. Principle of hyperfine filtering, here shown at the ^{85}Rb D_2-line: (a) hyperfine structure of the ^{87}Rb and ^{85}Rb D_2-line; (b) spectrum of the pumping lamp containing ^{85}Rb; (c) absorption spectrum of ^{87}Rb vapor in the cell with buffer gas; (d) spectrum of the pumping light on the cell's output. In a similar way radiation of ^{87}Rb can be filtered by means of a ^{85}Rb cell.

light, and an effective transfer of atoms to the other sublevel of the ground state (creating an orientation), accompanied by a gradual increase in the ^{87}Rb vapor transparency, is provided. The applied resonant radio-frequency field equalizes the population of the levels and hence reduces the vapor's transparency and the intensity of light, detected by a photodiode. When tuning the frequency of the hf field a line of absorption can be observed (Figure 7.3). This resonance signal is not large and for ^{87}Rb it is usually 0.001–0.01 from the background illumination, caused by lines of the buffer gas and nonresonant lines of the working element. It is difficult to receive a large signal/noise ratio by this method because of the fast increase in the spectral density of noise when reduction in the frequency takes place (Chapter 5). Therefore a phase modulation technique is usually used to increase the sensitivity. This modulation allows one to record the double resonance signal on the modulation frequency and thus to exclude low-frequency fluctuations of the light source, photodetector, and electronic circuits. Figure 7.4 shows the dependence of the first harmonic (the derivative) on the detuning of the radio-frequency field in relation to the absorption line. The sensitivity of this method is determined by the following relationship: $I_1 = S\delta\nu = I_n$, where I_1 is the radio-optical signal, $S = dI_1/d\nu$ is the slope of the discriminant curve, $\delta\nu$ the detuning of the radio-frequency field from the frequency of the atomic transition (from ν_0'), and I_n the noise voltage when $\delta\nu = 0$. The slope S is determined by the intensity of the pumping light, filling and size of the cell, characteristics of the radio-frequency field, and degree of the low-frequency modulation.

The frequency ν_0' of the reference spectral line is dependent on a number of disturbing factors—atoms of the buffer gas, the pumping light, the magnetic field,

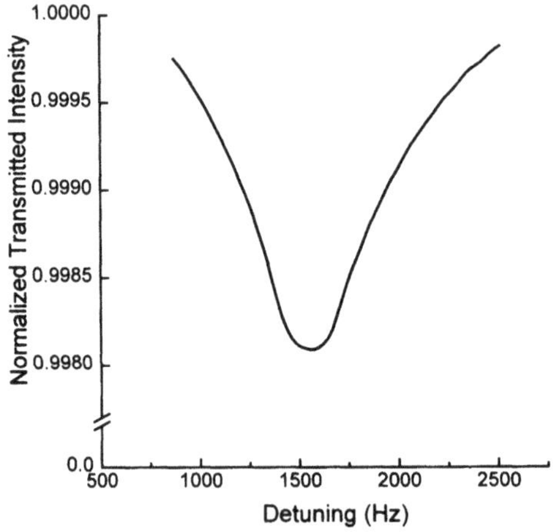

FIGURE 7.3. Radio-optical double resonance signal in a gas cell (volume 0.8 cm^3).

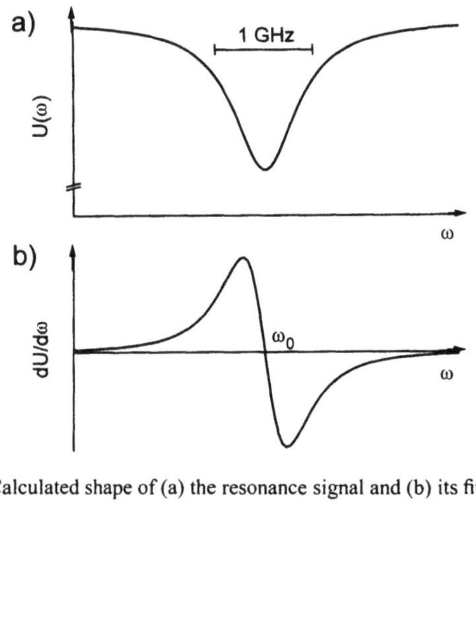

FIGURE 7.4. Calculated shape of (a) the resonance signal and (b) its first derivative.

— and is displaced in relation to the frequency ν_0 of the unperturbed transition:

$$\nu_0' = \nu_0 + n_\delta \int_0^\infty \left\{ \exp\left[-\frac{V(r)}{k_B T} \right] \right\} u(r) r^2 dr$$
$$+ \left[\int_0^l \beta(z) \frac{|\xi(z)|^2}{\gamma_1 \gamma_2} dz \right]^{-1} \int_0^l \beta(z) \frac{|\xi(z)|^2}{\gamma_1 \gamma_2} [\Delta_l(z) + \Delta_s(z)] dz + \alpha B^2 \quad (7.1)$$

where the second term on the right-hand side of the equation is the shift, caused by the atoms of the buffer gas with concentration n_δ, the third term is the light shift, the fourth term the shift, caused by a magnetic field; $V(r)$ is energy splitting of atomic levels, dependent on the distance between the colliding atoms; $u(r)$ is the energy of interaction of the atoms of the working element with the buffer gas; $\beta(z)$ is the dimensionless ratio of populations of the two hyperfine states of the ground level, induced by optical pumping; $\xi(z)$ is the distribution of the radio-frequency field in the cell, $\Delta_l(z)$ the local light shift, $\Delta_s(z)$ the spin-exchange shift, γ_1 and γ_2 are broadening, stimulated by longitudinal and transversal relaxation respectively. while z is the coordinate along the light path in the cell with length l.

High accuracy inherent to the process of frequency measurements (relative error 10^{-11}–10^{-13}) makes a precise determination of the real gas cell frequency possible. Using known values of the light shift, the magnetic field shift, and the shift caused by the spin-exchange interaction, it is possible to determine the shift, induced by the buffer gas, with high accuracy. Further on it is possible to calculate the gas pressure in the cell, using the value of the line shift and the known shift factor. Variants of such a method allow one basically to measure the pressure in the range from 0.001 to 250 mbar with very high accuracy ($\Delta p/p$ can achieve 10^{-6}). In certain cases the measurements can be carried out in closed volumes without destruction. Moreover, it is noteworthy that this method permits the partial pressures in a mixture of two or three gases to be determined.

The possibilities of such a double resonance technique for the measurement of the gas pressure were realized long ago, practically from the moment of creation of the first quantum frequency standards. This technique was used to calibrate gas cells and to control the gas pressure. In a patent [175] it was proposed to use a gas cell as a pressure gauge, having connected its volume with an investigated volume by a siphon, with an internal partition. A device for pressure measurements making use of the gas cell of a quantum generator as its sensitive element [176] was reported in 1969. In the beginning of the 1970s a method of pressure control in closed volumes (as well as in miniaturized cells) was described [67] and later on it was accepted in industry for manufacturing spectral devices. In 1975 a variant of the method for pressure measurement utilizing a flexible shell was described [177] and in [178] the partial pressure measurement technique was proposed. Nowadays these methods of pressure calibration and control find application basically in the fabrication of spectral devices for quantum frequency standards.

7.3. Effect of the Gas on the Frequency and Magnitude of the Double Resonance Signal

Collisions between atoms in the gas phase result in a shift and broadening of spectral lines. This effect, studied in detail for optical transitions [169, 191], influences also the energies of the hyperfine structure levels of the atoms. This fact was noticed already in the first experiments on observation of radio-optical resonance phenomena [192–194]. The physical reason for this effect are Van der Waals and exchange interactions. The perturbing atoms give rise to different line shifts: the transition frequency increases under collisions with light atoms ("blue" shift) and decreases under collisions with heavy atoms ("red" shift). The value of the shift is approximately proportional to the frequency of the studied transition. There exist important qualitative differences between optical and radio-frequency transitions: the broadening of optical lines is always larger than their shifts (approximately by a factor of 2.8), but the shifts of the radio-frequency transitions (at least, as are such caused by inert gases) are larger than their pressure broadenings. This difference can be explained by the following fact: radio-frequency transitions are magnetic-dipole transitions and optical ones are electric-dipole transitions. This specific behavior determines the high efficiency of the radio-optical resonance for the frequency stabilization and for the measurement of the foreign gas pressure. The pressure shift can be calculated using [195]:

$$\Delta\nu_p = n_\delta \int_0^\infty \exp\left[-\frac{V(r)}{k_B T}\right] u(r) r^2 dr \tag{7.2}$$

$\Delta\nu_p$ is proportional to the concentration of atoms up to pressures of 1–3 mbar [196]. Further increase in the concentration causes deviations from strict linearity due to the increase in the probability of triple collisions.

The temperature factor of the frequency shift can be estimated, taking the derivative of the expression for the shift with respect to temperature:

$$\frac{d\Delta\nu_p(T)}{dT} = \frac{n_\delta}{k_B T} \int_0^\infty \exp\left[-\frac{V(r)}{k_B T}\right] u(r) V(r) r^2 dr \tag{7.3}$$

For the description of the interatomic interactions the Lennard-Jones potential is used in the form [48]

$$u(r) = 4\varepsilon_1 \left[(\sigma_1/r)^{12} - (\sigma_1/r)^6\right] \tag{7.4}$$

or

$$u(r) = 4\varepsilon_2 \left[(\sigma_2/r)^{12} - (\sigma_2/r)^6\right] - D_1/r^8 - D_2/r^{10} \tag{7.5}$$

TABLE 7.1. Frequency Shifts of Radio-Frequency Transitions within the Ground States of Rb and Cs Caused by Inert Gases

Shift	Rb–He	Rb–Ne	Rb–N_2	Rb–Ar	Rb–Kr	Cs–He	Cs–Ne	Cs–N_2
By pressure [Hz/mbar] Experiment	541 ± 10	295 ± 6	390 ± 8	-38 ± 1	-436 ± 37	789	436	670
By pressure [Hz/mbar] Calculation	541	287	398	−158	−308	902	458	676
By temperature [Hz/mbar K] Experiment	0.75	0.20	0.45	−0.22	−0.38	1.13	0.075	—
By temperature [Hz/mbar K] Calculation	0.75	0.32	0.33	−0.11	−0.15	1.28	0.22	1.28

In Table 7.1 results of calculations performed in [48] are compared with experimental shift values.

However, in [48] adjustable parameters were used due to insufficient knowledge of all the essential characteristics of the colliding atoms. Later on a theory, allowing one to calculate the shift and broadening of a spectral line without any adjustable parameters with an accuracy of about several percents, was developed [197]. Using these results, one can in principle determine the gas-pressure by measuring the shift without any calibration experiments.

We speak about the measurement of pressure, but it is obvious from Eq. (7.2), that the frequency shift primarily characterizes the concentration of atoms in the volume. In those applications for which this method was developed and used, this question was not basic. As a rule, the pressure of a gas is more easily controlled and this parameter is usually used in experiments for measurement of the dosage of a gas.

For the analysis of the experimental dependence of the frequency shift on the concentration of atoms or on the pressure, let us expand expression (7.2) at a point T_1 in a Taylor series:

$$\Delta\nu_p(T_1) = Cn_\delta \sum_{k=0}^{\infty} (T - T_1)^k \left.\frac{\partial^k \varphi}{\partial T^k}\right|_{T=T_1} \tag{7.6}$$

where C is a constant; here

$$\varphi = \int_0^\infty \exp\left[-\frac{V(r)}{kT}\right] u(r)\, r^2 dr \tag{7.7}$$

Designating by p_0 the gas pressure in a sealed off cell we obtain

$$\Delta\nu_p(T_1) = p_0\left[c_0 + c_1(T-T_1) + c_2(T-T_1)^2 + c_3(T-T_1)^3 + \cdots\right] \tag{7.8}$$

where c_i is constant, $i = 0, 1, 2, \ldots$.

Therefore, the approximation of the experimental dependence can be expressed as a polynomial of the following form:

$$\Delta\nu_p(T_1) = p_0\sum_{i=0}^{\infty} c_i(T-T_1)^i \tag{7.9}$$

The shifts of the hyperfine transitions as a function of the temperature for Na, K, and Rb in He, Ne, and N_2 for temperatures from 100 to 850 °C have been studied elsewhere [198, 199]. A nonlinear function of the frequency as dependent on temperature was found, and could be approximated by the polynomial

$$\Delta\nu_p = p_0\left[c_0 + c_1x + c_2x^2 + c_3x^3 + c_4x^4 + c_5x^5\right] \tag{7.10}$$

where $x = T(°\mathrm{C})/1000$. Values of the factors c_0–c_5 are presented in Table 7.2.

The theoretical approximation and the experimentally determined factors presented in [199] seem to be rather reliable, but they were obtained only for three alkali elements and four gases. A determination of the shifts for ^{87}Rb is possible using the data for ^{85}Rb, multiplying the shifts by the factor 2.252, which is the ratio of the hyperfine splittings of the Rb isotopes [200].

The number of terms of the polynomial required for the approximation depends on the accuracy and temperature range. An approximation by a second-order polynomial was used in [200] and the coefficients for two gases have been determined:

$$\Delta\nu_p = p_0\left[c_0 + c_1(T-T_1) + c_2(T-T_1)^2\right] \tag{7.11}$$

where for argon $c_0 = -79.4$ Hz/mbar, $c_1 = -0.43$ Hz/(°C mbar), $c_2 = -0.00046$ Hz/(°C^2 mbar), and for nitrogen $c_0 = 728.7$ Hz/mbar, $c_1 = 0.69$ Hz/(°C mbar), $c_2 = -0.0017$ Hz/(°C^2 mbar), $T_1 = 60$ °C.

A comparison with the usually used approximation:

$$\Delta\omega = p_0\left[c_0 + c_1(T-T_1)\right] = \alpha p_0 \tag{7.12}$$

where p_0 is the pressure at temperature T_1, showed that neglecting the third term lead to an error of 0.08% in a vicinity of ±2 °C, of 0.5% for ±5 °C, 2% for ±10 °C, and 8% for ±20 °C.

Because the indispensable parameters of the approximations (7.10) and (7.11) for a majority of elements and gases are not available, one is compelled to

TABLE 7.2. Coefficients c_0–c_5 of the Polynomial (7.10) for Some Elements and Gases

Element	Gas	Range of temperature	c_0	c_1	c_2	c_3	c_4	c_5
Na	He	50–800	1035.62	2174.78	−2672.84	2130.28	−795.682	—
	Ne	85–850	654.073	1102.05	−1917.87	1516.80	−503.432	—
	Ar	100–700	−11.6738	502.190	−1098.32	515.151	—	—
	N_2	35–800	629.084	1764.03	−2620.53	2383.29	−998.774	—
K	He	−100–800	419.861	708.683	−1050.84	880.880	−334.209	—
	Ne	−65–750	237.893	244.247	−741.786	841.616	−384.843	—
	Ar	−80–750	−19.3265	−73.7039	−579.824	933.986	−534.416	—
	N_2	−100–800	294.677	498.461	−925.127	793.46	−303.578	—
Rb	He	−100–800	3043.58	4838.71	−7795.42	6928.33	−2665.42	—
	Ne	−70–500	1743.98	1385.74	−5625.24	9273.53	−6806.00	—
	Ar	−100–500	−196.621	−1168.26	−1801.90	1552.2	—	—
	N_2	−125–800	2270.92	3178.91	−7436.54	10989.0	−11219.1	4991.34

use the simplified approximation formulas for the shifts (7.12). A special analysis shows that this should be done cautiously, paying attention to the temperature, and other experimental conditions at which these characteristic parameters have been determined.

Further approximations are known [201]. In Table 7.3 the shift factors for formula (7.12) are presented. The sign (*) marks coefficients corresponding to the simplified form

$$\Delta\nu_p = c_0 + c_1 T \qquad (7.13)$$

The shift of a spectral line caused by a given gas is approximately proportional to its frequency (Table 7.4). The experimental dependences of shifts and broadenings of rubidium lines in cells coated with paraffine are given in Figure 7.5.

An inner coating used in some cells for the reduction of relaxations (Figure 7.6a) caused by collisions with walls also causes shifts in the spectral lines, usually of magnitude 1–100 Hz. The dependence of the shift in the Rb resonance line on the temperature for a cell coated with paraffine is shown in Figure 7.6b [17]. Similar results were obtained [190] for the material tetracontan.

Collisions of atoms of the element on which the resonance is observed with atoms of a foreign gas determine the shift and also the broadening of the transition. The sort and pressure of the gas determine the rate of the atomic diffusion as well, and therefore the rate of relaxation on the walls of the cell. These processes are characterized by the cross section of collisions and the diffusion coefficients. For a number of elements these values are presented in Table 7.5. They refer to longitudinal relaxation, except for those marked by a superscript '*a*,' obtained for transversal relaxation. Table 7.5 shows that the factor of diffusion is in the range 0.1 to 1.34 cm^2/s and decreases on taking heavier atoms. The cross section of collision varies within a larger range: from 10^{-25} cm^2 for light atoms and molecules to 10^{-18} cm^2 for heavy particles.

The dependence of the resonance signal and characteristic times of the longitudinal and transversal relaxations for Rb atoms on the pressure of a number of gases are presented in Figure 7.6. The pressure of gas strongly influences the signal resonance, but has no effect on the accuracy of measurements within certain limits. However, on increasing or decreasing the pressure the resonance signal becomes broader and the accuracy of the gas-pressure measurement begins to drop.

At optimal conditions the accuracy is determined mainly by an error incorporated in the factor of the line shift, which is 1–8%. By using the factors presented by other authors it is necessary to pay special attention to the type of approximation and temperature for which these parameters have been determined. The smallest error is presented in [201] for the ^{87}Rb isotope and is approximately

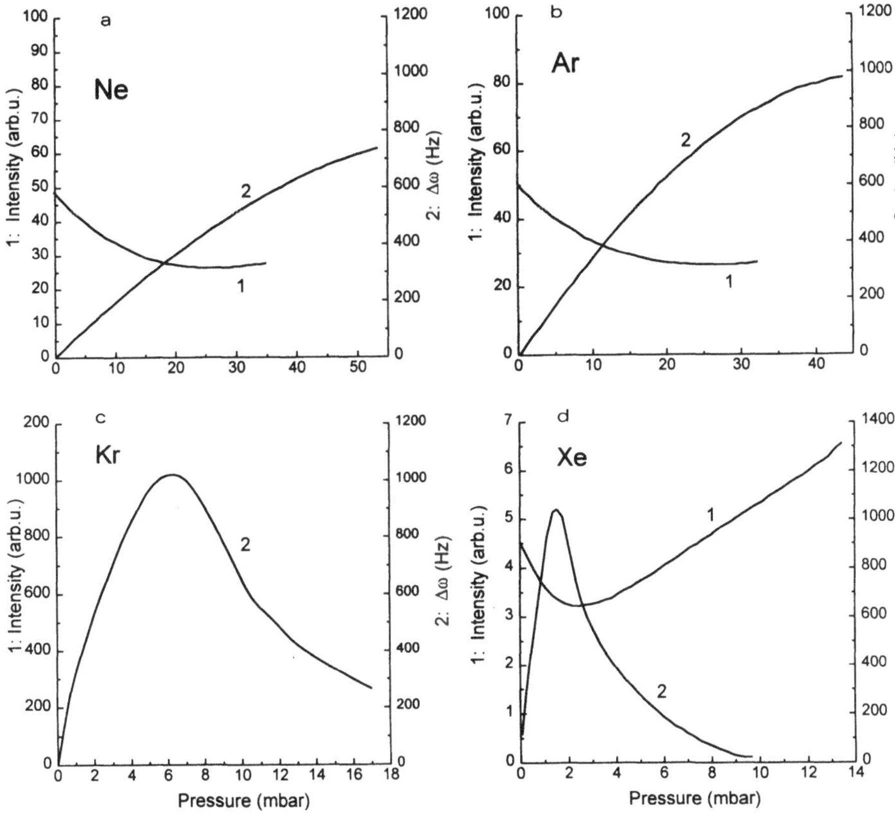

FIGURE 7.5. Pressure shift (2) and intensity (1) of the double resonance signal of Rb caused by: (a) Ne; (b) Ar; (c) Kr; (d) Xe.

0.2%. In order to attain such an accuracy, the error in temperature measurement should not be more than 1–2 °C. Besides, it is necessary to be sure about the purity of the studied gases. The precision of relative measurements can be significantly higher, but control of the gas temperature is also required for this purpose.

7.4. The Double Resonance Technique for Pressure Measurements

7.4.1. Variants of the Technique

There are two main variants of this method: (1) The gas under study is directly introduced into a gas cell, located in the radiospectroscope; in this device the change in the transition frequency of the working atoms under the effect of

TABLE 7.3. Frequency Shift Coefficients for Transitions between Hyperfine Levels of the Ground State for a Number of Elements Caused by Different Gases, Eq. (7.12)

Element	Gas	Factor of shift			Temperature	
		α [Hz/mbar]	c_0 [Hz/mbar]	c_1 [Hz/(mbar°C)]	T [°C]	Reference
Na	Ne	60			120–130	[95]
	Ar	0				
K	H	25 ± 2.2			65	[281]
	He	32 ± 2.2				
	Ne	18 ± 1.5				
	Ar	−0.3 ± 1.1				
	Kr	−32 ± 4				
Rb	He	246 ± 7.5			27	[17]
	Ne	135 ± 3.7				
	N	181 ± 6				
	Ar	−18 ± 0.6				
	N_2		637.7 ± 1.5	−0.780 ± 0.003	35–70	[201][a]
	Ne		529.5 ± 1.5	−0.751 ± 0.003		
	Ar		−55.9 ± 0.02	−0.1205 ± 0.0006		
	CH		571.4 ± 6.8	0.6681 ± 0.0022		
	Kr		683.0 ± 2.2	0.801 ± 0.003		
	Xe	−854			55	[282]
	H		496.2 ± 9.8	0.75	20–30	[194]
	D		503.7 ± 11.3	0.75		
	He		541.3 ± 10.5	0.75		
	Ne		294.7 ± 6	1.95		
	N		391.0 ± 7.5	0.45		
	Ar		−39 ± 1	−0.2		
	Kr		−436 ± 4	−0.4		
^{87}Rb	CH_4		−376	+0.45		
	n-pentane		−2105	+0.5		
	neo-pentane		−2180	−0.15		
	n-heptane		−3160	−0.15		
^{109}Ag	He	69.6 ± 2.4			597	[172]
	Ne	33.8 ± 0.3			624	
	Ar	18.7 ± 3.4			510–550	
	Kr	−13.2 ± 0.9				

TABLE 7.3. *Continued*

Element	Gas	Factor of shift			Temperature	Reference
		α [Hz/mbar]	c_0 [Hz/mbar]	c_1 [Hz/(mbar°C)]	T [°C]	
^{109}Ag	Xe	-84.2 ± 3.4				
	N_2	43.2 ± 3.4			620	
^{133}Cs	H_2	1382	—	—	15–22	[193]
		11430	—	—	22	[192]
	He	1200	—	—		
		788	—	—	15–22	[193]
		900	—	—		[287]
		—	891 ± 35	1.12 ± 0.11	0	[288][a]
	Ne	436	—	—	15–22	[193]
		488	—	—	22	[192]
	N_2	670	—	—	15–22	[193]
		700	—	—	22	[192]
		—	695.2 ± 5	0.468 ± 0.04	0	[288]
	Ar	−145	—	—	15–22	[193]
		−188	—	—	22	[192]
		−160	—	—	15–22	[287]
		—	-144 ± 2	-0.79 ± 0.04	0	[288][a]
	Kr	−981	—	—	15–22	[193]
		−977	—	—	22	[192]
		−992	—	—	15–22	[287]
	Xe	−1800	—	—	22	[192]
		−1766	—	—	22	[287]
		—	-1090 ± 40	-1.4 ± 0.4	0	[288][a]
	CH_4	—	-790 ± 22	1.1 ± 0.1		
	C_2H_6	—	-1392 ± 22	-1.41 ± 0.02		
	C_3H_8	—	2224 ± 28	-1.53 ± 0.06		
	C_4H_{10}	—	−3003	−1.5		
^{205}Tl	He	74 ± 22	—	—	323	[286]
	Ne	90 ± 7	—	—		
	Ar	-165 ± 7	—	—		
	Kr	-368 ± 7	—	—		
	Xe	-750 ± 60	—	—		

[a]Coefficients valid for Eq. (7.13).

TABLE 7.4. Normalized Shift Coefficients $\Delta\nu/(\nu_0 p)$ for a Number of Elements Caused by Different Gases

	Transition frequency	Normalized shift coefficient $\Delta\nu/(\nu_0 p)[10^{-9}\ \text{mbar}^{-1}]$							
Element[a]	ν_0 [Hz]	H_2	He	Ne	N_2	Ar	Kr	Xe	CH
^{7}Li	0.803×10^9	—	58.4	—	—	—	—	—	—
^{23}Na	1.772×10^9	—	55	26	—	0	—	—	—
^{39}K	0.462×10^9	71	54	38	—	−1	−68	—	—
^{85}Rb	3.035×10^9	81	44	60	−6	—	—	—	—
^{87}Rb	6.835×10^9	79	73	43	—	−6	−64	−125	−55
^{109}Ag	1.977×10^9	—	—	—	22	—	−7	−43	—
^{133}Cs	9.193×10^9	158	100	50	75	−17	−108	−20	—
^{205}Tl	21.31×10^9	—	4.6	4.2	—	−7.7	−17.3	−35.3	—
^{87}Rb[a]	3.771×10^{14}	—	9.2	0.9	−12.0	−9.8	—	−12.0	—

[a] The data of the shift factors for optical transitions are taken from [282] for Rb and from [197] for Li.

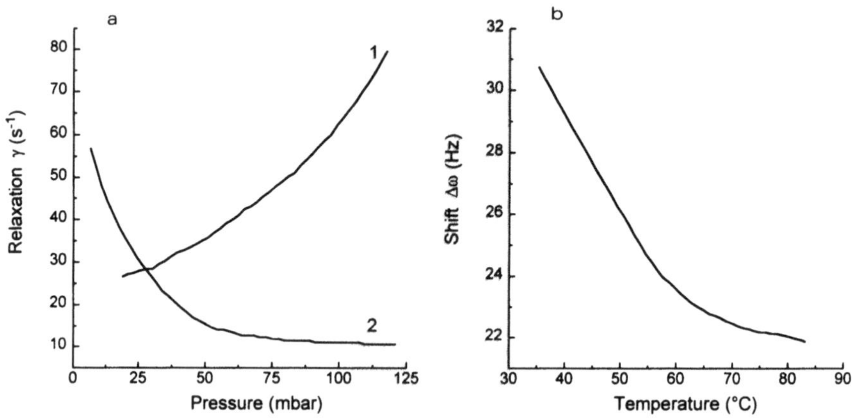

FIGURE 7.6. (a) 1, longitudinal and 2, transverse relaxation rate of Rb for a cell coated with paraffin. (b) shift of the radio-optical resonance as a function of the temperature for a coated cell.

TABLE 7.5. Diffusion Coefficients and Collision Cross Sections for Some Elements and Gases

Element	Gas	Coefficient of diffusion [cm^2/sec]	Cross section of collision α [cm^2]	T [°C]	Reference
Na	Ne	0.050 ± 0.17	$(1.8 \pm 0.6) \times 10^{-24}$	152 ± 2	[283]
	He	1.0 ± 0.3	$(3 \pm 4) \times 10^{-24}$	154 ± 2	
	Ar	0.2	5×10^{-23}	—	[284]
K	He	1.2	1.1×10^{-25}	85	[285]
^{85}Rb	He	0.32 0.42[a]	2.94×10^{-22}[a]	25	[200]
	Ne	0.16 0.20[a]	5.55×10^{-22}[a]		
	N_2	0.13 0.15[a]	8.0×10^{-23} 7.4×10^{-22}[a]		
	Ar	0.14 0.12[a]	4.9×10^{-22} 3.7×10^{-22}[a]		
^{87}Rb	He	0.425 ± 0.065	$(4.9 \pm 1.5) \times 10^{-25}$	—	[285]
	Ne	0.31	1.6×10^{-22}	—	
	Ar	0.22	9×10^{-22}	—	
	Kr	0.16	3.3×10^{-21} (3.1×10^{-20})	—	
Rb	H_2	1.34	3×10^{-24}	70	
	D_2	—	4.3×10^{-23}	35	
	N_2	0.33	5.7×10^{-23}	55	
	CH_4	0.5	8×10^{-24}	60	
	C_2H_6	0.3	3.8×10^{-23}	—	
	C_2H_4	0.24	1.3×10^{-22}	—	
	C_6H_6	—	$(6.0–7.5) \times 10^{-19}$	55	[285]
	C_6H_{12}	0.10	4.5×10^{-22}	50	
	CO	—	$(1 \pm 0.5) \times 10^{-22}$	55	
	$(CH_3)_2O$	—	$(3 \pm 1) \times 10^{-18}$	55	
	$C_4H_8O_2$	—	4×10^{-18}	—	
	NH_3	—	$(8 \pm 2) \times 10^{-18}$	55	
^{109}Ag	Kr	—	5.4×10^{-20}	680–730	[172]
	Xe	—	2.0×10^{-19}		
^{133}Cs	He	0.37	2.5×10^{-24}	26	[285]
	Ne	0.40(0.24)	$5.3(8.4) \times 10^{-24}$	44(26)	
	Ar	0.23(0.19)	$8.0(2.6) \times 10^{-23}$	44(26)	
	Kr	—	2.1×10^{-21}	44	
	Xe	—	4.6×10^{-20}		
	N_2	0.29	7.0×10^{-23}	52	
	$^{14}N_2$	0.22	4.7×10^{-18}	44	
	$^{15}N_2$	0.20	5.3×10^{-23}		
	C_4H_6	—	4.7×10^{-18}	20	
^{205}Tl	He	—	$(1.9 \pm 0.2) \times 10^{-20}$	430	[286]
	Ne	—	$(1.1 \pm 0.1) \times 10^{-19}$		
	Ar	—	$(3.4 \pm 0.5) \times 10^{-19}$		
	Kr	—	$(4.4 \pm 0.6) \times 10^{-19}$		
	Xe	—	$(1.4 \pm 0.2) \times 10^{-18}$		

[a] Valid for transversal relaxation; all other values are given for longitudinal relaxation.

this gas is measured. (2) The studied gas somehow affects the volume of the gas cell; this causes a change in the pressure of the buffer gas, leading to a shift in the frequency of the working transition. We shall name these variants the "contact" and "contactless" methods of gas-pressure determination. In the contactless method the cell of the radiospectroscope is used as a high-sensitivity pressure gauge. For this purpose a gas with a large frequency shift factor of the transition is introduced into the cell, such as heptane. The working cell is connected by its flexible side wall to a volume in which the gas pressure should be determined [177]. Sometimes, the flexible shell consists of a siphon with a partition [175].

The sensitivity of the double resonance effect to the change of pressure is rather high. For example, the use of ^{87}Rb and heptane with a pressure of about 13 mbar and a relative error in frequency measurement of about 10^{-12} could insure an error in the pressure measurement of less than 2×10^{-7} mbar, and a relative error less than 2×10^{-8} (if the error were related only to the frequency measurement). The achievable accuracy is determined mainly by the characteristics of the flexible shell of the cell. A measurement of higher pressures is possible when using a rigid shell. For a cell with a wall thickness of 2 mm, filled with a mixture of neon and argon, a pressure range from 250 up to 1050 mbar (full frequency shift, 4.8 kHz) with an accuracy of 0.25 mbar is possible. It is evident that the range can be widened, and that the accuracy of measurements can be increased by a reduction in the wall thickness of the gas cell. A disadvantage of this variant of the pressure-measurement technique is connected with the necessity to use certain elastic elements for transferring the outer pressure to the gas in the cell. Therefore a calibration by other manometers, enabling an absolute measurement of pressure, is necessary.

In the contact technique the gas, the pressure of which should be determined influences directly the working atoms. For a calculation of the pressure the sort of gas and the factor of the frequency shift must be known. The frequency shift depends linearly on the pressure, but the range of reliable measurements is restricted and depends on the sort of gas owing to differences in the disorientation cross sections. Therefore measurements with high accuracy up to 150–250 mbar in mixtures of neon and helium, up to 20–25 mbar in krypton, and up to 6 mbar in neon are possible. At pressures lower than 0.5 mbar the sensitivity of this technique decreases. One main problem of this method is the determination of the temperature of the outer volume and of the cell. The sensitivity of the method depends on the sort of gas and varies, for example, from less than 2×10^{-7} mbar in heptane to 1.5×10^{-4} mbar in argon. The absolute accuracy of the pressure measurement is determined by an error of the shift factors which, in the best case, is not less than 0.25%. The shift factor can be calculated in principle by means of atomic constants, but the achieved accuracy for Rb in such a way is not more than several percents.

The decrease in the accuracy of the contact method at low pressure is connected with the increase in collisional relaxation on the walls of the cell. To suppress this processes it is possible to use a cell with a special coating that prevents the disorientation. In such cells, under low pressure, the sensitivity of the double resonance is determined by on-wall relaxations. Assuming that the relative error of frequency determination is 10^{-12} it is possible to evaluate the achievable sensitivities of the pressure measurement, which will be the same as for the contactless method. However, in this case the lower limit of the pressure to be determined is given by the required accuracy. For example, with an error of 1% the pressure of heptane can be measured down to 2×10^{-5} mbar, nitrogen and neon up to 3×10^{-4} mbar, and argon up to 1.5×10^{-2} mbar. Thus, by the use of coated cells it is basically possible to overlap the range of pressures from less than 10^{-4} to more than 250 mbar.

Using the contact method it is possible to measure with high accuracy the partial pressures of a mixture of two or three known gases [178]. For this purpose it is necessary to measure frequency shifts of atomic transitions in the ground state of two or three isotopes (the number of isotopes should equal the number of gases in a mixture). For this case it is possible to write the following system of equations:

$$\nu_k - \nu_0 = \sum_i \alpha_{ik} p_i \tag{7.14}$$

where ν_0 and ν_k are the frequency of the atomic transition of the kth isotope in absence and presence of the gas respectively, while p_i are the partial pressures.

Resolving Eq. (7.14) for the partial pressures, we obtain

$$p_i = D_k / D \tag{7.15}$$

where

$$D_k = \sum_{i=1}^{n} \alpha_{ik} f_i, \qquad D = \det \alpha = \begin{vmatrix} \alpha_{11} & \cdots & \cdots & \alpha_{n1} \\ \alpha_{12} & \cdots & \cdots & \alpha_{n2} \\ \cdots & \cdots & \cdots & \cdots \\ \alpha_{1n} & \cdots & \cdots & \alpha_{nn} \end{vmatrix} \tag{7.16}$$

In this case the accuracy of determination of the partial pressures is also caused by an error in the shift factors. The relative pressure change can be recorded in as precise a manner as in the case of a one-component gas.

In the case of a two-component gas the technical realization of this method is not complicated. A cell has to be used containing a second isotope, and an additional radio-frequency must be applied, driving a hyperfine transition of this second isotope.

The equipment for measuring three components of gases seems to be more complicated and we have no information on the technical realization of this method. However, it seems possible to create a device for measuring the partial pressures of a three-component mixture using an alloy of Rb and Cs isotopes with known characteristics.

A particular advantage of the contact method is the pressure measurement in closed volumes. For this purpose the working element must be present in the volume, and be illuminated by the pumping light and irradiated by the radio-frequency radiation. Nowadays this method is effectively applied in manufacturing and investigating hf ELs and gas cells for quantum frequency standards. The volume in which the pressure is determined can vary over a wide range beginning from 0.5 to 200 cm^3, and even more. The accuracy of the pressure determination for smaller volumes and for pressure ranges from 1 to 20–250 mbar (depending on the sort of gas) is determined by the error in the frequency shift coefficient. The duration of measurement depends on the time to establish the working temperature and the required accuracy, and it varies from seconds to 30 minutes and more.

By comparing contact and contactless methods, we see that in both cases it is possible to ensure high accuracy in determining gas-pressure changes. For the contact method the accuracy depends on the shift factor of the gas, and in the contactless method on the characteristics of the flexible shell and on the accuracy of the calibration procedure by means of an absolute manometer. This calibration is not necessary for the contact method but, on the other hand, this method essentially involves knowledge of the value of the frequency shift factor of a transition. In the contactless method it is not necessary to know the composition of the investigated gases. The contact method requires this knowledge as well as a guarantee that the gas whose pressure is to be determined does not influence the working element in the gas cell. The contact method basically allows measurement of the partial pressures in a mixture of known gases. One advantage of the contactless method is the possibility to measure the pressure over a higher range from 250 to 1050 mbar. Thus, both methods have their specific benefits and their fields of application.

7.4.2. Experimental Setup for Gas-Pressure Determination

An apparatus for gas-pressure determination with the aid of radio-optical resonance is a quantum frequency standard in which the studied volume replaces the gas cell, or in which the volume of the gas cell is related to the volume under study.

The frequency of the radio-frequency field must vary over a rather wide range to allow double resonance observation. This can be achieved either by the use of a specially tunable crystal oscillator and frequency multiplier, or by the use of a crystal oscillator and a tunable synthesizer. Both variants are used in pressure-measurement devices. The radio-frequency oscillator operates with

FIGURE 7.7. Block diagram of the device for gas-pressure measurement in the spectral lamps: (a) electronic scheme: 1, discriminator with slope S_d; 2, frequency multiplier with a factor of multiplication N; 3, integrator with a factor of transfer $1/(1+\zeta T)$; 4, control element with slope S_y; 5, crystal oscillator; (b) block diagram: 6, setup for pressure measurement; 7, comparator; 8, secondary frequency source; 9, frequency meter; 10, recorder.

automatic frequency tuning. The frequency of the oscillator is compared with the output of a quantum frequency standard. We shall consider the relation between the parameters of the radiospectroscope and the system for automatic frequency tuning, with the instability of the quantum frequency standard, which determines the real sensitivity of the device for gas-pressure measurements. The device is presented in a block diagram (Figure 7.7) in which all the main control elements are present. The radiospectroscope is the block for comparing of the frequency detuning with the reference line.

7.4.3. *Recording Scheme*

The frequency of the oscillator, fulfilling the double resonance condition, must be compared with a reference frequency. The system consists of a comparator, which accepts the measured signal, and the signal of a reference oscillator. The output signal of the comparator is processed and stored by a small computer. As reference oscillator a quantum frequency standard of any type can be used (accuracy of pressure measurements, 0.1–0.01%). When performing limit measurements it is necessary to ensure an instability of the reference oscillator of about 10^{-12} and a corresponding accuracy of the frequency determination which depends on the measurement time. For high precision pressure measurements the limiting factor is the accuracy of the determination of the temperature of the gas or its gradient. Therefore the discussion should concern first of all the relative error of the pressure measurement as well as of the frequency. In the majority of cases it is possible to use the cheapest and most reliable quantum frequency standard, based on a gas cell, as reference oscillator.

For a required accuracy of the pressure measurement of about 1–5% the stability of the reference oscillator should be approximately 10^{-9}. That is provided by the crystal oscillator itself. Thus, the device requirements for the measurement of the gas pressure are reduced. In this case it is possible to record the output frequency directly by a frequency meter without comparator.

7.4.4. *Radiospectroscope*

The characteristics of devices for gas-pressure measurements are determined mainly by the parameters of the radiospectroscope. It consists of the following main parts: a light source, a gas cell with working element, a thermostat and a resonator, a magnetic screen, and a photodetector with preamplifier.

A hf EL filled with inert gas and vapor of the working element is used as light source, emitting spectral lines used for optical pumping in the test cell.

In the case of Rb isotopes, optical pumping is ensured by filtering the radiation of one isotope passing through a cell filter, filled by vapors of the other isotope. The principle of the filtering process is explained in Figure 7.2. In the case of Cs vapors the same effect is achieved using a cell filter, filled with vapor of the same isotope, but placed in a strong magnetic field, or using a cell filter with buffer gas at pressure 250–600 mbar [180].

By increasing the temperature of the cell filter it is possible to reduce its length to 2–5 mm [181], and place it in the thermostat. Then this unit becomes rather compact [59]. Detailed characteristics of filters with various fillings are presented for Rb isotopes in [181], for Cs isotopes in [180].

The gas cell is the part in which the radio-optical resonance signal is formed under the interaction of the ensemble of working atoms with the pumping light and the radio-frequency field. The resonance frequency is influenced by the pressure of the gas in the cell. The cell is usually a spherical or cylindrical bulb containing the saturated vapor of the working element and a buffer gas or a mixture of gases. The volume of the cell is connected to the volume, where the pressure should be determined.

In order to observe the radio-optical resonance, the gas cell should be placed in a resonator or in a waveguide for the radiation of the radio-frequency field. It allows one to choose the required polarization of the radio-frequency field and to reduce its power. The maximum strength of the magnetic component which causes transitions between hyperfine sublevels of the ground state of atom should be located at the center of the cell, where optical pumping is most effective and the influence of the cell walls is minimal. When using a resonator the radio-frequency field is distributed in the volume nonuniformly. Resonators of the type of wave H_{001}, which provide a maximum magnetic component strength, are usually used. The space distribution of the electric component is also important, since it determines the loss (the quality of resonator). A typical resonator has a large volume of 60 cm^3 and a homogenous spatial distribution exists only in its central part. It is used when there are no limitations on the dimensions of the radiospectroscope.

Silicon or GaAs photodiodes are used for detection. They have a quantum efficiency of 0.4–0.6 in the range of wavelengths of the head resonance doublets of Rb and Cs. The photosensitive area is usually $\approx$ 1 cm^2, but is sometimes much smaller. Criteria for choosing the photoreceiver and the preliminary amplifier in relation to noise reduction are reviewed in [179].

The frequency of atomic transitions depends on the magnetic field strength at the interactive region. Therefore, a magnetic screen should be placed in the radiospectroscope for realizing high sensitivity. It should also prevent a frequency change less than the allowable error when moving the device. As $\Delta f = \alpha B^2$, where $\alpha = 5.73 \times 10^4$ Hz/(mT)^2 for the isotope ^{87}Rb, we have $\delta(\Delta f) = 2\alpha B \delta B$, so the acceptable change of the magnetic field strength is

$$\delta B = \frac{\delta(\Delta f)}{2\alpha B_{int}}$$

Further on, the factor of shielding can be expressed as

$$k_s = \frac{B_{ext}}{\delta B} = \frac{B_{ext} 2\alpha B_{int}}{\delta(\Delta f)} \tag{7.17}$$

where B_{ext} is the external magnetic field and B_{int} the field inside the screen.

It is known that the factor of shielding of magnetic screens is determined by the magnetic permeability of the materials, the number of screens, and its sizes. In the case when the spacing between cylindrical screens is 5–10 times greater than their thickness the screening factor is

$$k_s = \left(0.5\mu\frac{d}{r_2}\right)^n \tag{7.18}$$

where μ is the magnetic susceptibility, r_2 the screen radius, d the thickness of the screens, and n is the number of screens.

The distance between the screens is recommended in practice to be approximately equal to the distance between the first screen and the nearest edge of the screened object. In designing screens it is necessary to use materials with maximal available magnetic permeability and to avoid joints and seams with a high magnetic resistance on the path of the magnetic power lines. Use of as many screens as possible with a maximum available distance between them increases the effectiveness. The use of screens as carrying devices is impossible. Two or three screens are usually used, but one screen is enough in cases when the accuracy can be lower than 10^{-10}.

For increasing the probability of magnetic dipole transitions it is necessary to place the cell in a constant magnetic field, directed parallel to the vector of the magnetic component of the radio-frequency field. The requirements pertaining to the uniformity of this field are caused by avoiding line broadening stipulated by its nonuniformity to be larger than the width of the radio-frequency line. As a nonuniformity of B causes a broadening $\delta(\Delta f) = 2\alpha B\delta B$, and the width of a spectral line is $\gamma = \sqrt{(W+\Gamma)^2+U^2}$, the allowable nonuniformity will be

$$\delta B \leq \sqrt{\frac{(W+\Gamma)^2+U^2}{2\alpha B}} \tag{7.19}$$

The field strength should be higher than the residual magnetism, but less than the value at which fluctuations in the magnetizing current give rise to an instability of the resonance frequency. An increase in the magnetic field strength in the cell increases also the requirements to screen external magnetic fields. A magnetic field strength is usually applied in the range from 2 to 10 mT.

7.5. *Sensitivity Limit*

By considering the sensitivity of double resonance methods it was assumed that the noise limiting the sensitivity has a physical origin. However, in the radiospectroscope a number of high-frequency and ultrahigh-frequency signals are applied. On decreasing their effect to the level of noise, it is expedient to take special precautions. As mentioned above, it is necessary to reduce stray signals from the thermostating system. When reducing the influence of the oscillator to the light source it is expedient to screen it accurately and release all conductors through filters. Similar precautions should be taken for the high-frequency channel and the crystal oscillator.

In recent years advances in the theory of the radio-optical resonance method have been made with regard to factors limiting its sensitivity. We will briefly analyze some problems of the sensitivity of the method relevant for pressure measurement of a foreign gas.

7.5.1. *General Aspects*

The signal at the output of the radiospectroscope will be

$$I_1 = S\delta\nu \tag{7.20}$$

where $\delta\nu$ is the detuning of the radio-frequency field frequency in relation to the frequency of the reference transition.

Assuming the disturbances to be small and the system to be linear, we introduce the noises of the radiospectroscope and crystal oscillator. If any fluctuations are statistically independent, the response of the system will be a sum of responses to any fluctuation effect. The transmission coefficient K_i of a closed system for the noise of the discriminator is expressed as [179]

$$K_1(p) = \frac{S_A/(1+\varsigma T)}{1+NS_yS_A/(1+\varsigma T)} \tag{7.21}$$

and for the noise of the crystal oscillator as

$$K_2(p) = \frac{1}{1+NS_yS_A/(1+\varsigma T)} \tag{7.22}$$

where S_y and S_A are transmission factors of the synchronous detector and the selective amplifier, N is the amplification factor of the multiplier, ς the noise factor, and T the time constant of the automatic frequency tuning system.

The dispersion of the fluctuations can be calculated by the spectral density of the processes according to the known formula

$$\sigma_i^2 = \frac{1}{\pi}\int K_i^2(\Omega)\, g_i(\Omega)\, d\Omega \tag{7.23}$$

where g_i is the spectral density of the process and K_i is the transmission coefficient.

By an algebraic transformation and integration we obtain the contribution from the radiospectroscope in the following form:

$$\sigma_i^2 = \frac{S_A^2 g_N}{2T(2+NS_yS_A)} \tag{7.24}$$

and, as $NS_yS_A \gg 2$, we have

$$\sigma_i^2 = \frac{g_N S_A}{2TNS_y} \tag{7.25}$$

The physical meaning of expression (7.25) is obvious: the parameter $g_N S_A$ is the sensitivity of the discriminator which determines the instability of the frequency standard.

The approximative autocorrelation function of the crystal oscillator, $B(\tau)$, is given by the expression

$$B(\tau) = \sigma_\nu^2 e^{-\Omega_0|\tau|} \tag{7.26}$$

where τ is the measuring time and σ_v^2 the dispersion of the crystal oscillator frequency, Ω_0 a constant. Using the formula of Wiener–Khinchin we can calculate its spectral density of fluctuations. Then we can compute the dispersion of the output signal frequency stipulated by the crystal oscillator:

$$\sigma_2^2 = \sigma_v^2 - \frac{\sigma_v^2 (1 + NS_y S_A)^2}{\Omega_0 T^2 +^2 (1 + NS_y S_A)} \tag{7.27}$$

When the automatic frequency tuning is switched off, the dispersion of the frequency of the device is determined by the dispersion of the crystal oscillator. When the automatic frequency tuning is working, the dispersion of fluctuations begins to decrease. The increase in the automatic frequency tuning constant causes a rise of the instability, because the rapid fluctuations stop its filtering by the automatic frequency tuning system. With allowance for the statistic independence of fluctuations, the dispersion of an output signal of the device will be

$$\sigma^2 \cong \frac{g_N}{2TN^2 S_A^2} + \sigma_v^2 \frac{\Omega_0 T}{\Omega_0 T + N^2 S_A^2 S_y^2} \tag{7.28}$$

The instability under the prevalence of one sort of noise was analyzed above. Now we have to find the optimal value for the automatic frequency tuning time constant T. We can calculate T by the rules for evaluate the function extremum. In doing so we shall obtain a condition for the slope of the radiospectroscope,

$$S \geq \frac{1}{\sigma_v N} \sqrt{\frac{g_N \Omega_0}{v^2}} \tag{7.29}$$

This condition should be satisfied when determining the instability by the radiospectroscope. For example, when $g \simeq 3.6 \times 10^{-9}$ V^2/Hz, $\sigma_\omega/\omega = 10^{-11}$, $\Omega_0 = 0.005$–0.02 s, the slope of the radiospectroscope should be more than 5×10^{-5}–10^{-4} V/Hz, and the sensitivity higher than 0.8–1.6 Hz^{-1}.

As was earlier mentioned, the sensitivity of the radio-optical resonance is determined by the relationship of amplitude of the resonance signal to the width and the amplitude of the noise. The value of the resonance signal is proportional to the number of participating atoms. The width of the signal profile is determined by the relaxation rate, which causes destruction of the pumped state. The main factors causing relaxations are the following: (1) spin-exchange atomic collisions; (2) collisions of Rb atoms with the wall of the cell; (3) collisions of Rb atoms with atoms of the buffer gas; (4) Doppler broadening.

The effect of the optical and radio-frequency fields, which also disturb the pumped state and cause optical and radio-frequency broadening, as well as collisional and Doppler broadening are analyzed separately as these factors contribute to the "dark" line width.

The spin-exchange collisions of the working atoms are determined by an effective cross-section of the interaction which, for the Rb atoms, is about (1.7–3.2) $\times 10^{-14}$ cm^2 (for other elements these values are presented in Section 7.3). The spin-exchange process depends on the atomic concentration of atoms N and the temperature T of the cell. Its contribution to the linewidth turns out to be

$$\Delta\nu_{\text{Rb–Rb}} = \frac{N\sigma_{\text{Rb–Rb}}}{\pi}\sqrt{\frac{3k_B T}{m_{\text{Rb}}}} \tag{7.30}$$

where k_B is the Boltzmann constant and m_{Rb} the mass of Rb atoms.

For a cell temperature of about 330 K, is $\Delta\nu_{\text{Rb–Rb}} \simeq 30$–$60$ Hz, and for $T = 320$ K it becomes less than 15–30 Hz. Thus, the reduction in temperature allows one to reduce the contribution of this effect to the linewidth.

The dependence of the linewidth on atomic collisions with the walls is determined by the time between collisions, depending on the shape of the cell. A broadening by collisions with the walls can be determined by the following formula:

$$\Delta\nu_{col} = \frac{S}{\pi V}\sqrt{\frac{RT}{2\pi m_{\text{Rb}}}} \tag{7.31}$$

where R is the gas constant, while S and V are the values of the surface and the volume of the cell, respectively. In a spherical cell of 50 mm diameter at a temperature of 330 K, broadening by collisions with walls will be $\Delta\nu_{col} \simeq 2.6 \times 10^3$ Hz.

Broadening caused by the Doppler effect is determined by Eq. (3.18). At a temperature of 330 K the Doppler broadening of the radio-frequency transition for Rb atoms will be $\Delta\nu_D \simeq 9 \times 10^3$ Hz. To avoid the Doppler effect the cell dimensions must be less than half the wavelength of the radio-frequency field.

Thus, the Doppler effect and collisions of atoms with the walls introduce the main contributions to the linewidth. Spin-exchange collisions play a considerable role when the temperature is higher than 350 K. Collisions with inert or molecular gases do not cause a decrease in the populations and loss of the coherency of states of active atoms. Such gases are usually introduced into the cell to reduce the broadening influence of the Doppler effect and decrease the collisions with the walls. The walls are coated by special layers for the same purpose. Reduction of line broadening by introducing the buffer gas occurs due to the increase in relaxation time and also due to the slow atomic diffusion during interaction with the radio-frequency field.

For a spherical cell, collisions with the walls will cause the following broadenings:

$$\Delta\nu_{col} \cong \frac{4\pi D_0}{r_0^2}\frac{p_0}{p}\sqrt{\frac{T}{T_0}} \tag{7.32}$$

and for a cylindrical cell

$$\Delta\nu_{col} = \left[\left(\frac{\pi}{L}\right)^2 + \left(\frac{2.045}{r_0}\right)^2\right] D_0 \frac{p_0}{p}\sqrt{\frac{T}{T_0}} \tag{7.33}$$

where r_0 and L are the radius and length of the cell, respectively; D_0 is the diffusion coefficient for the working atoms in the atmosphere of the inert gas with pressure p_0 and temperature T_0, while p and T are the gas pressure and temperature in the cell, respectively.

The value of collisional broadening in a spherical cell of diameter 50 mm with 13 mbar nitrogen pressure is reduced by a factor of 1000 and would be about 1 Hz. The Doppler width for nitrogen as a buffer gas is

$$\Delta\nu_D = \frac{4\pi D_0}{\lambda_0^2}\frac{p_0}{p}\sqrt{\frac{T}{T_0}} \tag{7.34}$$

where λ_0 is the wavelength of the reference transition. The Doppler width of Rb will be decreased from 9×10^3 down to 20 Hz and lower. On the other hand, however while filling the cell with buffer gas B there exists a broadening due to collisions with atoms of the buffer gas:

$$\Delta\nu_{\text{Rb-gas}} = p\sigma_{\text{Rb-gas}}\frac{1}{\pi}\sqrt{\frac{3}{\mu_{\text{Rb-gas}}kT}} \tag{7.35}$$

where $\sigma_{\text{Rb-gas}}$ is the effective cross section of interaction of the working atoms with the atoms of the inert gas; $\mu_{\text{Rb-gas}} = \frac{m_{\text{Rb}}m_{\text{gas}}}{m_{\text{Rb}}+m_{\text{gas}}}$ is the reduced mass of the system compressing Rb and buffer gas atoms.

The collisional broadening increases when increasing the buffer gas pressure, while other broadening factors are decreasing. Thus, an optimum pressure of the buffer gas should be ensured in a cell. This pressure is determined by the condition of minimal contribution of all sorts of broadening, and it depends also on the dimensions and shape of the cell and the sort of gas. As mentioned above, the frequency of atomic transitions is perturbed by atoms of the buffer gas and the light. The first effect is used when measuring the buffer gas pressure, and the second must be excluded or at least be taken into account. The light shift is dependent on the distribution of radiation in the volume of the cell. As for observing the radio-optical resonance, the optical thickness is about 1, the light intensity decreases in the cell, while the shape of the contour of the pumping light changes. As a result, the optical pumping rate and the resonance signal are different throughout the volume of the cell. The situation is more complicated under the inhomogeneous distribution of a radio-frequency field in a cell. As a result there exist complicated relations between the light shift, the value of the signal, the frequency shifts from the radio-frequency field, the light intensity, the filling of the cell, the temperature and the cell dimensions. The most direct method of accounting for the light shift (but suitable only for rather large cells) is the reduction in light intensity and operation under a small optical thickness. If the limit sensitivity is not required, it is expedient to use small dimension cells [182], of 0.5–5 cm^3. On the other hand, it is worthwhile to use high values of the optical density and intensity if the maximal wide range of the measured pressures is needed.

For effective operation, the cell must be warmed up to a temperature corresponding to the maximal signal. It is achieved at an optical thickness of about 1. The temperature has to be sustained with high accuracy. In the case of nitrogen at about 130 mbar the temperature shift is about 46 Hz/(mbar K). Thus, to determine the shift with an accuracy of 0.01%, the error in the temperature measurement should be less than 0.1 °C.

The thermostating system, consisting of a thermostat with a heater and an electronic thermocontroller, enables one to maintain the temperature mode. A similar system is used in thermostating the spectral lamp. For the gas cell the range of temperatures is 35–55 °C, and for the lamp it is 90–140 °C. For a reduction in dimensions, the filtering cell is placed in the thermostat of the gas cell or in the thermostat of the lamp. For high sensitivity, it is necessary to reduce the effects caused by the heating current. For this purpose bifilar heating coils and smooth temperature modes of the thermoregulating circuit are used.

7.5.2. Theoretical Limits of Sensitivity

The accuracy of the buffer gas-pressure measurement and the limits of this technique are determined by the sensitivity of the radio-optical resonance method. By the sensitivity we understand in this case the relationship between the slope of the discriminant curve and the noise voltage within a band of 0.25 Hz. Another important parameter is the threshold of sensitivity, which is the reciprocal value of the sensitivity. There are two specified types of sensitivity: the real and the limit sensitivity. The limit sensitivity is calculated for the case of the shot noise and ideal conditions of observation of the radio-optical resonance. The real sensitivity is obtained in the real experiment or calculated on the basis of the real noises at the output of the radiospectroscope, assuming real conditions for the optical pumping process.

At first we shall consider the limit sensitivity of the radio-optical resonance for an optically thin ideal cell. The current of the photoreceiver, i_Φ, as a function of the detuning of the frequency of the radio-frequency field in relation to the resonant frequency of the reference atomic transition, $\delta\nu = \nu - \nu_0$, can be expressed as [65]

$$i_\Phi = i_\Phi^0 \{1 - F(\Gamma, W, \delta\nu, u)\} \tag{7.36}$$

where $F(\Gamma, W, \delta\nu, u$ is the profile of the contour, Γ the "dark" width of the reference transition, W the optical pumping rate,

$$\delta\nu = \delta - \varepsilon\cos\Omega t, \tag{7.37}$$

(with δ the discriminated part of the detuning), ε the deviation of the frequency, Ω the frequency of the modulation ($\Omega < \Gamma$), and u the amplitude of the radio-frequency field.

The double resonance signal, which is formed at the output of the photodetector is used for tuning the frequency of the crystal oscillator by means of the frequency of the atomic transition. This signal is proportional to the amplitude of the first Fourier harmonic, which can be calculated by performing a Fourier transformation [52] with the initial line profile [183] as follows:

$$I_1 = G_1 \frac{2}{T} \int_{-T/2}^{T/2} \frac{\alpha^2 \cos\Omega t dt}{1+(k-\beta\cos\Omega t)^2+\alpha^2} \tag{7.38}$$

where

$$G_1 = i_\Phi^0 \beta z \frac{I(W+\Gamma)W}{[I(I+1)W+(2I+1)\Gamma]^2 \xi}, \qquad \alpha = \frac{2n^2}{W+u}\frac{2I+1}{I}\xi, \qquad \beta = \frac{2\varepsilon}{W+\Gamma}$$

$$x = \frac{2\delta}{W+\Gamma}, \qquad \xi = \frac{4I^2+I-1}{I+1} + \frac{(2I+1)(\Gamma+W)}{(2I+1)\Gamma+(I+1)W}$$

I is the spin of nucleus. After some calculations the amplitude of the first Fourier harmonic is obtained in the form

$$I_1 = \frac{\alpha^2 G_1}{2A} \frac{1}{AC+AB} \tag{7.39}$$

where

$$A = 1+\alpha^2+x^2-2\beta x+\beta^2, \qquad B = 2(1+\alpha^2+x^2)-\beta^2$$

$$C = 1+\alpha^2+x^2+2\beta x+\beta^2$$

These equations enable the optimal parameters of the radio-frequency field and the deviation of frequency, and also the conditions for which the signal amplitude is maximal, to be found. The slope of the function, $S = (\partial I_1/\partial\nu)$, at the point $\nu = \nu_0'$ is extremely significant. By calculating derivatives α and β and equating them to zero, we receive a system of equations

$$\frac{\partial}{\partial\alpha}\left[\left.\frac{\partial I_1}{\partial x}\right|_{x=0}\right] = 0 \quad \text{and} \quad \frac{\partial}{\partial\beta}\left[\left.\frac{\partial I_1}{\partial x}\right|_{x=0}\right] = 0 \tag{7.40}$$

The solution of Eqs. (7.40) is $\alpha^2 = 2$ and $\beta^2 = 3/2$, which do not differ from results received elsewhere [52].

With allowance for conditions (7.40) we find

$$u = (W+\Gamma)\left[\frac{I}{(2I+1)^2}\right]^{1/2} \quad \text{and} \quad \varepsilon = \frac{3}{2}\left[\frac{W+\Gamma}{2}\right] \tag{7.41}$$

For a small detuning of the amplitude of the first harmonic of the signal we have

$$I_1(x) = \frac{\alpha^2 x B' G_1}{2^{1/2} F'^{3/2}(A'^{1/2}+D'^{1/2})} \tag{7.42}$$

where $A' = 1+2\alpha^2-2\alpha^2\beta^2+2\beta^2+\beta^4+\alpha^4, B' = 1-2\beta^2+2\alpha, F' = 1+\alpha^2+\beta^2$, and $D' = 1+\alpha^2-\beta^2$.

Assuming optimal parameters for the radio-frequency field, the deviation of frequency, and the intensity of light, the amplitude of the first Fourier harmonic will take the form

$$I_1 = \frac{2xG_1}{27} \tag{7.43}$$

or, with allowance for Eq. (7.38),

$$I_1(\delta\nu) = G_1\frac{4}{27}\frac{\delta\nu}{W+\Gamma} \tag{7.44}$$

On the other hand, we recall that $I_1 = S\delta\nu$; so we now obtain the slope of the discriminant curve S in the form

$$S = \frac{4}{27}G_1\frac{1}{W+\Gamma} \tag{7.45}$$

For determining the sensitivity one should equate the amplitudes of the first Fourier harmonics of the signal and noise. We employ an expression for the spectral density of the amplitude fluctuations of the signal output of a radiospectroscope [184] with account for the transformation of the radio-frequency field fluctuations within the amplitude–frequency characteristic of a spectral line from the band $\Omega/2\pm\Delta\Omega$ into $\Omega\pm\Delta\Omega$ [185], on the assumption of absence of amplitude–phase fluctuations. This yields

$$g(\Omega,n) = g_L(\Omega)+g_{hf}(\Omega)+4\frac{Q_k}{\omega_0}\frac{n^2}{(1+n^2)^2}I_0^2 g_\nu(\Omega)+S^2 k_\nu g_\nu(\Omega) \tag{7.46}$$

or

$$g(\Omega,n) \cong g_L(\Omega)+g_{hf}(\Omega)+4\frac{n^2}{(1+n^2)^2}S^2 g_\nu(\Omega)+S^2 k_\nu g_\nu(\Omega) \tag{7.47}$$

where g_L is the spectral density of the photocurrent fluctuations caused by fluctuations in the light flow of the source of light, g_{hf} the spectral density of photocurrent fluctuations dominated by the amplitude fluctuations of the radio-frequency field, $g_v(\Omega)$ the spectral density of the phase fluctuations, I_0 the amplitude of the resonance signal, Q_k the quality of resonance, $n = (\omega - \omega_0)/\gamma$, γ the width of resonance, $S = Q_k I_0/\omega_0$ with ω_0 the resonant frequency of the reference transition, $g_v(\Omega/2)$ the spectral density near frequency $\Omega/2$, while k_v is the factor of transformation of fluctuations from the band $\Omega/2 \pm \Delta\Omega$ to the band $\Omega \pm \Delta\Omega$, depending on the amplitude and the width of a spectral line.

In the case of predominance of the shot noise, we have

$$g(\Omega,n) \simeq g_L(\Omega) = 2ei_\Phi \tag{7.48}$$

Assuming that the pumping light is monochromatic and does not contain background lines, namely, $i_\phi \simeq i_\phi^0 \simeq c_1 W + c_2 \simeq c_1 W$ (c_2 is the photocurrent caused by background lines), then the optimal intensity of the pumping light is

$$W = \Gamma \frac{3(2I+1)(8I+7)}{8(I+1)^2} \tag{7.49}$$

and the threshold of sensitivity, δ', can be calculated as

$$\delta' = \frac{27\,[I(I+1)W + 2(I+1)\Gamma]^2}{4\sqrt{C_1}W^{3/2}kl} \tag{7.50}$$

where k is the absorption coefficient and l is the length of cell. For the ^{87}Rb isotope we obtain $W = 4.6\,\Gamma$, $\varepsilon = 3.4\,\Gamma$, and $u = 1.4\,\Gamma$.

Figure 7.8a shows the limit sensitivity as a function of the radius of the gas cell. When decreasing the size of the cell the threshold quickly increases, but for a radius of about 3–5 mm it is still rather small. Figure 7.8b illustrates the dependence of the threshold of sensitivity on the optical pumping rate. This dependency is weak: for a change in the pumping rate from 1 up to 10, the threshold of sensitivity varies by only 25%.

The calculation of sensitivity was carried out under the assumption of an optically thin cell that allowed one to obtain and analyze the analytical expressions, relating the parameters of the pumping light to the sensitivity of the resonance. The assumption of an optically thin cell is not fulfilled for the maximal sensitivity. The problem of sensitivity has been investigated [186] with regard to the process of radiation transfer in the cell. In this work the integrated intensity transformed by the photoreceiver was expressed in the form

$$U = U_0 \int j_0(\nu) \exp[-\beta(\nu)\psi(\tau)]\,d\nu \tag{7.51}$$

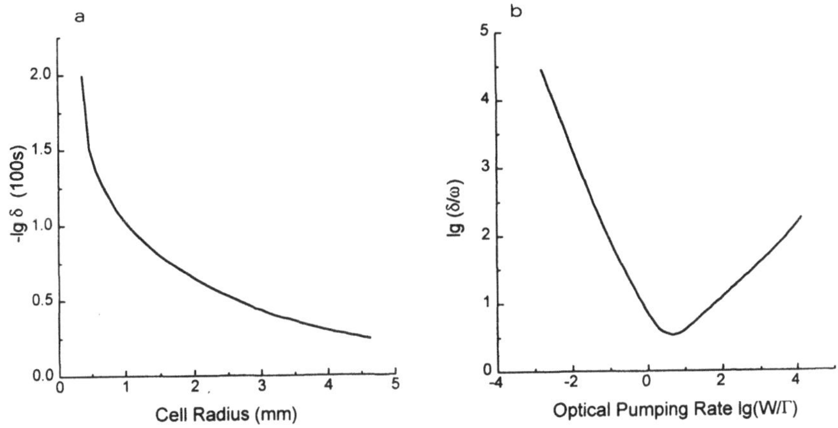

FIGURE 7.8. Dependence of the sensitivity limit on (a) the radius of the cell and (b) the optical pumping rate.

where $U_0 = U(\tau)$ is the intensity at the input of the cell, $\beta(\nu)$ the profile of the absorption line of the optical pumping light, $j_0(\nu)$ the normalized spectral density, while $\psi(\tau)$ is the nonlinear optical length, which is calculated from the solution of the differential equation for the resonant radiation:

$$d\psi/d\tau = \rho, \qquad \psi(0) = 0 \tag{7.52}$$

where $\tau = h\nu_0 n_0 \sigma_0 z$ is the linear optical length, n_0 the concentration of working atoms, ρ the population of the hyperfine sublevel from which pumping is undertaken (it is obtained from the solution of the system of kinetic equations for the elements of the density matrix), and z the coordinate along the cell.

As a result of the calculations, for the case of a cell with a window of 10 cm^2 and $\Gamma = 100$ s^{-1} for ^{87}Rb, we obtain $\sigma(\tau) = 5 \times 10^{-14}\ \tau^{-1/2}$, which is less than the quantitative assessment of the threshold sensitivity of the optically thin cell by approximately a factor of 10. The conditions of maximal signal observation are also different in this case and correspond now to the value $W = 13.3\Gamma$.

7.5.3. *Real Sensitivity*

Calculations give extremely high estimates for the limit sensitivity, which are not achieved in the experiments. This is connected with the lower efficiency of the optical pumping and the increased real level of noise. Actually, spectral lamps emit background resonant lines. With regard to this factor the expression for the sensitivity threshold of an optically thin cell will take the form

$$\delta' = \frac{27}{4}\frac{[I(I+1)W + 2(I+1)\Gamma]^2}{W^2 kl}(C_1 W + C_2)^{1/2} \tag{7.53}$$

As was noted above, the signal makes less than 0.01 of the background and therefore the threshold sensitivity can be higher than the calculated one by a factor of 10. It is possible to use interference filters or selective photodetectors for suppressing the background intensity, but usually also great losses in the pumping light intensity are inevitable, which is not always acceptable. An effective way to increase the sensitivity may be to use narrow-band laser radiation, well suited for optical pumping due to its high monochromaticity. But most present-day lasers have a significant level of noise and are therefore not competitive. Radio-frequency field fluctuations are another reason, limiting the threshold sensitivity. Assuming, according to [185], that the main process is the transformation of fluctuations from a band $\Omega/2 \pm \Delta\Omega$ into a band $\Omega \pm \Delta\Omega$, which is proportional to the slope of the discriminant curve, it is possible to write the following relation:

$$g(\Omega, n) = S^2 k_v g_v(\Omega/2, n) \tag{7.54}$$

If now, as well as before, we equate the first Fourier harmonics to the noise voltage, we find that in this case the threshold sensitivity depends only on the spectral density of the phase fluctuations of the radio-frequency field:

$$\delta'_v = \sqrt{k_v g(\Omega/2, n)\Delta\omega} \tag{7.55}$$

Practically this means the independence of the sensitivity on parameters determining the slope of the discriminant curve, i.e., on the size of the cell, on buffer-gas pressure, and on the intensity of the pumping light. This fact is indirectly used in designing small-sized radiospectroscopes with a cell volume of 0.8–1.2 cm^3 [182], in the creation of precision radiospectroscopes [187], as well as in gas-pressure measurement over the range of observation of the radio-optical resonance of (1–2 mbar and less).

The real sensitivity depends on the characteristics of the radio-frequency field and is, e.g., 70–80 Hz^{-1} t$^{1/2}$, for the radiospectroscope used in [36]. Therefore the required sensitivity can be achieved by an increase in the measurement time. The relative threshold sensitivity obtainable is less than 10^{-12} for a sensitivity value of about 20 Hz^{-1} t$^{1/2}$ and a measurement during a time of about $\tau = 100$ s. For $\tau = 1000$ s the relative threshold sensitivity could reach even less than 3×10^{-13}. However, further increase in observation time does not in practice increase the sensitivity, because for measurement times $\tau \gg 10^4$ s the contribution of the light intensity fluctuations and instabilities of the surrounding temperature prevails on the frequency regulation capacity. When determining the gas pressure, the process of temperature establishment can practically predetermine the achievable accuracy. Best accuracies can be realized only if the volume, in which the pressure of the gas is measured, is thermostated. Even in this case the environmental temperature can limit the maximal achievable accuracy because the temperature effect on the frequency of the radiospectroscopes is usually in the

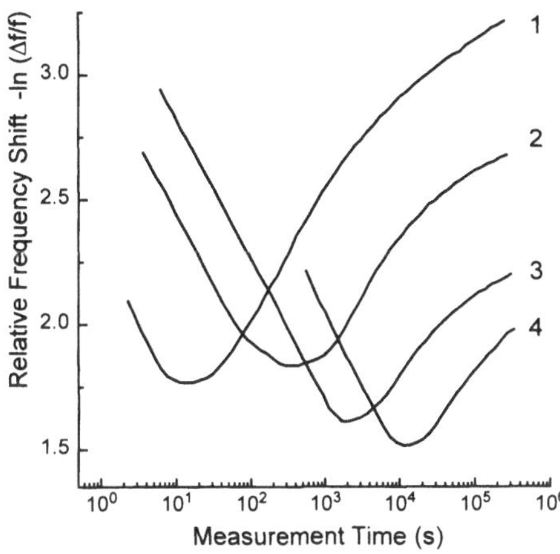

FIGURE 7.9. Dependence of the threshold sensitivity of radio-optical resonance on the measurement time and on the light shift: 1, 150 Hz; 2, 30 Hz; 3, 10 Hz; 4, 5 Hz. Obtained sensitivity of the radiospectroscope: 1, 3×10^{-9}; 2, 10^{-9}; 3, 3×10^{-10}; 4, 10^{-10}.

range from 5×10^{-13} to 2×10^{-12} 1/°C. If an error in the gas-pressure determination of 0.1–1% can be accepted, the measurements can be performed in about 1 s and other requirements pertaining to the experimental conditions also are reduced. Optimum modes can be found using diagrams constructed by an experimental fluctuation spectrum of the light source for different values of the light shift [188] (Figure 7.9)

The sensitivity of the double resonance method when measuring the atomic transition frequency, and consequently the precision of gas-pressure determination, is limited by the light shift (perturbation of the energy of atomic states by the pumping light). Under unfavorable conditions the light shift can reach dozens of Hz.

Quantum electrodynamic calculations lead to the following dependency of the shift on the parameters of optical pumping [4]:

$$\Delta\nu_{op} = \int_0^{\infty} A_k^2 \frac{\nu - \nu_0}{(\nu - \nu_0)^2 + \gamma^2} I(\nu) d\nu \qquad (7.56)$$

where A_k is a parameter depending on the wave functions of the atom, ν are frequencies of resonant and nonresonant light waves, and $I(\nu)$ is the spectral distribution of the optical radiation.

The value of the light shift is proportional to the light intensity and reaches its maximum for a detuning of the pumping line when the absorption line equals the half-width. The light shift is absent when the frequencies of the radiation and the absorption lines coincide.

Actually, the intensity and shape of the pumping line vary while passing through a cell with the vapor of the working substance. Together with the nonuniform distribution of the radio-frequency field in the volume it results in a nonlinear dependence of the light shift on the intensity incident to the cell [see Eq. (7.56)].

For a reliable determination of the true value of the light shift, it is necessary to perform measurements employing different intensities with the help of neutral filters, as well as to repeat these experiments at different temperatures of the filtering cell. Obviously, it is necessary to maintain the minimal intensity of light, which ensures the required sensitivity of the radio-optical resonance. Using a cell of volume 140 cm^3, it is possible to decrease the temperature of the cell, and ensure an optical thickness of $kl = 0.2$ in order to get a linear dependence of the light shift on the intensity of the light [189]. By operating with minimal intensity of the pumping light (photocurrent about 1 μA) it is possible to lower the light shifts to a level of 1×10^{-12}. In this case a high sensitivity of more than 100 $Hz^{-1}\, t^{1/2}$ could be realized.

In order to increase the sensitivity it is possible to use cells with a special coating on the inner surface, hence lowering the rate of relaxation of atoms under collisions with the walls [190]. Nevertheless, these types of cells were not used in the frequency standards because they show a significant frequency drift. However for pressure measurements coated cells can be used also. In the last case the low pressure range (10^{-3}–10^{-5} mbar) becomes accessible for the given method, for which ordinary cells are not applicable.

7.6. *Some Applications of the Double Resonance Techniques*

7.6.1. *Measurement of Pressure in High-Frequency Electrodeless Lamps*

The correct choice and control of the buffer gas pressure within hf ELs with vapor of metals is an important application. The pressure of the gas determines the main characteristics of these lamps. Therefore it is necessary to provide a correct choice of gas pressure and to insure accurate manufacture to improve the characteristics of spectral lamps. Using ordinary control methods the accuracy of pressure determination in the lamps is not high (10–15%). It is even lower because of uncontrollable gas-pressure variations while sealing off the spectral lamp and removing it from the vacuum system. Therefore a special setup has been constructed to allow pressure measurements in spectral lamps during manufacture and during their research. The characteristic feature of this device is the use of a crystal oscillator of frequency 4,996,112 Hz, tunable at ±2 Hz, and a frequency multiplier. The studied lamp is inserted in a special mounting, installed directly on the photodetector. This part (lamp–photoreceiver–preamplifier) is easily disconnectable from the main device, which is important for rapid replacement of tested lamps. The investigated lamp is held at a temperature of 55°C and is irradiated by the light of the pumping lamp and affected by a radio-frequency field, arriving through a waveguide. Lamps and cells of larger sizes are placed directly in

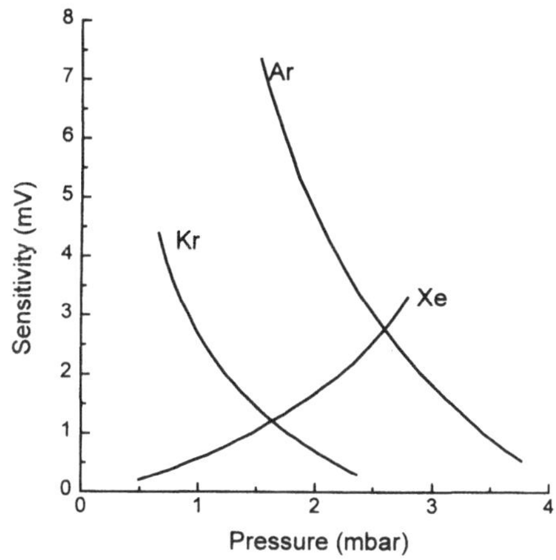

FIGURE 7.10. Dependence of the sensitivity of the pressure measurement technique for Ar, Kr, and Xe.

the thermostat of the radiospectroscope.

The lamps are filled with Rb vapor and inert gas, in most cases with krypton, allowing one to attain a stable discharge with rather high intensity of the metal spectral lines. Typical pressures are between 2.4 and 4.2 mbar.

The sensitivity of the technique as a function of buffer-gas pressure is indicated in Figure 7.10. Pressure measurements with an error less than 0.01 mbar over the full range of pressures from 1 to 13 mbar are possible.

In the case of spectral lamps, the accuracy of this method is limited by possible penetration of gas into the lamp bulb during testing and research. However, chemically reactive gases react with Rb and may reduce its quantity. A frequency shift is also possible if gas which is not reacting chemically with Rb penetrates through the walls. A high penetration rate through the glass is typical for hydrogen [65]. H_2 molecules have a high factor of frequency shift. Special spectroscopic studies did not indicate the presence of hydrogen to more than 0.005%, even in lamps exploited more than 8000 h. Probably, free hydrogen is bound by Rb in the discharge. He and H_2 diffusing through the glass of the bulb from the atmosphere could be the cause of an error. After equalizing of the partial pressure in the atmosphere and in the lamp, the frequency shift due to this reason will be

$$\Delta\nu_{H_2,He} = \alpha_{He} p_{He} + \alpha_{H_2} p_{H_2} \simeq 3 \text{ Hz} \tag{7.57}$$

where p_{He} and p_{H_2} are the partial pressures in the atmosphere for He and H_2.

Penetration of other gases from the atmosphere may be ignored. Thus, the error in the determination of the level shift due to penetration of gases is less than 0.4%.

Using the data on the temperature dependence of the shift factor for krypton, one could estimate the required accuracy of the temperature establishment. The temperature should be known better than within ±1 °C for an error less than 0.2%, and approximately ±5 °C for an error of about 1%.

Application of this method to lamps with Rb vapor enabled one to detect differences in pressures of the gas before detachment from pumping equipment, and the pressure after the sealing procedure. The following significant pressure variation was observed: when filling 2.1 mbar of krypton the final pressure in the lamps was found in most cases to be between 1.7 and 2 mbar.

7.6.2. Remote Sensing of Gas Pressure in Gas-Filled Objects

The gas pressure determines the major characteristics of many devices: instability of the frequency in absorption cells, reproducibility of the characteristics of cell filters, utilization characteristics of gas-filled lasers, etc.

The pressure in the cell of a quantum frequency standard completely determines the real value of the frequency and the temperature factor of the frequency; the constancy of the pressure in the lamp determines the long-term frequency stability. In order to provide frequency stability during time τ better than 1×10^{-13}, the pressure (i.e., the pressure of krypton or nitrogen) should not vary more than 6×10^{-6} mbar during the specified time of measurement.

The requirement for stability of filling of the filtering cells is significantly lower. For an instability of less than 1×10^{-13}, the pressure change should not be more than 0.006 mbar during the time of measurement. However, the filling should be undertaken with a rather high accuracy and the double resonance method is effectively applied for control. Pressure measurement in the filtering cells has its own characteristic features, because in this case the vapor of ^{85}Rb isotopes is used which contain less than 0.6% of ^{87}Rb isotope. Nevertheless, the sensitivity of the method was comparable with that of typical cells [181]. This fact is explained by the optimum value of the Rb-vapor density and the high efficiency in the use of the light. Owing to this fact an error in pressure measurement of less than 1% was achieved. This method of pressure measurement by traces of Rb isotopes allowed one to investigate certain kinds of irreproducibility of cell filters and to propose a new approach to its reproducible production [181]. A method of pressure control for the study of the longevity and reliability of the gas cells was also developed [202]. Gas diffusion out of the cell causing the frequency shift in the cell was detected. Analysis of measurements allowed one to propose a technique for manufacturing gas cells with increased reliability. Other applications of the method are also possible. For example, the time dependence of the pressure of He in a cell has been presented [109]. The gradual reduction in pressure is caused by the helium permeability of the glass.

7.6.3. *Dosage of Gas in Gas-Filled Objects*

One characteristic feature in the technology of fabricating spectral devices in quantum electronics is the high requirement of accuracy of the gas dosage. Before the creation of tunable frequency synthesizers, the value of the frequency in quantum standards was determined by the real value of the gas-cell resonance frequency. Therefore, since the creation of quantum devices, attention was paid to the problem of calibration, i.e., manufacturing cells with a well-determined filling. Frequency standards were applied to control the gas filling. The error in the calibration of a cell in the earliest works was 5×10^{-9}, later on 2×10^{-10} and even 10^{11} [179], corresponding to an error in the pressure for different cells of about 4×10^{-4} mbar.

Nowadays, two variants of manufacturing cells are used with increased reproducibility of filling. The first method consists in control by use of a reference cell, placed in the thermostat of the gas cell which is connected with the volume of the buffer gas. In this case the errors are caused by the difference of temperatures in the volumes, in which the manufacturing and reference cells are placed. Therefore, pipelines connecting the cells with an evacuated volume of the equipment should be calibrated and satisfy the requirements of providing an effective pumping out and minimum pollution under the sealing procedure. An error of about ± 6 Hz is achieved; this corresponds to an error in the pressure of less than 0.04 mbar for a full pressure of 25–30 mbar.

A detailed analysis of the filling process is required for increasing the accuracy of the filling. As shown above, the method of control and measurement of pressure with the aid of double resonance provides the best accuracy, but the process of disconnecting the cell from the vacuum equipment can change the pressure. It happens owing to heating of the pipelines and release of a part of the gas into the cell or from the cell during its overheating. The pollution of the cell by gas emissions from the glass during its heating is also an important factor [203].

In order to ensure a relative error in the pressure determination of $\Delta p/p \sim 10^{-4}$ practice, the radius of the capillary should be chosen to be a maximum of 0.25 mm. While filling with metallic Rb, the channel radius cannot be made smaller than 1–3 mm for technological reasons. Therefore, to provide high accuracy of the cell filling, a two-channel system is utilized. A wide channel is used for the metal filling and a capillary for filling with noble gas [179]. A wide channel is used also for the vacuum thermal cleaning. After cleaning and filling in the metal the lamp is prepared in the following way: (1) the gas is pumped out, (2) noble gas is filled through the capillary, (3) finally, dot sealing out of the cell is performed.

For precise calibration of gas cells the latter are fabricated with a supplementary volume. The cell is placed in the radiospectroscope and the supplementary volume into the thermostat. The dosage of gas in the cell is prepared by changing the temperature of the thermostat while controlling the frequency of the radiospectroscope. A detailed description of such a gas cell pressure calibration procedure, utilized in industrial cell production for quantum frequency standards, is presented elsewhere [179].

8
Creation of Highly Stable and Reliable Electrodeless Spectral Lamps

In this chapter various methods to increase the service-life time and the reliability of spectral lamps are considered. Manufacturing technologies of the spectral lamps and their connection with longevity and reliability are discussed in detail for the first time. Original methods for tests in real and accelerated time scales are described for estimating the longevity and reliability.

8.1. Cleaning of Spectral Devices

8.1.1. Physics of Cleaning Processes

In order to obtain stable parameters of spectral devices, it is necessary to ensure the smallest possible rate of change in the quantity of rubidium and buffer gas. This is a function of the temperature of the bulb, the concentration of particles, the material of the bulb, and the degree of pollution of all the materials. The concentration of atoms in the bulb, $n(T)$, and on its surface, n_0, is described by the Langmuir isotherm

$$n(T) = n_0 \exp\left(-\frac{A}{k_B T}\right) \tag{8.1}$$

where A is the work done by atoms leaving a surface in thin films as a function of the number of atoms on the surface.

The physics of degassing consists in the displacement of the balance of the solid state–gas boundary into the gas phase and the subsequent removal of the evaporated atoms in a mode of viscous and molecular flow (deep stage of degassing). The temperature of degassing is clearly restricted by the values above which softening or destruction of the glass takes place. Duration of the vacuum-thermal degassing process cannot be too long because the evaporation of atoms from the surface as well as the diffusion and subsequent evaporation of atoms from near surface layers of glass will form vacancies. After the working substance is inserted into the spectral devices its atoms would fill these vacancies, giving rise to a slow frequency drift of the device.

The high-frequency discharge itself is also an effective method of degassing. Under a correctly chosen power density, a hf discharge cleans the surface of the bulb of a spectral device.

The final stage of manufacturing spectral devices is separation of the device from the pumping equipment. The object can be polluted at this final stage, too. In the following sections we shall examine the main stages of manufacturing spectral devices and their relation to the reliability of the product.

8.1.2. Vacuum-Thermal Degassing

The manufacturing process of a spectral device is conventionally divided into the following steps:

(1) Pumping without heating.

(2) Pumping under increased temperature (vacuum-thermal degassing).

(3) Atomizing of the metal inside a device.

(4) Buffer gas filling.

(5) Sealing-off the spectral device and separation from the vacuum system.

All stages except the fourth will be analyzed below.

Basic measurements have been undertaken using equipment to provide the vacuum-thermal degassing of work pieces of absorbing cells and filling with a gas mixture (Figure 8.1). The high vacuum was obtained by employing a diffusion pump that used silicon organic oil. All the pipes were of 20 ± 1 mm diameter and the length of the cells was 60 mm. The work pieces of the cells and pipes were cleaned mechanically, rinsed with rectified spirit, and fused to the pumping equipment by a hydrogen burner. The total pressure was measured together with the qualitative spectral structure of the gases and the partial pressures. The omegatron gauge used and the connecting tube were degassed by both vacuum-thermal and ion–plasma processing. The latter was carried out by excitation of an electrodeless high-frequency discharge in argon at a pressure of 2.6 mbar. The studied volume with the absorbing cells was only vacuum-thermally processed at a temperature of 400 °C. Finally, after cooling the system a vacuum no worse than

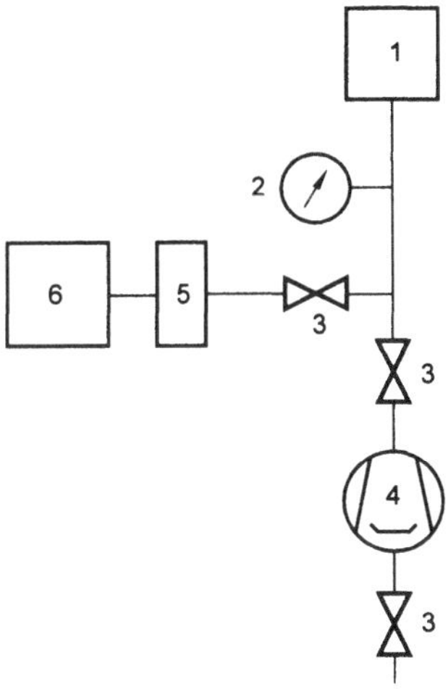

FIGURE 8.1. Experimental setup. 1, volume, pumped out; 2, manometer; 3, vacuum valve; 4, diffusion pump; 5, RMO-4C gauge; 6, UPDO-2A omegatron.

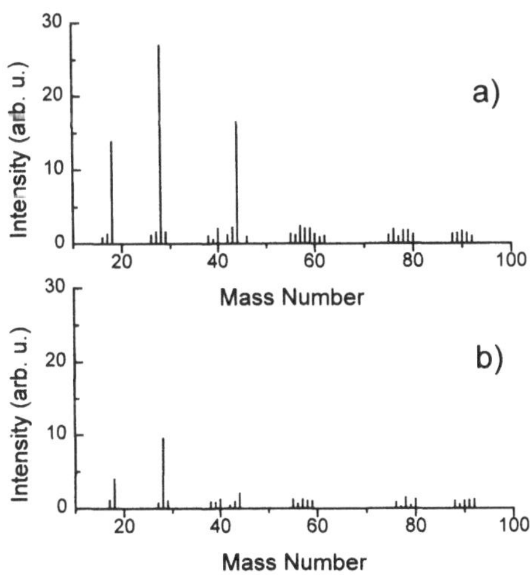

FIGURE 8.2. Typical mass-spectrum of the rest gases. (a) after pumping out; (b) after 60 h degassing at 400 °C.

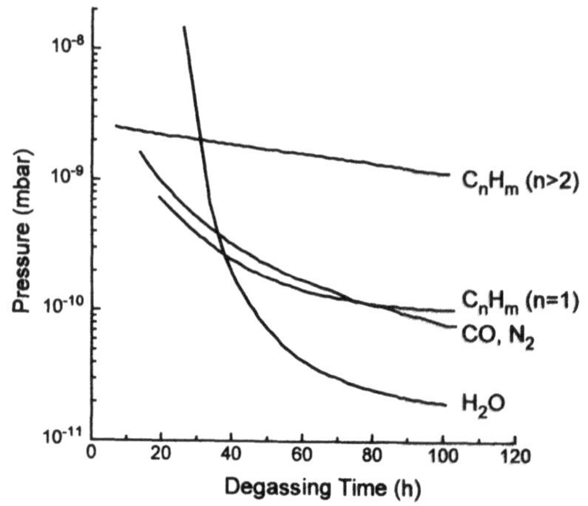

FIGURE 8.3. Dependence of the residual gas pressure on the duration of vacuum-thermal degassing (T = 400 °C).

2.6×10^{-7} mbar was achieved after 60 h of degassing. A typical mass spectrum of the rest gases is shown in Figure 8.2.

Moreover, vapors and molecules characteristic of electric vacuum devices, such as hydrogen and water vapors, molecules of oxygen, carbon dioxide, nitrogen and carbon monoxide, and molecules with masses from 40–150 atomic units, have been observed. Heavy hydrocarbons which are vapors of the oil used in the diffusion pump, as well as decomposition of these vapors, correspond to these atomic weights. This assumption is confirmed by the independence of vapor density on the time of baking-out and pumping. Therefore, it is evident that for further improvement of the spectrum of the residual gases in spectral devices it is necessary to develop and use an oil-free system of pumping. In order to choose an optimum degassing mode, the dependence of the rest-gas pressure (derived from mass spectra) on the heating time has been investigated (Figure 8.3).

Baking at a temperature of 400 °C for 20–30 h completely cleans the surface of the glass cells, and subsequent elevation of temperature by a further 100 °C does not lead to further essential outgassing. In the first hours of the baking out (with simultaneous pumping) vapors of water and carbon dioxide are most effectively outgassed from the glass. This is quite natural, as the water is dissolved in the glass, and carbon dioxide is a product of the decomposition of carbonates, which stem from the glass. The mass spectrum of Figure 8.2 shows clearly the light hydrocarbons which are removed during the first hours of pumping under heating. After 30 h, the pressure of the residual vapor becomes much less than the vapor pressure of the heavy hydrocarbons. It is therefore expedient to conduct the bake-out process for no more than 30 h, otherwise the inner surface of the bulb will be oiled. To estimate the effect of the degassing process on the long-term stability of spectral devices, the pressure increase due to inleakage of gases and

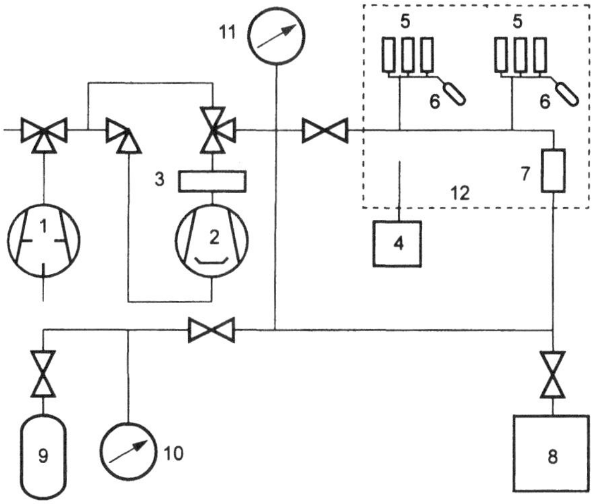

FIGURE 8.4. Vacuum setup. 1, forevacuum pump; 2, diffusion pump; 3, nitrogen trap; 4, thermocouple devices; 5, work pieces of spectral devices; 6, ampoules with Rb; 7, cell of the mass spectrometer; 8, mass spectrometer; 9, bulb with Ar; 10, membrane manometer; 11, ionization vacuummeter: 12, high-temperature furnace.

outgassing from the surface into the studied volume was investigated. It could be shown that after reaching 10^{-7} mbar during degassing, the pressure increases during some hours to no more than 3×10^{-7} mbar.

8.1.3. Gas Emission at Various Stages of Spectral-Device Manufacture

The increase in accuracy of the mass-spectrum determination was studied using modified equipment (Figure 8.4), which allowed more careful examination of the degassing process. In this case the qualitative composition of gases was measured by a mass spectrometer, which could be attached to various points of the circuit.

Before measurements, the analyzer was degassed at a temperature of 200 °C for 5 hours, so that the residual vacuum in the cooled analyzer was less than 1.3×10^{-7} mbar. A mass spectrum of the residual gases after pumping of the circuit during 10 h without heating is shown on Figure 8.5a.

Outgassing while heating the system up to 200 °C is shown in Figure 8.6. Active outgassing of vapor of water, carbon monoxide, nitrogen, and hydrogen, decreases gradually with the warming-up. The evident outgassing decrease at a temperature of about 170 °C is typical for glasses of the molybdenum group. The time dependence of outgassing at a temperature of 300 °C is shown in Figure 8.7. After 30 h of pumping the content of the residual gas components became reduced to H_2, $CO_2 + N_2$, and a group of gases typical for the vapor of the diffusion

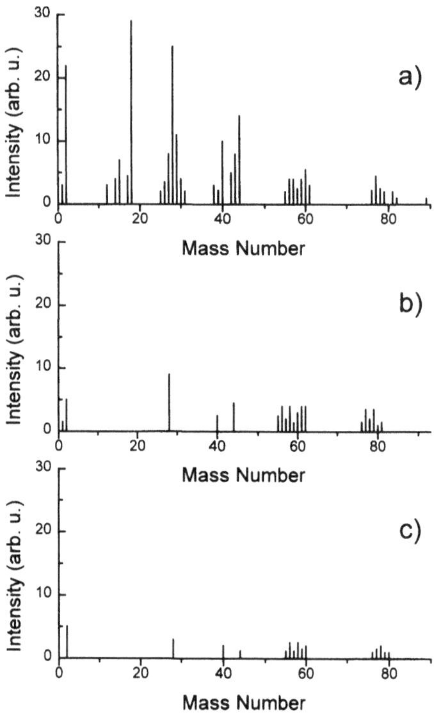

FIGURE 8.5. Mass spectra of the residual gases at various stages of manufacture. (a) mass spectrum after 10 h of pumping out without heating, $p = 1.6 \times 10^{-6}$ mbar: (b) mass spectrum during outgassing after 48 h of heating to $T = 400$ °C, at $p = 2.6 \times 10^{-5}$ mbar; (c) mass spectrum after atomizing rubidium, $p = 10^{-7}$ mbar.

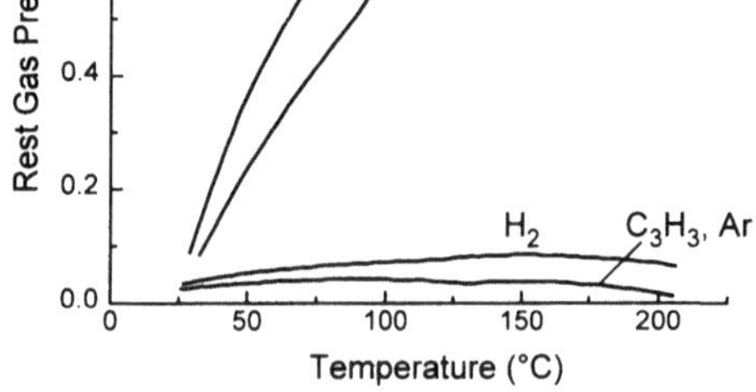

FIGURE 8.6. Temperature dependence of the outgassing process during pumping.

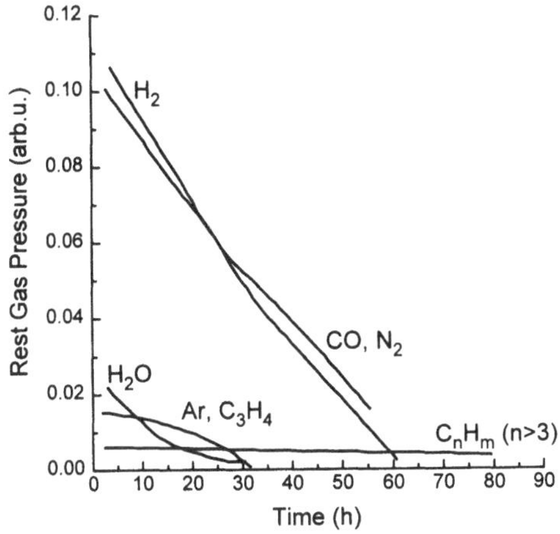

FIGURE 8.7. Temporal dependence of the outgassing process ($T = 300\ ^\circ\mathrm{C}$).

pump oil. A mass spectrum after the thermal degassing is shown in Figure 8.5b. Comparison with Figure 8.5a, shows that the spectrum has become purer and the main components of the residual gases have changed. If previously it was mainly water vapor, now it consists of molecular hydrogen ($m = 2$) and carbon monoxide and nitrogen (both $m = 28$) [149, 204]. Insertion of rubidium atoms causes a decrease in carbon monoxide, which flows from the glass surfaces. The presence of carbon on the surface of glass cells is confirmed by special Auger-spectrometric studies. Carbon monoxide, as is known, can react with rubidium, reducing the quantity of molecules of mass number 28.

A study of the sealing-off process was carried out by a special block consisting of three cells, one of which was connected by a tube to the analyzer of a mass spectrometer. The cells were vacuum-thermally degassed at $T = 400\ ^\circ\mathrm{C}$ during 30 hours, then cooled (the pressure in the cooled system was about 5×10^{-7} mbar), and then subsequently sealed off one by one. An increase in the gas components of masses 2, 14, 16, 28, 44 (molecular hydrogen, atomic nitrogen or CH_2, atomic oxygen or CH_4, carbon monoxide or molecular nitrogen, carbon dioxide) was found.

Table 8.1 indicates the experimental parameters of outgassing when sealing-off the cells from the vacuum system. The sealing-off of every cell causes appreciable pollution of the others. This can be explained by the fact that sealing-off with a hydrogen burner is carried out at temperatures exceeding the temperature of the glass transformation. It means that gases absorbed by the glass at the expense of diffusion are released, resulting in an increase of residual pollution. Taking into account that the volume of the connecting pipes is 10–12 times bigger than the volume of a cell, the real pollution in a cell also increases by the same factor and

TABLE 8.1. Outgassing Pressure p_i [10^{-6} mbar] during Sealing-off of the Cells

Operation:	Mass [atomic units]					$(\Sigma_i p_i)$
sealing-off	2	14	16	28	44	[10^{-6} mbar]
First cell	0.7	2	2	1.3	2	8
Second cell	0.9	2.7	4.4	1.3	5.2	14.5
Last cell	47	31	42	62	6	188

affects the vacuum in the cell by up to 2.6×10^{-4} mbar. At such a pressure within the volume, a monolayer of polluting substance corresponding to a pressure of about 1.3 mbar should exist on the walls of the cell. The active pollution may result from interaction with rubidium, located in the volume, and it may cause frequency shifts in the cells or unstable burning of the spectral lamps. In order to reduce gas emissions during sealing-off, which limit the service life and long-term stability of the spectral devices, it is necessary to avoid a hydrogen burner and to seal-off by some other techniques, such as cold sealing. An electrical discharge is sometimes used to reduce pollution when disconnecting the cells. However, a significant voltage at the location of sealing-off is required. Sealing-off by the discharge, carried out at increased temperatures up to 400–500 °C inside the thermostat, improves the situation. This method seems to be promising and convenient in the industrial production of spectral lamps and also allows one to carry out the dosing of a working substance in the lamp. Thus, the mass spectrum of the residual gases at various stages of manufacture of the spectral devices shows that the most significant pollutions occur at the sealing-off step by the gas burner.

8.1.4. Degassing in a Discharge

Methods considered in Sections 8.1.2 and 8.1.3 provide the purification of the surfaces, while the studied volume is connected to the pumping equipment. In the course of commercial mass production of spectral devices the long-term duration of the vacuum-thermal degassing process is a serious disadvantage. Other cleaning methods are required and particular interest is evoked in using a gaseous discharge [71]. Obviously a high-frequency electrodeless discharge is most suitable for discharge degassing owing to the possibility of creating a plasma without inserting electrodes into the volume. As inert gases are used in spectral devices, it is natural to use the same that are supposed to be filled in the product. The pressure of the gas should provide effective cleaning, which is expected when a majority of ions recombine on the wall of the bulb. It can be achieved in vessels of 2 cm radius under a pressure of less than 4 mbar [205]; in the range from 4 to 0.1 mbar ambipolar diffusion is effectively realized. In the electrical field arising under the ambipolar diffusion the ions are accelerated up to 3–7 eV, and the electrons, at a temperature of $(1.2–1.7) \times 10^4$ K inside the plasma are moving more slowly [64]. A mixed (E and H) discharge is most effective for this purpose due to the simultaneous creation of a significant potential and a high density of the charged particles. Therefore degassing processes were investigated in more detail in mixed discharges under argon pressure less than 4 mbar.

The problem of degassing was investigated experimentally, using an arrangement similar to that presented in Section 8.1.3. The preliminary vacuum was created by a forevacuum pump, and the high vacuum by an oil pump with silicon organic oil. The studied bulbs were manufactured from borosilicate glass C52-1 and C50-4. The total pressure and partial pressures of gases were measured by a vacuummeter and the mass spectrometer respectively. Before the measurements the analyzer of a mass spectrometer was degassed at a temperature of 200 °C during 5 h. As a result, the vacuum in the cooled analyzer was no worse than 10^{-6} mbar. For excitation of the discharge, semiconductor generators of frequency 100 MHz were used. The measurements were performed as follows: the vacuum system with the gas bulbs was pumped (without heating) to a vacuum of 2.6×10^{-6} mbar, then filling with argon was carried out, and a high-frequency discharge in the bulbs and inside the vacuum system was excited. After the degassing by the discharge the generator was switched off, the gas was pumped out and after reaching the initial pressure, a mass spectrum of the residual gases was recorded. The following typical degassing modes were studied in more detail: (1) power of the discharge $P = 15$ W, time of degassing 20 and 40 min; (2) power of the discharge $P = 100$ W, time of degassing 40 min. The power $P = 15$ W was not enough to clean the bulbs (in 20 as well as in 40 min); the mass spectrum after degassing was identical to the spectrum before degassing. At $P = 100$ W a change in the mass spectrum was observed: the oil content of the vapor was considerably decreased. That was associated with the decomposition of complex vapors in the discharge, and the removal by pumping of H_2, O_2, CO, and CO_2 as the oxygen- and carbon-containing gases.

Thus, the surface of the glass was cleaned from organic pollutions by the discharge. Besides, cleaning by the discharge was also studied after vacuum-thermal degassing at 615 °C during 48 h. When a pressure $p = 3 \times 10^{-7}$ mbar was attained argon was filled into the system and the discharge was excited. At power $P = 15$ W the effect of cleaning was not observed, and at $P = 100$ W the color of the discharge varied from bright orange (probably connected with pollution) to light violet. The intensity of coloring depended on the diameters of the capillaries through which the pumping was carried out. In bulbs of spectral lamps and in other small devices in which the pumping was conducted through a capillary of 0.8–1.5 mm diameter and 10–15 mm length, orange coloring was always observed at the beginning. In bulbs of larger dimensions, pumped through pipes of 3–4 mm in diameter, light-violet luminescence was observed at the beginning. In this case an outgassing by the discharge could be characterized by the increased time of pumping. Thus, even after careful vacuum-thermal degassing an additional outgassing was observed during excitation of the discharge. This can be explained by the fact that during the discharge, additional cleaning is carried out by electron and ion bombardment of the surface and the near-surface layers of the glass. Electron bombardment affects only a very thin layer, strongly saturated with gases. Ions penetrate into the near-surface layers and expend energy on the displacement of atoms or ions from equilibrium positions. When this energy exceeds the binding energy of atoms in the glass, and if the momentum of ions is directed perpendicular to the surface, evaporation will take place. The evaporation induced by the ion bombardment differs qualitatively from usual thermal evaporation. The energy of the interaction process is in this case much higher, and particles may leave not only from the surface (as during heating) but from the near-surface layers of the glass of 10–20 atomic layers in thickness. Thus, during the additional degassing in the discharge in an inert gas, deeper cleaning of the glass surface is implemented.

The discharge degassing processes at two argon pressures of 0.7 and 1.6 mbar were studied in more detail. Significant differences in the mass spectrum of the residual gases were not observed after cleaning by the discharge at these two pressures. At a reduction in pressure on the one hand, the energy of the bombarding particles is increased, but on the other hand their concentration decreases. It is also significant that during the use of high-frequency degassing there is an opportunity to clean unheatable parts of the vacuum system right up to an internal surface of vacuum valves. The use of additional discharge cleaning of spectral devices is expedient in manufacturing spectral devices.

8.1.5. Study of the Internal Surface of Spectral Devices

In order to study more comprehensively the effect of the quality of cleaning on the properties of vacuum devices, research on their internal surface has been undertaken. It is known that the aging process of vacuum devices is related to the processes of sorption and desorption, which depend on the conditions of the internal surface of the glass of the bulbs. The current state of the inner surface, which varies for different samples, is changing throughout the work, and effects the real durability and the reliability of the parameters of a device. These changes are caused by uncontrollable outgassing and adsorption, directly connected to the characteristics of spectral devices, especially for gas cells, and reduce the long-term stability. For this purpose X-ray spectroscopic investigations (Figure 8.8) of the internal surface of gas cells have been carried out [206]. A technique of sequential removal of layers of the surface by a high-energy argon–ion gun was applied. The energy of the etching beam was about 4 keV. The residual pressure in the vacuum chamber of the spectroscope was $p = 10^{-10}$ mbar. The samples were produced from the end faces of the cells (size 10×15 mm). While a sample was irradiated by X-rays, the number of electrons emitted by the sample was recorded by an analyzer.

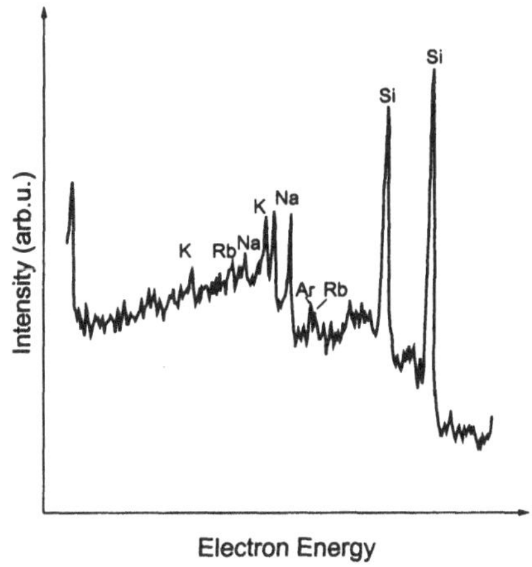

FIGURE 8.8. Typical spectrum of X-ray released electrons, taken from the C50-4 glass surface.

Surfaces of gas bulbs of hf ELs, made of C50-4 glass were studied. The lamps were filled with Rb vapor and a mixture of N_2 and Kr with a total pressure of about 13 mbar, and subjected to an artificial accelerated "aging" at elevated temperatures (200 °C during 150 h). Other spectral devices in which, due to "aging," the double resonance signal had disappeared, as well as other devices showing various drifts of frequency were selected for investigations, too. It was established by processing the spectra that these samples were characterized by their content of rubidium on the surface and by the depth of its diffusion into the glass. For the bulbs in which the signal had disappeared, increased content of rubidium at some depth inside the glass and the absence of Rb on the glass surface were observed (Figure 8.9). This means that complete diffusion of rubidium into the glass took place. The rubidium penetrates into glass and interacts with components of the glass, namely, with Na_2O and B_2O_3; thus there was a redundant concentration of Na and B in near-surface layers.

The depth of rubidium as a function of boron concentration was found, and an increase of boron concentration in the glass reduced the rate of diffusion of rubidium atoms deep in the glass of a bulb. In spectral devices in which a significant frequency drift was observed, increased content of carbon on the surface was found. The latter can be connected with insufficient cleaning or with deposition of products of the cracked oil fragments. On interacting with oxygen, carbon atoms form molecules of mass 28 within the volume of a bulb, clearly observed during mass-spectrometric analysis (see Figure 8.5). In turn carbon monoxide can interact with rubidium and cause a frequency drift. Thus for removal of the frequency drift, which limits the longevity of the spectral devices, one must prevent the presence of carbon-containing substances on the internal surface. This implies

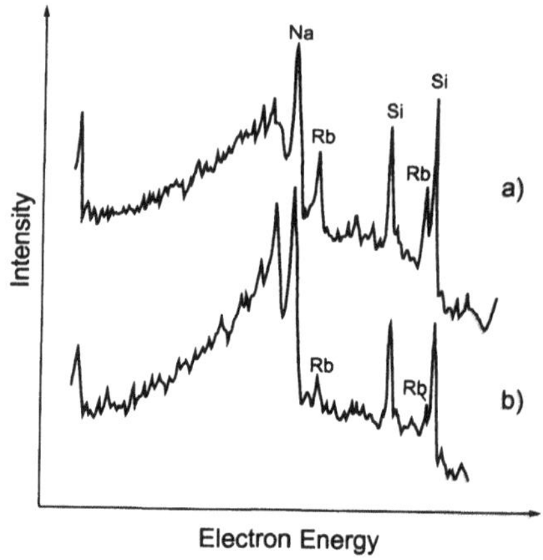

FIGURE 8.9. Spectra of X-ray released electrons from the internal surface of a cell in which, after enforced utilization (150 h, 200 °C), the fluorescence signal disappeared: a, from the cell surface; b, after removing a surface layer of 14 nm.

the necessity to use an oil-less high-vacuum pumping, degassing by discharge and more careful cleaning of work pieces. In case of small-size bulbs (diameter 10–13 mm), typical for spectral lamps, X-ray spectroscopic studies of the internal surface were not carried out, as the equipment did not one allow to investigate samples with a bent surface. However, the conclusions on the correlation between the boron concentration and the depth of diffusion of rubidium seemed especially important in this case, as the lamps are produced from glass with a decomposition temperature of 650 °C. During fusing of glass boron is evaporated from the glass, therefore it is necessary to enrich the flame of the burner with boron. However, this process is uncontrollable and could be the reason for the variation in the reliability factors of different spectral lamps. The main requirements in the process of manufacturing to insure reliability and durability of lamps are cleanliness, the use of oil-less pumping, degassing by discharge, boron enrichment, correct choice of degassing time, and its depth.

8.2. Methods for Determining Reliability of the Spectral Lamps

8.2.1. Main Reasons for Aging of Spectral Lamps

The failure of a spectral lamp is usually caused by one or more of the following reasons: the depletion of metal; the decrease in buffer gas pressure; devacuumization of the bulb of the lamp; extinction of the discharge or relaxation due to mismatch of the lamp and oscillator.

The emission of products of the glass destroyed by ion bombardment or emission of dissolved gases from the glass, is also possible. However, in the case of hf ELs with rubidium vapor we were unable to reliably observe such processes.

An indication of the depletion of metal in the bulb is the absence of luminescence of atoms of the metal. At a decrease in pressure the discharge becomes unstable, and what is required is an increased power of the discharge for stabilizing ring-shaped luminescence, i.e., for the excitation of the H-discharge.

Devacuumization of the bulb of a lamp is possible in the case of poor quality of the manufacturing process. Another reason may be the use of unsuitable cement, used for fastening a lamp. It can be avoided by the use of glue-free fastening of the lamp or the use of glues of elastosil type.

Sometimes metallization of the bulb of the lamp occurs, leading to problems with ignition of the discharge. This kind of failure was described elsewhere [110], where it could occur because of an insufficient temperature gradient on the surface of the bulb of a cylindrical spectral lamp. In order to exclude this effect, periodic operation of a lamp at increased temperature was proposed. In the case of correct utilization of the light sources, metallization was not observed and is not considered here when estimating the reliability of spectral lamps.

Extinction of the discharge or the onset of relaxation, due to mismatch of the lamp and oscillator is possible as a result of the pressure decrease in the gas, for a redundant quantity of metal in the lamp or ineffective tuning of the oscillator. This kind of failure is specific to the light source as a whole and in the present section it will be disregarded.

The main reason for a failure of electrodeless spectral lamps with rubidium vapor is the depletion of metal and the consecutive change of the lamp's filling.

Up to a certain level a disappearance of metal has no effect on the characteristics of a spectral device, as its parameters are determined by the concentration of atoms of the metal in the saturated vapor, the density of which is a function of temperature.

The specific evaporation energy q is described by the Clausius–Clapeyron formula

$$q = \frac{dp}{dT}(V_2 - V_1) \tag{8.2}$$

where V_1 and V_2 are the volumes of two coexisting phases, while dp/dT is the slope of the vapor pressure curve $p(T)$.

If there exists a condensed phase, the vapor density should not depend on the quantity of metal at a free state within the spectral device. Therefore the lifetime of the lamp, τ, is in this case equal to

$$\tau = \frac{N_1}{dN_1/dt} \tag{8.3}$$

where dN_1/dt is the rate of vapor disappearance from the volume of the device and N_1 the initial quantity of substance in the device.

TABLE 8.2. Dependence of the Quantity of Rb [10^{-3} mg] on the Utilization Time for Eleven Investigated Lamps

Number	Utilization time [h]								Lifetime
of lamp	0	100	200	400	500	1000	2000	3000	[h]
1	75	62	52	42	39	36	34		$> 10,000$
2	50	43	38	35	34	32			$> 10,000$
3	7	5.5	3.6	2.8					900
4	7.2	6.8	4.8	3					700
5	43	35	31	20					—
6	19	18	17						700
7	30					18	17.5	16	$> 10,000$
8	21								3100
9	57								$> 10,000$
10	31								$> 10,000$
11	52								$> 10,000$

A change in the pressure of the buffer gas causes a direct change in the lamp parameters. Thus the serviceable lifetime of a lamp is linked to acceptable changes in radiation intensity, impedance, and voltage.

8.2.2. Studies in the Real Time Scale

Studies on variations in working parameters of the lamp were carried out for spherical spectral lamps of 13 mm diameter. In these lamps a H-discharge with power $W = (3 \pm 1)$ W at a surface temperature of 170 ± 10 °C was excited.

The change in the quantity of rubidium was studied with the help of methods described in Chapter 6. Some results are presented in Table 8.2. The most significant decrease in the quantity of Rb takes place during the first 500 h of operation, after which a decrease in the disappearance rate of Rb was observed. This fact is probably connected to the saturation of the glass with the metal (Table 8.2, lamps 1, 2). When inserting less than 0.01 mg Rb into the lamp (Table 8.2, lamps 3, 4), all the metal disappears as a result of interaction with the glass during less than 1000 h. When inserting more than 0.03 mg Rb the lamps work more than 5000 h as a rule (Table 8.2, lamps 7, 9–11). The lamp with 0.021 mg of Rb (lamp 8) has worked only 3100 h. In some special cases (Table 8.2, lamps 3–5) an even faster decrease in Rb was observed before failure. The reason probably consists in poor pumping and the presence of pollution, as well as in the uncontrollability of the chemical structure of the glass (lamp 6).

For eleven similar lamps the connection of longevity with the change in the quantity of Rb during the first 100 h of operation was studied. Some results are presented in Table 8.3. For some of the lamps (1, 5, 9–11) a decrease in the relative Rb quantity $\Delta M/M$ over 100 h by 2–5% was observed and corresponds to an estimated lifetime of more than 3000–5000 h at similar discharge conditions,

TABLE 8.3. Change in the Quantity of Rb during the First 100 h of Operation

Number	Quantity $[10^{-3}$ mg]			$(\Delta M/M)$	Longevity
of lamp	m_0	M_1	$\Delta M = m_0 - M_1$	100[%]	[h]
1	31	29.5	1.5	4.8	> 10,000
2	17.5	15.5	2	11.4	800
3	7.2	6.3	0.9	12.5	700
4	7	5.5	1.5	21	900
5	36.5	35.5	1	2.8	> 10,000
6	18	14	4	22	1500
7	21	9	12	56	300
8	44	32	12	27	900
9	19	18	1	5	5000
10	37	36	1	3	> 10,000
11	43	41.5	1.5	3.5	> 10,000

taking into account stronger absorption of Rb in the first 100 h. For some lamps the variation in the quantity of rubidium during this period is stronger 10–50%. Such lamps (numbered as 2–4, 6–8 in Table 8.3) fail rather rapidly.

With the help of the method developed for measuring the metal quantity, the absorption of rubidium by the different glasses was investigated for the various manufacturing technologies of the work pieces. It should be noted that for spectral lamps made of pyrex glass, during the first measurement the bulb of the lamp darkened and Rb was completely absorbed by the glass, so that further measurements became impossible.

From Figure 8.10 we see that C51-1 glass, enriched with boron (curves 1, 2), absorbs rubidium much more passively. We conclude that it is necessary to use glass with enhanced boron content to achieve the required longevity.

An analysis of these results yields the following conclusions:

(1) There is a tendency for the glass to be saturated with rubidium and stabilization of its quantity in the lamp, therefore the use of a quantity of rubidium from 0.005 mg up to 0.02 mg is necessary at least.

(2) Spectral lamps containing more than 0.03 mg Rb subject to accurate manufacture can operate more than 5000 h.

(3) There is an opportunity to forecast roughly the operating lifetime of a lamp after the change in the quantity of Rb in the first 100 h of operation. In this connection it is recommended to reject lamps with an initial quantity of Rb less than 0.03 mg.

(4) In order to attain a long service life of a spectral lamp while avoiding metallization, it is recommended to produce it from C51-1 glass enriched with boron.

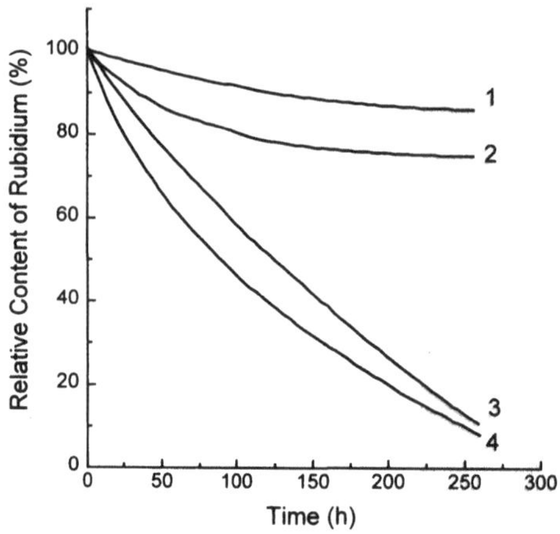

FIGURE 8.10. Change in the relative content of Rb in lamps made of C51-1 glass for subsequent measurements: 1, 2, for lamps fabricated from boron-containing glass; 3, 4, from glass without boron.

The most important parameter of a lamp is the pressure of the buffer gas introduced into the lamp to form the glow discharge and stabilization of the operating modes. This parameter directly determines the ignition voltage and the intensity of radiation. It is known that a decrease in gas pressure is possible while the lamp is working. This is the main reason for the failure of luminescence lamps filled with argon.

The change of gas pressure in lamps during their operation was studied by means of a method described in [67]. Figures 8.11 and 8.12 indicate the pressure of Kr in lamps of 13 and 8.5 mm diameter as a function of the duration of their operation. In the first 100–200 h a rapid decrease in pressure was observed, after which the rate of decrease in gas pressure stabilizes to a value of 0.0022–0.001 mbar/day. Since the optimum pressure for the ignition of discharge in lamps of 13 mm diameter was equal to $p = 2$ mbar, for an increase in longevity up to 5000 h, taking into account the decrease in gas pressure, it is recommended to insert *ca* 2.4 mbar gas into the spectral lamp. By these means the voltage of ignition changes no more than 10% and the light sources can be operated up to 5000 h.

The transparency of the C51-1 glass in 8000 h of lamp operation decreased from $69 \pm 4\%$ down to $50 \pm 3\%$. The pressure drift of Kr as well as of Rb is caused by two main reasons: (1) chemisorption of the glass and subsequent diffusion; this process depends on the temperatures of the gas and the glass. (2) Ionic bombardment of the glass. The rate of drift of atoms of Kr and Rb at the expense of this process should depend on the intensity of the discharge and be proportional to the degree of ionization in the discharge. Thus for any further essential increase in lamp longevity it is expedient to reduce the power of the discharge, for example, by transition to the E-discharge mode.

Therefore a longevity of more than 5000 h can be achieved by introducing

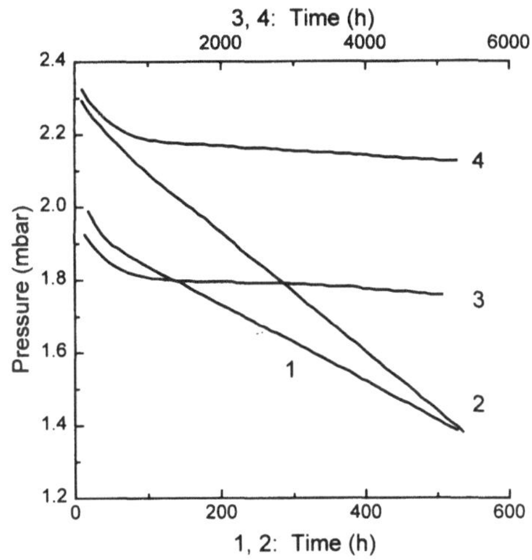

FIGURE 8.11. Dependence of Kr pressure (60 °C) in spectral lamps as a function of operating time. Lamps 1, 2 have been utilized 500 h, lamps 3, 4 5000 h.

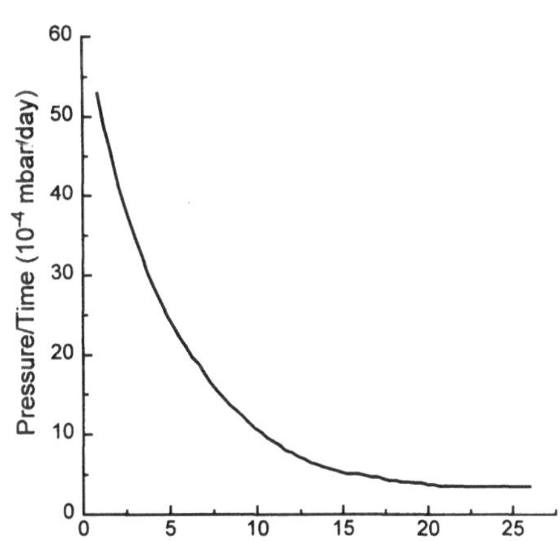

FIGURE 8.12. Rate of Kr pressure change in a lamp of 8.5 mm diameter as a function of the time of equivalent operating age for a specific power at a discharge of 1 W/cm^3.

into a spectral lamp made of C51-1 glass more than 0.03 mg of Rb and observing its aging in the first 100 h, while filling Kr at a pressure of 2.4 mbar. Further studies have shown the possibility of achieving longevity of more than 10,000 h. In the 1980s these results were confirmed elsewhere [125, 127, 207] where it was also recommended that 0.03–0.2 mg of Rb be inserted into spectral lamps with a volume of about 1 cm^3.

In order to determine the reliability of the spectral lamps and to more systematically study the different aforementioned factors that cause their aging, groups of identical light sources (10–13 pieces) were tested. Altogether more than 60 light sources have been tested up to their failure. They worked in a H-discharge mode and were installed on a test bench together with a temperature controller, which maintained a given temperature in the thermostat. Before and after the beginning of tests the internal resistance of the light sources was measured. During each operating term the integrated intensity of emission was recorded as well. Data for the tested group of spectral sources, distinguished by the power of the generator consumption, the diameter of the lamp, and the initial pressure of krypton, are given in Table 8.4. The length of the vessel for an excess quantity of metallic rubidium in the first group of lamps was about 2 mm, containing 0.03–0.07 mg of Rb, and in the second group 4 mm, containing 0.08–0.2 mg of Rb.

The lamps were made of alkali-resistant C51-1 glass. The operating mode of the light sources provided spectral lines of the alkali metal, as well as the line spectrum of the buffer gas. The intensity of the rubidium lines was 2–3 times higher than the intensity of the brightest krypton line $\lambda = 811.2$ nm.

The averaged values of the pressure changes in the inert gas and the contents of metal are indicated in Table 8.4. The pressure changes in the lamps ranged from 0.1 to 0.6 mbar, and the contents of metal varied between 0.07 and 0.15 mg. The averaged changes in the light emission of the lamps are also indicated in Table 8.4.

The power consumed when the discharge glows did not vary in practice but when a H-discharge is excited the power increases significantly. An additional experiment showed that the degree of the discharge power extinction was proportional to the change of gas pressure in the lamp.

The voltage–current characteristics of the light sources also show some characteristic variations. Such a dependence is given in Figure 8.13 for a source in which the pressure did not change for 5000 h of operating. In this case the difference between these characteristics at the beginning and end of the tests (curves 1 and 2, accordingly) indicates the change of parameters of the generator circuit. These features have been confirmed by subsequent tests of the generator without a spectral lamp. It was shown that in this case the parameters of the transistor generator have changed.

It follows from Section 3.2 that a decrease in the krypton pressure in a lamp

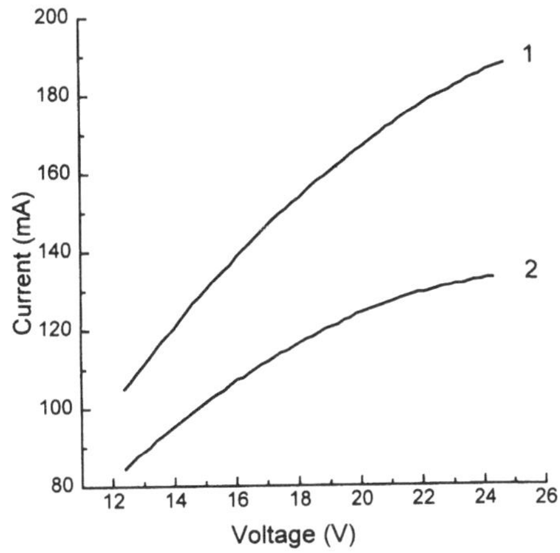

FIGURE 8.13. Voltage–current characteristics of the light source: 1, before beginning the durability test; 2, after 5000 h of operation.

should result in an increase in radiation intensity. The decrease of light intensity observed in the current tests was explained by the decrease of power transmitted into the discharge owing to a change in the generator parameters. However, after working for 5000 h the generators have retained their serviceability with the lamps of the third and fourth groups. The data on the dimensions and filling of the lamps, as well as on the characteristics of light sources from these lamps are indicated in Table 8.4.

The lamps of the third group, with dimensions and filling similar to lamps of the first group, required after less than 1000 hours of operation an increase in generator power owing to extinction of the H-discharge. The radiation of these lamps, having operated at the H-discharge mode in this group, was ten times more stable than for the first group. The changes of pressure in the lamps during operating period was about 0.1–1 mbar. In connection with studies on the aging processes and generator tests of lamps, samples with increased pressure up to 4 mbar having a bigger additional vessel and larger contents of metallic rubidium were tested. The generator power consumption was 2.5 W. The maximal change in krypton pressure during the testing period was 0.1 to 1.1 mbar, but in the majority

TABLE 8.4. Change in the Mean Parameters of the Lamp ($t = 5000$ h)

Number of lamp	ø [mm]	Initial and final pressure [mbar] p_i	p_f	Initial and final quantity of rubidium [10^{-3} mg] m_i	m_f	Rate of intensity change $\Delta I/(I\tau)$ [1/h]
1	10	2.8	2	650	500	5×10^{-5}
2	13	2.4	2.2	70	63	8×10^{-7}
3	10	3	2.5	60	50	—
4	13	4.9	4.3	75	68	10^{-6}

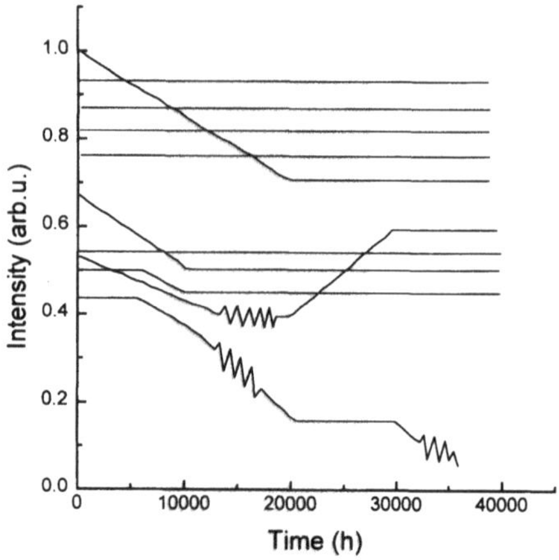

FIGURE 8.14. Change in the radiation intensity of different lamps during 40,000 h of operation.

of cases it was only 0.6 to 0.8 mbar. The change of intensity was essentially smaller than for the first two groups, and for three lamps it was not observed at all.

For the third and fourth groups of tested sources the generators were assembled from electronic elements made in the same year and having had the same period utilization. The deterioration which took place in these generators during the operating time was identical. Therefore the distinction in the studied modes of operation of the light sources can be explained mainly by the differences in the initial fillings of the lamps.

The results of periodic records of the radiation intensity of light sources, during working for 25,000–40,000 h, are presented in Figure 8.14. For the first 4500 h of light-source operation the intensity of radiation was constant within an error of 2%. A similar result was observed for the majority of lamps of the fourth group. For the interval of 25,000–40,000 working hours the stability of radiation of the light sources was within the error limits of the measurement. Figure 8.14 shows that for some light sources periodic changes in radiation intensity were observed over long periods of time. In another light source, stronger variations in intensity, up to 3% for 1000 h, were observed, while for other sources the change of intensity was less than 1% for 1000 h, and for yet other sources the intensity change was smaller than the errors of measurement, i.e., the relative change of intensity was less than 2×10^{-7} h^{-1}. Data on the pressure in the studied light sources are presented in Table 8.5.

By comparing the radiation intensity variation with the change in the gas pressure within the spectral lamps, we conclude that sources with a small change in the pressure worked more stably than the others. Results of studying the qualitative

TABLE 8.5. Results of Pressure Measurements [mbar]

Time of	Number of the light source								
measure [h]	1	2	3	4	5	7	8	9	10
0	4.5	4.5	4.4	4.4	4.5	4.5	4.4	4.5	4.5
10,000	4.3	4.3	3.8	4.2	4.25	4.3	4.1	4.3	4.2
25,500	—	4.1	3.2	4.1	4.0	4.2	3.9	4.0	4.0
34,500	—	4.1	—	4.1	4.0	4.2	—	4.0	3.7
40,000	—	4.1	—		3.7	4.1	—	4.0	3.0

composition of the glass imply that the relaxation of the intensity fluctuations of sources 1 and 3 should be related not only to change of parameters in the transistor of the oscillator, owing to which the latter could attain an unstable working point, but also to the effect of substances emanating from the glass on the parameters of the radiating plasma. Later these substances were probably partially absorbed by Rb. The character of the radiation intensity change in light source 3 can be explained by the same process of plasma contamination.

Tests of a light source during 80,000 h (from January 1979 to December 1990) have also shown a 5% change in radiation intensity for the first 2000 hours, but thereafter the intensity was constant to within an error of less than 2%.

The reliability of two groups of lamps was determined by operating and testing them during 12,000 and 24,000 h at a discharge power of 1.7 ± 0.2 W (Table 8.6). It was shown that the increase in buffer gas pressure allows one to reduce substantially the aging of the spectral lamps. Changes in buffer gas pressure also were not observed in the lamps in which the pure rubidium discharge was excited without the excitation of krypton (lamps 2 and 10 in Table 8.6).

It is interesting to note that a significant change in gas pressure (lamps 2, 6, and 10 in Table 8.7) was observed in the darkened lamps while in the other lamps the bulb color remained practically unchanged. The dark coloring disappeared under repeated excitation of the discharge in such a lamp and the gas pressure increased directly after excitation of the discharge, returning to the initial value. Spectral analyses of the filling of darkened lamps with an accuracy of 0.01% have not yielded any impurity inside them.

As an explanation of this effect it is possible to assume a connection between the adsorption of krypton and the state of the internal surface of lamps, which depends on the boron enrichment in the glass-blowing flame by which the lamp work pieces have been produced.

Evaluation of the failure rate and the probability of nonfailure during the guaranteed durability $P(t)$ for the studied lamps is shown in Table 8.8. The decrease of the gas pressure in a lamp for $p < 2.4$ mbar or the decrease in the quantity of metal for $m < 0.01$ mg were regarded as a failure in the calculation. The limit boundary of the error in the failure rate is given in the 6th line of Table 8.8.

TABLE 8.6. Pressure and Metal Quantity in Lamps (Group 1) for Different Moments during Their Lifetime

Number of lamp	τ [h]					
	0	0	1000	5000	12,000	12,000
	p_0 [mbar]	M_0 [mg]	p [mbar]	p [mbar]	p [mbar]	M [mg]
1	2.40	0.09	1.95	2.15	1.94	0.08
2[a]	2.40	0.13	2.20	2.20	2.19	0.11
3	2.35	0.12	1.90	2.15	1.80	0.12
4	2.30	0.16	2.15	2.25	2.0	0.15
5	2.45	0.04	2.0	2.15	2.0	0.04
6	2.50	0.07	2.40	2.50	2.35	0.05
7	2.40	0.09	2.10	2.25	2.0	0.06
8	2.55	0.14	2.15	2.30	2.10	0.12
9	2.40	0.15	2.20	2.20	2.14	0.13
10[a]	2.15	0.14	2.15	2.15	2.04	0.12
11	2.60	0.04	2.10	2.40	2.0	0.03

[a] Only ^{87}Rb emission was observed.

TABLE 8.7. Pressure and Metal Quantity in Lamps (Group 2) for Different Moments during Their Lifetime

Number of lamp	τ [h]						
	0	0	5000	10,000	15,000	20,000	24,000	24,000
	p_0 [mbar]	M_0 [mg]	p [mbar]	p [mbar]	p [mbar]	p [mbar]	p [mbar]	M [mg]
1	4.50	0.10	4.25	3.90	3.60	3.40	3.20	0.07
2[a]	4.30	0.09	3.90	3.40	2.80	2.25	1.80	0.06
3	4.20	0.10	3.95	3.70	3.50	3.25	3.00	0.08
4[b]	4.45	0.09	4.45	4.45	4.10	3.80	3.40	0.07
5	4.45	0.11	4.40	4.10	3.80	3.40	3.15	0.08
6[a]	4.35	0.08	4.00	3.80	2.70	2.25	1.90	0.06
7	4.10	0.15	3.90	3.40	2.95	2.45	2.15	0.10
8	4.00	0.07	3.75	3.40	2.95	2.40	2.00	0.05
9	4.20	0.09	3.85	3.45	3.10	2.50	2.10	0.08
10[a]	4.20	0.10	3.60	3.10	2.55	2.05	1.70	0.07
11	4.20	0.14	3.90	3.65	3.35	3.10	2.80	0.09
12	4.30	0.17	3.95	3.70	3.40	3.10	2.70	0.13
13	4.45	0.12	4.10	3.85	3.55	3.20	3.00	0.10
14	4.10	0.11	3.80	3.55	3.35	3.10	2.80	0.09
15	4.00	0.15	3.65	3.30	3.00	2.75	2.55	0.10

[a] The lamp would be blue.
[b] Only ^{87}Rb emission was observed.

TABLE 8.8. Reliability Tests of Spectral Lamps

Characteristics	Number of series 1	2
Number of lamps	11	15
Krypton pressure [mbar]	2.4 ± 0.25	4.2 ± 0.25
Operation time [h]	12,000	24,000
Number of failures	< 1	< 1
Intensity of failures	1.8×10^{-6}	0.5×10^{-6}
Guaranteed operation durability	5000	10,000
$P(\tau)$	0.9910	0.9975

The increased reliability of the lamps, as follows from the results described in Chapters 2 and 3, is possible under the higher increase in pressure of the buffer gas — up to 6–13 mbar — and under the reduction in discharge power. The expected longevity should be about 100,000–200,000 h, i.e., 10–25 years. However, to conduct comprehensive tests in a real time scale, is extremely time-consuming and expensive. Therefore, at the beginning of the 1980s, methods for accelerated tests of spectral lamps were developed to correct the technology employed in their manufacture.

8.2.3. Accelerated Aging of Electrodeless Spectral Lamps

The long duration of tests on a real time scale makes an operative quality check of spectral devices impossible, but such a check is necessary to improve the technological process of their manufacture. Further increase in the durability of these devices up to 10–20 years cannot be confirmed practically without accelerated tests of the lamps. A pulsed mode of operation of the light sources has been used to accelerate the aging process.

The light sources were modulated by introducing at the base of the transistor of the hf circuit a sinusoidal voltage produced by a sound generator. The amplitude of the pulses and the lamp radiation intensity were observed by means of an oscillograph. The emission intensity was estimated with a photodiode, connected to a microamperemeter. The pressure of the buffer gas was used as an aging parameter which was measured by the double resonance technique with an error less than 1%. The results of the experimental research are shown in Figure 8.15 and Table 8.9.

A decrease in gas pressure with an increase in pulse frequency has been detected. The acceleration of krypton ions penetrating into the glass is characteristic at the initial stage of discharge breakdown having increased the average energy of electrons and ions. On increasing the pulse frequency, the earlier stages of the discharge are introduced and the rate of absorption of krypton grows accordingly. The pulsed mode provides an increase in the absorption rate 10–40 times more than

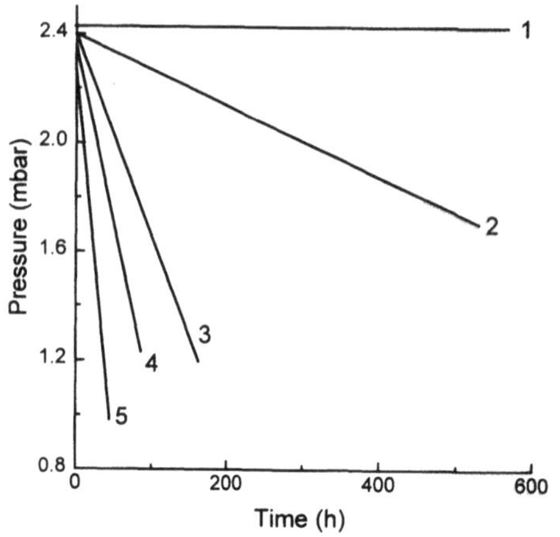

FIGURE 8.15. Variation of the gas pressure in spectral lamps during accelerated aging as a function of the pulse generator frequency: 1, stationary mode (DC discharge), $P = 1.7$ W; 2, a frequency of 10 kHz; 3, 20 kHz; 4, 30 kHz; 5, 50 kHz.

for the normal mode. Therefore modes with a pulse frequency of 20–30 kHz and a discharge power providing the photocurrent of the FD-7K photodetector equal to 600–800 μA on irradiation from a distance of 3 cm could be recommended.

The disadvantage of this method is the difficulty in establishing a connection between the rate of acceleration of aging and the real modes of discharge, the design and sizes of the lamp. Therefore, another method based on the use of increased discharge power was proposed.

As the gas pressure and metal concentration change, the general aging of a bulb results from the interaction of the ionized components with the glass. For acceleration of the aging processes it is necessary to increase the degree of ionization of the discharge plasma. This is possible by increasing the power introduced into the discharge.

For the realization of tests at increased discharge power, a number lamps of 13 ± 0.5 mm in diameter, filled with rubidium vapor and krypton at a pressure of 4.2 ± 0.3 mbar, were investigated.

In order to avoid overheating an additional radiator was installed on the spectral source that allowed tests to be conducted at powers up to 30 W. The

TABLE 8.9. Gas Pressure Variation Rate for Different Pulse Frequencies

p [mbar]	W/cm^3	f [kHz]	$\Delta p/\Delta t$ [10^{-3} mbar/day]
2.4	2.0 ± 0.2	10	60
2.4	2.0	20	115 ± 20
2.4	2.0	30	330 ± 60
2.4	2.0	50	200

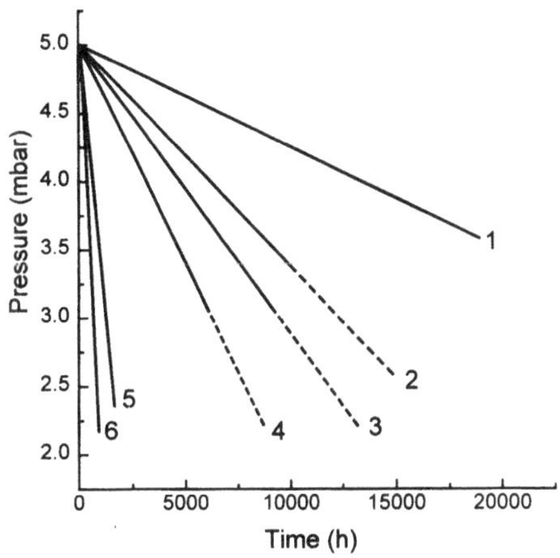

FIGURE 8.16. Variation of the gas pressure in spectral lamps during accelerated aging as a function of the discharge power: 1, $P = 1.7$ W; 2, 3 W; 3, 3.5 W; 4, 5.8 W; 5, 9 W; 6, 10 W.

frequency of the oscillator was 90–100 MHz. The spectral lamps were fixed at a heat conductive holder. At a test mode the ring-shaped luminescence of rubidium from the near-surface layers was selected, requiring an increase in thermostat temperature up to 135–145 °C. The power introduced to the discharge was determined by a calorimeter. For this purpose the power, consumed by the thermostat when switching on and off the discharge in the lamp, was measured. The difference in the consumed power was attributed to the power expended in the discharge itself. Measurement of the rate of gas-pressure variation are shown in Figure 8.16 and listed in Table 8.10.

The increased power results in the acceleration of the gas absorption (by acceleration factor A), which could be approximated by the power law

$$A = W^{\alpha} \tag{8.4}$$

with $\alpha = 1.5$ at $0.3 < W$ [W] < 2 and $\alpha = 2$ at $2 < W$ [W] < 7. This experimental result was compared with computations of the longevity of lamps as a function of the discharge power, carried out on the basis of the developed theory of high-frequency spectral lamps. In such a manner, the flow of ions on the wall of the bulb was chosen as a parameter for determining the rate of aging. Hence the longevity of the lamp is determined by Eq. (8.4)

Analysis of Eq. (8.4) allows one to reach important conclusions about the

TABLE 8.10. Observed Rate of Gas-Pressure Variation

	Number of lamp							
Characteristics	1	2	3	4	5	6	7	8
Power [W]	1.7	10	3.6	7	5	5.8	7.5	8.5
$\Delta p/\Delta t$ [10^{-3} mbar/day]	1.1 ± 0.1	70	5	43	21	30	43	40

TABLE 8.11. Expected Durability of Spectral Lamps (Calculated)

	Power [W]					
Spectral source	1.7	1	0.7	0.3	0.2	0.1
SS-1	5000[a]	11,000	19,000	67,000	124,000	350,000
SS-2	10,000[a]	22,000	38,000	135,000	248,000	701,000

[a] Guaranteed durability.

opportunity for significantly increasing the longevity of spectral lamps by decreasing the consumed power. Calculations of the expected longevity of lamps are given in Table 8.11.

In view of the obtained results it is possible to calculate the probability of the nonfailure operation of spectral lamps, indicated in Table 8.8, at the decreased power of 0.3 W. Thus during 50,000 h, for a pressure of krypton of 2.40 ± 0.25 mbar the probability of failure-free operation will be 0.9933, and for a pressure of 4.20 ± 0.25 mbar it will be 0.9982.

The long-term longevity of lamps of 10 and more years will be limited by the properties of the glass and by the plasma effect.

This method of accelerated aging under increased power was applied to the longevity of lamps 8.5–9 mm in diameter. The results of the tests are presented in Table 8.12.

As a result it was established that the aging process could be described by the relationship

$$\tau \sim \left(\frac{W}{S}\right)^{-1.72} \tag{8.5}$$

where S is the surface of the lamp (see Chapter 3).

Hence in order to maintain high level of the lamps while reducing their dimensions, it is necessary to decrease the power in the discharge by a factor of r_0^2. For example, in a lamp 9 mm in diameter the level can be maintained by reducing

TABLE 8.12. Aging of Lamps during Tests with Increased Power

No.	W [W]	Rb [mg]	p [mbar]	t_1 [h]	p [mbar]	t_2 [h]	p [mbar]	$\frac{\Delta p(t_1-t_0)}{(t_1-t_0)}$ [mbar/day]	$\frac{\Delta p(t_2-t_1)}{(t_2-t_1)}$ [mbar/day]	$\frac{\Delta p(t_2-t_0)}{(t_2-t_0)}$ [mbar/day]
2	5.3	0.0270	4.0	393	4.0	2358	3.6	0.016	0.0052	0.0070
3	4.1	0.0265	3.7	264	3.4	2097	2.7	0.024	0.0034	0.0106
4	4.3	0.0097	5.3	84	—	—	—	—	—	—
5	5.6	0.011	3.5	141	3.3	2106	—	0.045	—	—
6	5.1	0.013	4.1	393	3.9	1907	3.7	0.012	0.0032	0.0051
7	5.2	0.0265	3.4	393	3.4	2358	2.7	0	0.0081	—
Mean value	—	—	—	—	—	2165	—	0.019	0.0063	0.0068 ±0.0016

TABLE 8.13. Variation in Lamp Pressure after 2160 Hours of Operation at 4.5 W Power

	Number of lamp				
p [mbar]	1	2	3	4	5
Initial	7.2	4.9	4.3	12.6	4.5
Final	6.2	3.5	4.0	12.2	4.0

the power in the discharge by a facto of 2 in comparison with lamps of 13 mm diameter.

8.3. *Upper Limits of the Lifetime*

The complex studies described above show how to increase the lamp longevity up to 10 years and more. On the other hand, the method for accelerating lamp tests up to 35-fold at 10 W discharge power, allowed the operational supervision of the quality during the manufacture of the spectral lamps.

The method for accelerating tests also enabled lamps with increased pressure (Table 8.13) to be also checked. These lamps were operated for 90 days at a discharge power of 4.5 W/cm^3 and 130 days at 2.5–3 W/cm^3. A calculation for a discharge power of 1 W/cm^3 yields a mode of operation is equal to 55,000–60,000 h of real work.

The average change of pressure in lamps for 40,000–60,000 h at a discharge power of 1 W/cm^3 does not exceed the average change in pressure of the lamps of the fourth group (see Table 8.5), though the time of their operation is greater. On comparing Table 8.12 with Table 8.13 one should bear in mind that in Table 8.12 the power consumed by the generator is indicated, and the efficiency of power transfer into discharge is approximately 50%.

On the whole these studies show that the service life of spectral lamps in the H-discharge mode with power 1 W/cm^3 exceeds 50,000 h.

Taking into account the observed change in the pressure of gas in lamps, it is expedient to increase the initial pressure of the inert gas up to 6–8 mbar in order to attain a service life of 10 years and more.

The increase in the lifetime of the source as a whole up to such high values requires appropriate durability of the elements of the oscillator circuit. Besides, the setup generator of the light source should have a rather large excess of power above the threshold at which the discharge fails.

Thus, the developed technology for manufacturing spectral lamps for appropriate longevity of the service life of a generator, decreases discharge power, and increases gas pressure of the permits one to ensure a service life of more than 100,000 h.

9

Some Problems in Designing Light Sources Based on High-Frequency Electrodeless Lamps

9.1. Choice of the Design of the Light Source

9.1.1. Mode of the High-Frequency Electrodeless Lamp

The choice of mode of the lamp and of its design begins with the choice of the bulb dimensions. The same intensity can be achieved at different dimensions of the lamp by adjusting the gas pressure and temperature of the thermostat. To increase the longevity and reliability of the lamp and reduce the power consumed by the thermostat, it is expedient to select modes with the smallest possible power in the discharge, with higher gas pressure, and with possible lower temperature of the lamp. Thus the thermostat temperature should be chosen to provide the highest intensity of radiation for a given discharge power. From Figure 9.1 it is seen that the highest density of radiation of rubidium resonance lines is about 10–25 mW/cm^2 in the H-discharge mode. When a power of about 1–10 mW/cm^3 is required, it is expedient to use a lamp of dimensions necessary to obtain the required optical power using the E-discharge mode. In specific experimental conditions, the lamp dimensions are chosen so that the power consumed in heating the tested objects should be at a reasonable level. For example, to obtain a rather low power (0.50 mW) an E-discharge of 0.4 W, possessing the highest integrated optical density (2–5 mW/cm^2) can be used. For a reduction in light source dimensions it is possible to reduce the size of the lamp with little increase in oscillator power and thermostat temperature in order to conserve the light flux. For the E-discharge mode a reduction in the size of a spherical lamp is possible down to 8 mm diameter. Further reduction in lamp dimensions is inexpedient as the lamp

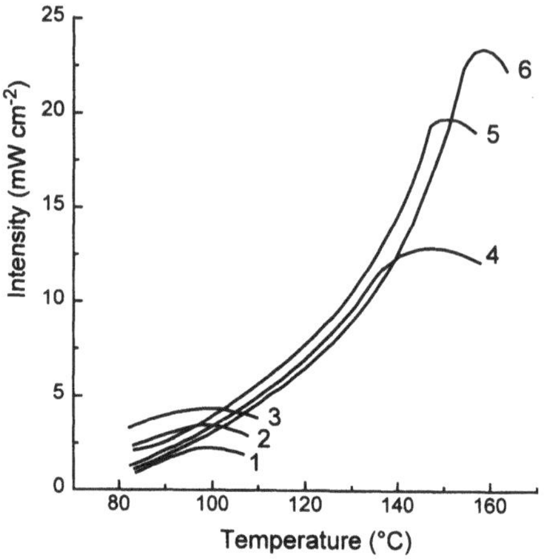

FIGURE 9.1. Radiation intensity of rubidium lines as a function of discharge power and thermostat temperature: 1–3, E-discharge; 4–6, H-discharge; discharge power: 1, 0.1 W; 2, 0.2 W; 3, 0.3 W; 4, 1.5 W; 5, 2.7 W; 6, 7 W.

reliability becomes lower, in which case measurement of the content of rubidium, aging at increased temperature with control of the rate of metal depletion, etc., would become difficult.

9.1.2. Design of Light Sources

In Chapter 2 we considered some problems of light-source design, in particular the principles of selecting inductors for excitation of the H-discharge. Some designs of light sources with diameter from 10 mm to 40 mm are described in the literature [22, 52, 146]. Light sources utilizing the E-discharge mode were not frequently used. The design of one commercial variant (light source No. 3, in Table 9.5) in use since 1977 is shown in Figure 9.2. Its characteristic feature is the use of two electrical conductive holders, touching the lamp at diametrically opposite sites of its surface. In practice the light source represents a spectral lamp located inside the grounded electrode, which is simultaneously the thermostat in which the potential electrode is placed. The light source is located in a Dewar vessel and consumes no more than 1.5 W power, while the oscillator consumes no more than 0.8 W in the temperature range from −60 up to 55 °C. Thus at a thermostat temperature of about 95 °C it radiates about 1 mW/cm^2 in the resonance lines of rubidium. A modified design of the light source is described elsewhere [59]. This modification consists in the installation of a small cell filter in the lamp holder, i.e., in the thermostat. The high temperature of the thermostat allows us to decrease the cell-filter dimensions from 30–50 cm down to 0.4 cm (see also [77]). Further modification of the light source allows one to increase its stability to external mechanical effects and simplify the setup of the oscillator, and is described in [60]. In order to increase the stability under

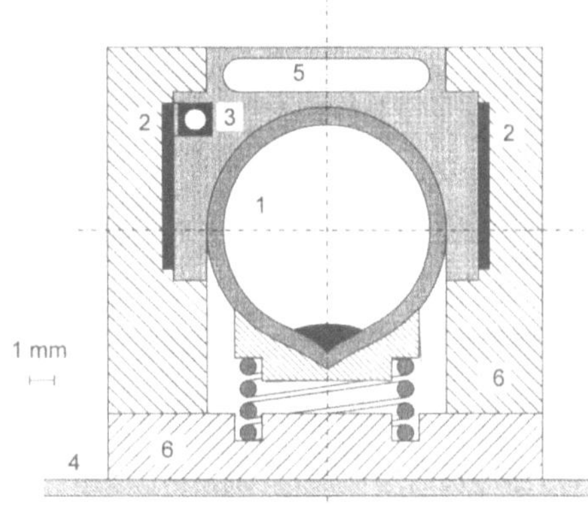

FIGURE 9.2. Light source in the E-discharge mode: 1, electrodeless lamp; 2, electrodes; 3, thermostabilizer; 4, plate of generator's printed circuit; 5, cell filter; 6, thermoisolation.

external temperature variations, the following design was proposed: In this light source the lamp was located in an evacuated vessel containing some droplets of easily vaporizable liquid. The vapor density of the filling varied as a function of the environmental temperature [53]. In this case we expanded the temperature range up to 65 °C and decreased the requirements of the system thermostating. In view of its importance we shall consider this design in detail in Section 9.2. In order to reduce the dimensions, light sources without a Dewar vessel (light source No. 5) were also developed. Their volume is 60 cm^3 and the consumed power was 1–6 W as a function of the environmental temperature, while the oscillators consumed power from 0.5 up to 1.5 W depending on the required intensity. Tests have shown the necessity of careful thermostating of the light source to provide the required stability at environmental temperatures of up to 60 °C. Such a light source had double thermostating of a lamp of 8 mm diameter and a cell filter (light source No. 6). The main problem of the given design is the maintenance of the required reliability of the lamp of smaller size than usual and prevention of overheating upon increase in environmental temperature. For this purpose the power of the oscillator should be less than 1.0 W. Actually we chose a mode with power of about 0.5 W that was necessary for stable working of the thermostating system. In order to decrease heat losses the thermostat was surrounded by a layer of thermoisolation. The light-source dimensions were decreased by locating all radioelements vertically on a board close to the external thermostat, providing lowered working temperature and increases reliability. Table 9.1 indicates the parameters of a number of light sources.

Experience in developing alkali metal vapor light sources may be used to create light sources with other filling metals, too. On the basis of light source No. 5 we have developed and investigated a light source emitting highly-stable radiation of the resonance lines of mercury (light source No. 7). The light source consists of the spectral source, the stabilizer of the power supply, and the temperature

TABLE 9.1. Parameters of a Number of Light Sources (LS) Developed and Produced Industrially

Characteristics	Light source (LS) number							
	1	2	3	4	5	6	7	8
Development date	1972	1978	1975	1985	1986	1990	1992	1995
$2R$ [mm]	10	13	13	13	13	8	13	8
Voltage supply [V]	19	22	10	15	10	8	8	6
Current supply [mA]	125	140	80	100	80	70	70	30
Temperature of thermostat [°C]	75	120	95	100	90	100	60	90
Frequency [MHz]	120	96	140	155	230	260	230	400
Longevity[a,b] [h]	10^4	$5 \cdot 10^4$	$5 \cdot 10^4$	$5 \cdot 10^4$	$5 \cdot 10^4$	$> 10^5$	—	$> 10^5$
Radiation intensity[c] [mW]	120	200	50	80	35	15	4	60
Spectral density of noise[d]: $5 < \omega < 10^4$ Hz, shot noise	+	+	−	+	−	−	−	+
$30 < \omega < 10^3$ Hz, shot noise	+	+	+	+	+	+	+	+
Relative change of intensity during observation time, $1 < \tau < 10^{-6}$ s, $\tau^{1/2}$	—	$< 7 \cdot 10^{-6}$	—	$< 7 \cdot 10^{-6}$	—	$5 \cdot 10^{-6}$	$< 10^{-5}$	$5 \cdot 10^{-6}$
Volume[e,f] [cm^3]	200	300	100	180	60	30	100	15
Batch [number]	850	1300	100	50	25	20	100	—
Range of working temperature [°C]	−5–50	0–50	−60–60	−5–60	−60–70	−60–80	−50–50	−60–??
Intensity change for 1 °C [%]	0.5	0.15	0.02	0.05	0.1	0.05	0.2	0.03
Vibration resistance	—	—	10g	5g	10g	15g	10g	15g

[a] By longevity we mean the technical service life of sources LS-1 and LS-2, and the guaranteed service lifetime for others.
[b] There is an opportunity to manufacture lamps with a resource of more than 25 years.
[c] Radiation intensity is indicated for rubidium resonance lines in the LS-1–LS-6, LS-8 sources and for the mercury resonance line in the LS-7 source.
[d] See Chapter 5.
[e] The dimensions of LS-3 are indicated without Dewar vessel 6 in which it is located.
[f] The LS-3–LS-6, LS-8 units contain miniaturized thermostated cell filters, while LS-7 and LS-8 contain a small-sized temperature controller and voltage stabilizer.

controller, working in a smooth mode. The electrical circuit of the source is shown in Figure 9.3. The spectral source contains the discharge activator, spectral lamp, spectral-lamp thermostat, and chassis. The discharge activator represents a transistor generator assembled on the basis of a three-point capacitor circuit and creating an electromagnetic field at a frequency of about 230 ± 10 MHz. The discharge activator is assembled on the board fixing the transistor and the constructive capacitor of the oscillating loop, created by the potential inductor and the internal surface of the thermostat. The transistor is located on the chassis of the spectral source. An electrodeless spectral lamp with spherical bulb and lengthened reservoir for the surplus of metal is located within the constructive capacitor. The lamp is filled with mercury vapor and krypton (1.5 Torr). The discharge in the bulb is excited and supported by the electric field of the capacitor. A thermoresistor used as a temperature gage of the lamp surface is inserted in the chassis of the lamp thermostat. The temperature is fixed by a smooth temperature controller with an accuracy of 0.2 °C. The voltage stabilizer and discharge activator are switched on to ignite the discharge. The connection of the spectral source with the temperature controller and voltage stabilizer is effected through passing filters to reduce the high-frequency pickups. The light source is designed as a complete constructive unit and can be used in equipment of different apparatus. Its technical characteristics are given in the caption of Figure 9.3.

Comparison of the characteristics of the presented source with that described in [80] demonstrates essentially higher stability, smaller consumed power, and higher reliability. Our source was used successfully in atomic-absorption analyzers.

9.1.3. Designs of High-Frequency Electrodeless Spectral Lamps

There are many designs of spectral lamps. The first electrodeless light sources used the lamps described in [7–9] (Figure 9.4). They were excited by an electrical field of a capacitor with a frequency of a few MHz and are only of historical interest. The first easily producible and convenient design of a hf spectral lamp was proposed in [22]. These spectral lamps had diameter 10 mm and were filled with krypton (about 1.6 Torr). A H-discharge was excited at a frequency of 100 MHz; their service life could reach 10,000 hours. At the same time spherical lamps of 30 mm diameter [24] and cylindrical ones of 18 mm diameter and 55 mm length have been created [208]. The perfection of these lamps was attained via research on their parameters and some modifications in design. It was found that to provide high stability and reproducibility it is better to use lamps with a lengthened reservoir for metallic rubidium, which should be separately thermostabilized. To provide high radiation stability it is necessary to avoid discharge burning in this reservoir [209]. According to [68, 69] the lamps should be filled with 1.8 Torr krypton and 0.03–0.2 mg rubidium, in order to

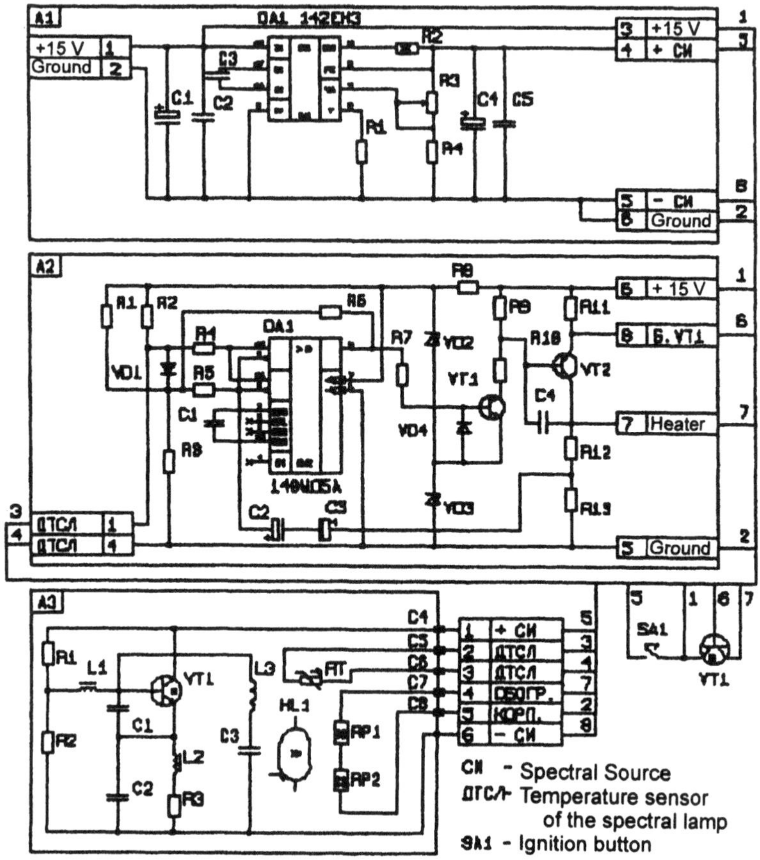

FIGURE 9.3. Basic electrical circuit of a source in the E-discharge mode. Power supply voltage $V = 15$ V. Consumed power of the voltage stabilizer $W = 1.5$ W. Consumed power of the temperature controller: in the heating mode $W = 5$ W, in a stationary mode $W = 1$ W. Power of resonant radiation not less than 2.7 mW. Resonant light flux in the complete solid angle not less than 3.5×10^{15} photons/s. Outlet resonant light How not less than 1.5×10^{14} photons/s. Entering into mode with instability of complete radiation not more than: 1%, 20 min; 0.3%, 40 min. Drift of the resonant radiation intensity during 1 h not more than 0.15%. Dispersion of the complete radiation intensity: 1–10 s, $(1.0 \pm 0.3) \times 10^{-5}$; 10–100 s, $(5.0 \pm 2.0) \times 10^{-5}$; 50–500 s, $(1.3 \pm 0.7) \times 10^{-4}$. Temperature range -50 to $+50$ °C. Instability of the resonant radiation within this temperature range, $< 30\%$. Diameter of lamp, 13 mm. Volumes: spectral source 60 cm^{-3}, temperature controller 32 cm^{-3}, voltage stabilizer 22 cm^{-3}. Weight 190 g.

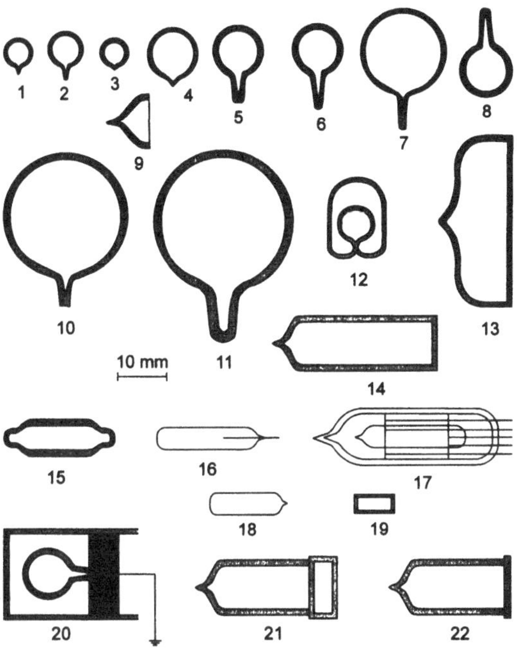

FIGURE 9.4. Designs of spectral lamps: 1, from [68, 126, 211]; 2, [211] (HP5965 A); 3, [68, 211]; 4, [211]; 5, [165]; 6, [211]; 7, [32]; 8, [292]; 9, [196]; 10, [63]; 11, [211]; 12, [32]; 13, [135]; 14, [110]; 15, [117] (HAM-111); 16, CH1-50*; 17, CH-69*, CH1-73*; 18, [7, 9, 22, 293]; 19, sapphire lamp, [119]; 20, [196]; 21, [291]; 22, [213]. The asterisk denotes Russian types.

provide longevity of more than 5000 h and ensure the highest radiation intensity. At low-power modes these lamps can provide a guaranteed longevity of more than 30,000 h up to 100,000 h.

Lamps with a quantity of Rb from 0.08 up to 0.2 mg are used in light sources with the highest longevity [68, 69, 125–127, 210, 211]. Analysis of the "service life" of these lamps and the development of methods for accelerated tests has resulted in an increased buffer-gas pressure of up to 4–5 mbar [68, 126, 211].

The use of lamps with increased pressure (6–13 mbar) [117, 211] allows one to combine highest radiation intensity and a resource of 100,000 h and more.

For small-dimension devices it is better to use lamps in the E-discharge mode [69, 211]. In this case the discharge power could be decreased down to 0.1–0.3 W and, as a result, the lifetime of lamps could reach 100,000–200,000 h and even more. Due to the low power of the discharge and the oscillator, the creation of an integrated unit including the light source becomes possible.

For further reduction in light-source dimensions the lamp diameter should also be reduced, but this will entail a decrease in the intensity of radiation and the reliability of the spectral source. The voltage of the breakdown field will increase. On reducing the dimensions of a lamp it is necessary also to increase the gas pressure (see Chapter 2).

Light sources with cylindrical bulbs of 8 mm diameter and 35 mm length have been developed and manufactured in Nizhni Novgorod (Russia) [110, 179]. The hf field within the capacitor evaporates metal from the surface of the lamp and excites it in an E-discharge. Using a vacuum jacket the power consumption could be reduced to 0.8 W. Despite the extremely low power in the discharge, until recently the guaranteed longevity was no more than 5000 hours, stipulated by the other types of failures. Characteristics for the design are given elsewhere [110].

Another design was developed and produced in Moscow [26]. The lamp has a cigar shape of 9 mm diameter and 15 mm length. Owing to the shape of the lamp it is easily installed in a light source. These lamps have an operating time of more than 14,000 h [90].

Russian industry produces spherical spectral lamps (10 mm and 20 mm diameter) for spectral analysis. The lamps have various fillings, including a complex mixture of several elements. The characteristic longevity is about 1000 h and the intensity of the emitted radiation lies in the range from 1 to 8 mW/(cm^2 sterad) for different elements. The designs of these lamps as well as modifications in their vacuum jacket are described in [32, 36]. Tests have shown a high radiation stability of this type of spectral lamp, which was used in the quantum frequency standards delivered by "Varian." Nowadays Hewlett–Packard devices employ spherical lamps of 10 mm diameter, filled with rubidium and krypton at a pressure of 4.5 mbar, with a guaranteed lifetime of three years.

The firms Rohde & Schwarz, Frequence Electronics, and Efratom use cigar-shaped lamps. Extensive studies on the optimization of the manufacturing process of lamps were carried out by the firm Efratom in the 1980s. In particular this research was devoted to controlling the content of metal and establishing a connection between longevity and stability. These results confirmed earlier results obtained at the Leningrad Radio Technical Institute [68, 69, 111]. The data have been used in the production of lamps since 1974. A decrease in the power consumed by the oscillator down to 0.4 W has allowed the high reliability of these light sources to be achieved.

Extensive investigations of spectral lamps of 10 mm diameter with various fillings were carried out in Japan in the 1970s. Many important results concerning the characteristics of lamps and their operational parameters were obtained. In 1981 a spectral lamp with a miniaturized cell-filter lamp located at the front part was developed (compare with [60]). This cell-filter lamp was used in a number of light sources by the firm NEC. In 1991 the service lifetime was specified to be about 15 years.

In France, a lamp of cylindrical form with reservoirs on both sides for the surplus metal was used in the rubidium frequency standard HAM-111. The design was rather large, the consumed power was 2 W, and the temperature was 110 °C.

A comparison of lamps working in the high-frequency range with lamps operating in the ultrahigh-frequency range was carried out [81]. It was found that the latter had the better stability. Therefore these lamps are widely used in various fields of spectral analysis [50], emitting rather intensive lines of the second resonance doublet of rubidium. To obtain intensive emission of the second and third resonant doublets of sodium it was recommended [135] to add some argon (0.1 Torr) to the buffer gas.

Spectral lamps containing certain elements, covering practically the whole periodic table, as well as lamps with double or threefold alloys have been developed at the Latvian University in Riga [5, 41, 42, 44, 45, 105]. Especially noteworthy is the high-intensive and long-term light source filled with helium [5]. This type of spectral lamp usually has a diameter of 25 mm or 40 mm. The problem of achieving a high intensity of mercury ion lines while suppressing the resonance lines of Hg was solved [5] using an absorbing cell filled with vapor of special mercury isotopes at the front part of the lamp.

High-frequency electrodeless sapphire spectral lamps with rubidium vapor and an expected service life of more than 100 years were elaborated elsewhere [119]. The characteristics of these lamps were similar in all aspects to those made from glass excluding the fact that the ignition voltage of the discharge was higher than for glass bulbs. In order to prevent ionization of the surrounding air by ultraviolet radiation emitted by such lamps, an additional cover of the bulb by a layer absorbing the ultraviolet emission is required. Sapphire was chosen for the following reasons: (1) it is a simple crystal, optically transparent after polishing; (2) it is chemically neutral; (3) it has good optical, thermal, and electrical properties; (4) it has a rather high melting temperature (2040 °C), despite all the technical difficulties of producing and sealing-off the sapphire bulbs. Also, the material itself is much more expensive than ordinary glass, but the long service life of these lamps stimulated the development of such a light source. The design of a sapphire lamp produced by the Aerospace Corporation is shown in Figure 9.4 [119]. Experimental sapphire electrodeless high-frequency lamps are now also fabricated and tested in Nizhni Novgorod (Russia) and at Zenit in Zelenograd (Russia).

9.2. *Thermoisolation of the Light Source*

The stability of the radiation intensity of a light source depends on the quality of the thermostating system. Passive and active thermostats are employed in the temperature stabilization of the electronic units. Materials with low thermal conductivity, such as glass fiber and foam plastic, as well as vacuum isolation are used for thermal isolation in thermostats.

With respect to the efficiency per unit of thickness, vacuum isolation exceeds other kinds of thermoisolation by some dozens of times. However, it is not always expedient to maintain a high degree of thermoisolation over the range of environmental temperature changes.

Passive ovens with thermoisolation, having a thermal conductivity which does not depend on the external temperature, allow one to thermostabilize electronic equipment under short-term temperature jumps. Concerning slow changes in the environmental temperature, these thermostats cannot provide effective thermostabilization.

Therefore, until now active systems of thermostating were used to suppress slow changes in the environmental temperature. However, in this case a special circuit comprising a temperature controller and its power supply is required. This requirement increases the dimensions of the lamps, makes the design more complicated, and results in increased energy consumption.

Thermoisolation possessing a thermal conductivity dependent on the temperature has been proposed [210]. The results using such a system in the development of light sources will be discussed in the following section.

9.2.1. Systems with Temperature-Dependent Thermoisolation

In order to obtain a thermal conductivity of vacuum isolation that depends on the environmental temperature, a saturated vapor can be used. The concentration of atoms (or molecules) of this vapor should vary within required limits, depending on the environmental temperature [61, 210, 212]. Vapors of hydrocarbons, silicates, vacuum oils, and smearings or liquid metal can be used as a heat carrier. The main physical properties of such heat carriers are given elsewhere [149].

For a light source with a Dewar vessel, the temperature of its thermostat is always higher than the environmental temperature. The pressure of the saturated vapor in the space of the vessel between the inner and outer surface corresponds to the temperature of the coldest part. Thus the pressure of the heat-carrying saturated vapor is a function of the temperature of the external wall, and therefore a function of the environmental temperature. Within the rather narrow range of the considered temperature change, the pressure of the saturated vapor of the substance has an exponential character (see Section 6.2).

At low pressure of a vapor, the mean free path of molecules of the heat carrier exceeds the distance between the walls. Molecules can fly from one wall to the other without collisions, and transfer energy directly from wall to wall.

If the mode of a gas flow is expressed in terms of the Knudsen number $K_n = l_0/L$, where l_0 is mean free path of molecules and L the characteristic size, then free molecular flow corresponds to $K_n > 1$. When $K_n < 10^{-2}$, collisions between molecules prevail over the collisions with the walls. The area of gas flow located between these values, $10^{-2} < K_n < 1$, is referred to as the intermediate area.

In the free molecular flow of gas, energy is transferred by each molecule from one wall to the other, and the heat flow is proportional to the concentration of vapor in the heat carrier.

The energy exchange of the vapor in the heat carrier with the walls of the Dewar vessel is characterized by the accommodation factor, determined by the following expression:

$$\alpha = \frac{T_i - T_r}{T_i - T_s} \tag{9.1}$$

where T_i is the temperature of the particles, colliding with the surface, T_r the temperature describing the average energy of particles reflected from the surface, T_s the temperature of the surface.

Within the considered range of temperatures of the environment (−60–60 °C) the temperature of molecules, falling onto a surface of condensation slightly exceeds the temperature of the vapor condensation. All the falling molecules will be condensed in this case. Thus, in the ratio (9.1), by considering the heat exchange on an external wall the molecules of the evaporating substance will have the temperature of the surface. The value α, characterizing the energy exchange of molecules with the wall on which evaporation and condensation take place, is rather close to one in Eq. (9.1). The factor α on a hot wall is less than one. However, within the simple physical picture we shall assume that the reflected molecule also has the temperature of the wall.

In the stationary mode the number of molecules leaving the hot wall should equal the number of molecules leaving the cold wall. The number of such molecules is proportional to the concentration of molecules with the temperature of the cold wall, and to their average collision rate.

Therefore the flow of energy from the hot to the cold wall is equal to the difference of the counterflows of energy, and proportional to the value

$$q \sim n_2 v_2 T_2 - n_1 v_1 T_1 \tag{9.2}$$

where n_1 and n_2 are the concentrations of molecules at temperatures T_1 and T_2, respectively; v_1 and v_2 are the velocities while T_2 and T_1 are the temperatures of hot and cold walls, respectively.

Taking into account the fact that the average rate of molecules is proportional to the square root of the absolute temperature, we obtain the thermal flow as a function of the pressure and temperature under free molecular heat transfer:

$$q \sim \frac{p_1}{T_1}(T_2 - T_1) \tag{9.3}$$

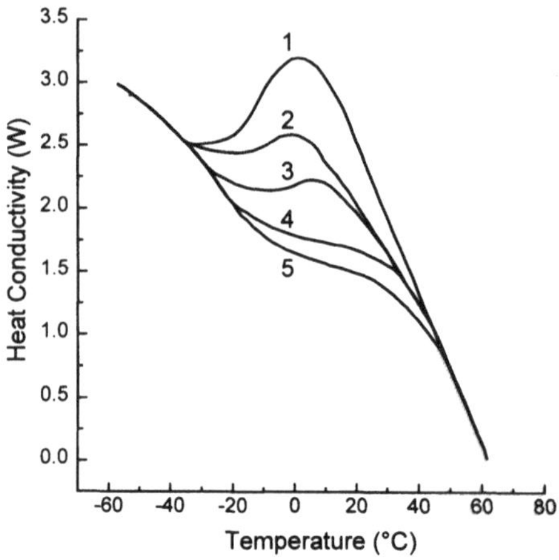

FIGURE 9.5. Heat conductivity of a Dewar vessel filled with different substances as a function of temperature: 1, naphthalene; 2, isoamilbenzoate; 3, cupron acid; 4, selen-4-chloride; 5, caprylaldehyde.

where p_1 is the pressure of the vapor at the temperature of the cold wall, T_1.

The ratio (9.1) yields an expression for the thermal conductivity of the thermoisolating vapor: $q \simeq p_1/T_1$.

Following Eqs. (9.3) and (6.3), as well as the condition governing the choice of heat carrier, the value of q obeys an exponential law, practically beginning from zero.

Actually, when $p > 0.1$ mbar, the dependence of the thermal conductivity on the environmental temperature becomes weaker, after which the heat transfer becomes independent of pressure.

9.2.2. Choice of Heat Carrier

The heat carrier is chosen so that at the minimum environmental temperature the vapor pressure of the heat carrier is below the pressure of residual air in the empty Dewar vessel (about 10^{-5} mbar). In this case the flow of heat transferred by the molecules of the heat carrier is negligibly small and has practically no effect on the heat conductivity of the vessel. On the other hand, heat transfer by the heat carrier at maximum temperature of environment should exceed heat transfer through the empty thermoisolation.

Application of the temperature-dependent isolation allows one to expand the temperature range and considerably stabilize power losses. When the environmental temperature range is rather broad, further growth in environmental temperature leads to the independence of the thermal conductivity on pressure.

The thermal conductivity of the heat carrier, $\sigma(T)$, is connected to the

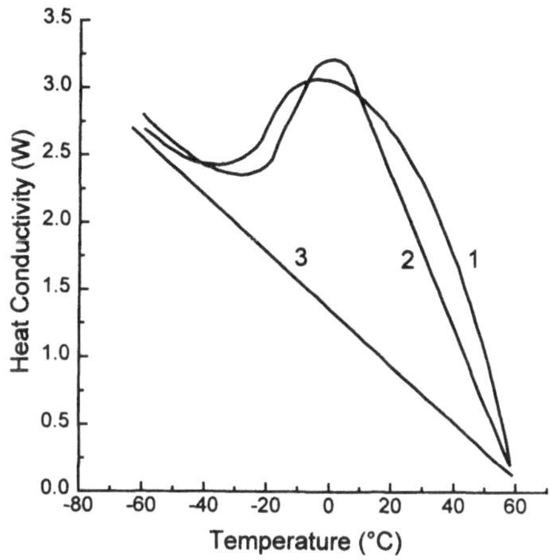

FIGURE 9.6. Dependence of the heat conductivity of a Dewar vessel filled with naphthalene on the environmental temperature: 1, calculation; 2, experiment; 3, power dispersed by thermal stabilization inside a Dewar vessel without an additional heat carrier.

equivalent thermal conductivity λ_n^{eq} the following relation:

$$\lambda_n^{eq} = \frac{\sigma(T)}{q} \tag{9.4}$$

Knowing the temperature dependence of the equivalent thermal conductivity of the cold wall of the Dewar vessel, the substance to be used for the heat-transport medium can be selected. Approximately 140 substances have been analyzed. Figure 9.5 shows the power dissipated through the Dewar vessel, as a function of the environmental temperature. The calculated dependence of the power losses on the environmental temperature is in good agreement with experimental data (Figure 9.6).

A comparative analysis carried out for various substances has shown that the peak of the maximum dissipative power decreases with decreasing gas kinetic cross section. The maximum is also reduced with an increase in the molecular weight of the substances.

The maximum is displaced to the right with an increase in the density of the heat carrier. The influence of the convective heat emission factor α on the external wall of the Dewar vessel was also found during these studies. With the growth in α the maximum of the dissipated power is displace to the right, i.e., the increase in the heat emission factor provides an additional opportunity for the temperature range to expand.

9.2.3. The Thermostat with Temperature-Dependent Thermoisolation

The vessel under study had the following dimensions: external diameter 68 mm, internal diameter 58 mm, length 195 mm, distance between the walls

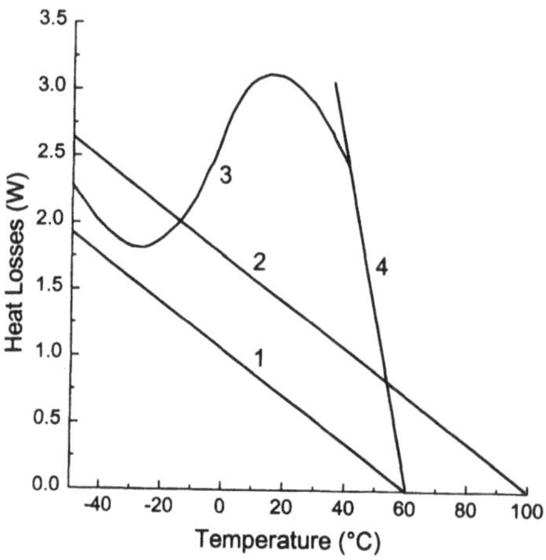

FIGURE 9.7. Comparative characteristic of heat losses for thermostats with different thermoisolation: 1, vacuum isolation for thermostat temperature of 60 °C; 2, vacuum isolation for thermostat temperature of 100 °C; 3, vacuum isolation with naphthalene vapor; 4, polythene foam.

3 mm. Naphthalin as working heat-carrier substance was initially cleaned and introduced into the interwall space of the Dewar vessel. The technology of filling assured conservation of the high thermoisolation properties at low environmental temperatures.

The average temperatures were taken as temperatures of the Dewar vessel. The sealing off temperature was determined as a sum of the environmental temperature (determined with an accuracy of 0.05 °C) and the temperature drop, determined by the difference between the readings of copper–constantan thermocouples, the error of which was also no more than 0.05 °C.

Thus the error in determining the coldest point's temperature was not more than 0.1 °C, which allows the vapor density of the heat conductor to be determined with an error of no more than 2.5%.

In order to study the thermal conductivity of the Dewar vessel as a function of the environmental temperature, a heater with a temperature controller was placed in the vessel. The heater temperature was kept constant ($T = 60 \pm 0.2$ °C). During the experiment, the power consumed by the heater thermostat upon variation in temperature over a range of −60 to 60 °C was measured. The measurements were carried out in the stationary mode. Figure 9.6 shows the results of experiment. Comparing the heat losses for the filled and empty vessel we see that, at low temperatures, they are practically identical. Heat losses through the isolation of the filled vessel grow with increased environmental temperature. Experimental data (Figure 9.6) show that at a total power of 0.8 W dissipated in the thermostat, the thermostating system with vacuum isolation fails even at a temperature of 25 °C. For temperature-dependent thermoisolation, the maximal temperature for the same dissipated power is 50 °C. Employing usual vacuum isolation is practically excluded for a power of 1.5 W, while the filled system allows one to use the effective vacuum isolation at temperatures up to 40 °C.

In Fig. 9.7, different kinds of thermoisolation are compared.

9.2.4. Temperature Stabilization by Temperature-Dependent Thermoisolation

The temperature of the internal wall of the Dewar vessel can be calculated by the formula

$$T_W = T_e + \frac{Q+Q_n}{\sigma_e} + \frac{Q+Q_n}{\sigma_i^V + \sigma_c(T_1)} \tag{9.5}$$

where T_W is the temperature of the internal wall of the Dewar vessel, the change of which is regarded as the stabilization criterion, T_e is the temperature of the environment, Q the power dissipated by the active elements of the thermostated unit, Q_n the power of the additional heater, σ_e the thermal conductivity from the external wall of the vessel into the environment, σ_i^V the thermal conductivity of the vacuum isolation, including the thermal conductivity of gas inlets and the other constructive elements, while σ_c is the thermal conductivity of the condensing heat-transport medium determined by the external wall temperature T_1.

The external wall temperature T_1 is determined by the two first terms of Eq. (9.3):

$$T_1 = T_e + \frac{Q+Q_n}{\sigma_e} \tag{9.6}$$

In spite of the fact that the value σ_e is considerably higher than the values of σ_i^V and σ_c, a difference in the temperatures between the external wall and the environment should not be neglected because the dependence $\sigma_c = \sigma_c(T_1)$ is very steep.

The power of the spectral lamp or the power of the light source as a whole can be used to stabilize the temperature. When necessary, an additional heater can be installed. Narrowing of the temperature range can be evaluated by analyzing relation (9.6). We assume that the maximum and minimum values of function $T_W = T_W(T_e)$ are attained at the ends of the environmental temperature interval, noting that, in principle, the extreme values are possible within this interval. The difference of the wall temperature ΔT_W could be calculated using relation (9.6) for the maximal (T_1^{max}) and minimal (T_1^{min}) temperatures:

$$\Delta T_W = \Delta T_e - \frac{(Q+Q_n)\left[\sigma_c(T_1^{max}) - \sigma_c(T_1^{min})\right]}{\left[\sigma_e - \sigma_c(T_1^{max})\right]\left[\sigma_e - \sigma_c(T_1^{min})\right]} \tag{9.7}$$

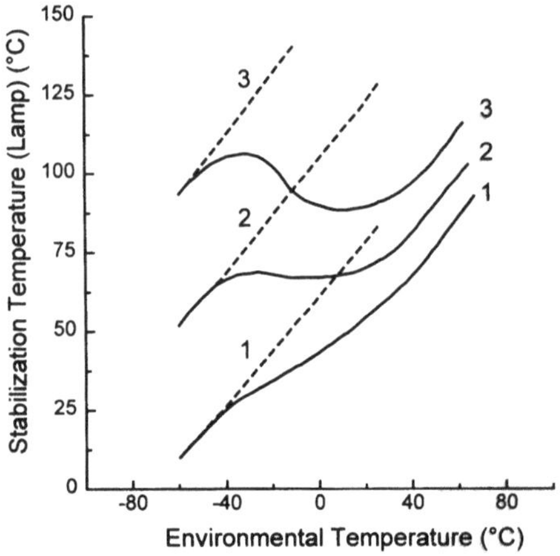

FIGURE 9.8. Stabilization temperature as a function of the temperature of the environment for various values of the total power: 1, 1 W; 2, 2 W; 3, 3 W. Solid lines refer to naphthalene vapor in the space between the walls. Dotted lines refer to vacuum isolation.

Analysis of the expression shows that the right part of the equation is always positive and therefore $\Delta T_W < \Delta T_e$. Thus it is possible to reduce ΔT_W by selecting the magnitude of Q_n. Obviously, a decrease is possible within certain limits determined by this formula.

The experiment on the temperature stabilization was carried out with the same Dewar vessel described before.

A heater simulating the discharge and an additional heater were placed inside the vessel. The power was varied over the range from 1 up to 4 W. The Dewar vessel was placed in the climatic chamber. The temperature in the chamber was changed from −60 to 60 °C at steps of 10 °C. The temperature of the air inside the Dewar vessel was measured under stationary conditions. Figure 9.8 shows the results of the experiment. An analysis of the plot shows that in this range of −60 to 60 °C the best temperature stabilization is attained at a dissipated power of $Q = 3$ W. The maximum difference of temperatures in this case is 23 °C, i.e., less than 20% of the complete range of the environmental temperature change (120 °C). In the range −5 to 40 °C the maximal temperature difference is 10 °C, or 22% of the total range of temperature variation (45 °C).

At $Q = 2$ W in the temperature range −40 to 20 °C the maximum difference of temperatures is no more than 6 °C, or 10% of the complete range. The range of temperatures for which a constant temperature was maintained with an error no more than 1 °C was 30 °C for 1.7 W. Selecting the material of the heat carrier, it is possible to shift the working point into the required area.

9.3. Schemes of Light Sources

9.3.1. Oscillators for Exciting Discharge in High-Frequency Electrodeless Spectral Lamps

High-frequency oscillators working at a frequency of 10 to 300 MHz are used to excite the discharge. In the first light sources, two-stage tube-type oscillators were used. Later, in the mid-1960s, transistor oscillators assembled on a three-point capacitor circuit, or a Clapp circuit, were utilized [142].

A large part of the investigations presented here were carried out with different variants of this circuit. This circuit was applied in many commercial light sources of the Leningrad Radio Technical Institute (since 1990, Russian Institute of Radionavigation and Time), as well as of different companies (Microdevice, Hewlett–Packard, Rohde & Schwarz, Frequence Electronics, etc.). The Clapp generator allows one to obtain a high voltage on the inductor and provides reliable ignition of the discharge. The circuit is sensitive to the spread of transistor parameters and spectral lamps and requires, as a rule, tuning to achieve stable ignition and stable work of the operating mode. Calculations concerning the Clapp oscillator circuit have been published [142]. The oscillator, loaded with a lamp, provides an instability of high-frequency power of less than 10% in the range from -15 up to 60 °C. The instability of the intensity in an E-discharge is of the same order. For the H-discharge, it depends on the scheme of the thermostat and can be reduced down to 5%. When no thermostat is used the instability of intensity is much higher, owing to fluctuations in the environmental temperature acting directly on the lamp. The selection of the lamp design can reduce the intensity instability only down to 30% in the range from -5 to $+45$ °C. Therefore thermostabilization is an important problem in the construction of spectral sources.

The oscillator problem is solved by special selection of the electronic elements used. In a series of tests no change was observed in the voltage–current characteristic, and failures in tests during 40,000–70,000 h in the intense mode were not caused by the oscillator.

Only a few works have examined oscillators. Most publications were devoted to special oscillators used in magnetometry. Their general feature is that these oscillators are loaded on a cable of length 2–6 m, at the end of which there is an inductor where a spectral lamp is located [213, 214]. The main problem was to set up correctly the generator–cable–spectral lamp circuit [215].

In quantum frequency standards developed at the Research Institute in Nizhni Novgorod, light sources with extremely low power consumption (from 0.3 to 1 W) were utilized in various devices. An additional generator, creating pulses to facilitate discharge ignition, has been utilized [179].

The stability of the light source in many respects depends on the stability of the feeding voltage. The requirements pertaining to this characteristic have been considered in Chapter 4. It was found that for light sources operating in the E-discharge mode a more stable power supply is required. Usually the supply voltage is stabilized, while the stabilization of the current is more difficult and does not increase the intensity stability.

9.3.2. Systematics of Thermostating

Careful thermostating is required for lamps containing saturated vapors. For this purpose the lamp is placed in a cavity, the surface temperature of which is maintained constant by an electronic temperature controller. Modern temperature-dependent electronic elements provide a sensitivity of 0.001 °C. The real sensitivity is smaller and is determined by the design of the thermostat. For maintenance of stable intensity of the resonance lines of rubidium with an accuracy of 0.1%, it is necessary to provide a temperature stability no worse than 0.016 °C.

The main difficulties to ensure the required temperature stability inside the source when the environmental temperature is changing are caused by the exit aperture of the optical radiation, worsening the thermostat quality, the small size of the thermostated surface, and by uncontrollable sources of heating.

It should be noted, that when using thermostats in a pulsed mode, it is difficult as a rule to remove electric inductions interfering with the working conditions. The values of these interferences depend on a number of factors, including the scheme of the oscillator setup. Therefore temperature controllers in a smooth mode of operation are favorable and they are more frequently applied. However there are problems of work maintenance over a range of environmental temperatures resulting from the necessity to dump the surplus of power.

During recent years, temperature controllers using large matrix crystals have been utilized with high-reliability temperature stability over a wide temperature range. Their extensive use allows one to change the conventional shape of the light source. The use of stabilizers and oscillators developed on the basis of integrated electronic circuits will widen the application of spectral sources.

10
Measurement of the Optical Line Shift and Broadening

10.1. Methods of Measuring Line Shifts and Broadenings Based on the Zeeman Effect

10.1.1. Introduction

In this part we consider the shift and broadening of the lines of a working element of the electrodeless discharge spectral source. For ignition of the discharge and intensity stabilization of the working element, buffer gas (mainly Kr) is usually used in the spectral lamps. Therefore the emitted radiation results from optical transitions of the studied atoms placed in a foreign gas medium. Atoms of the foreign gas perturb the energy levels of the working element atoms, resulting in a shift and broadening of the emission and absorption lines. Therefore, for an exact description of the spectral distribution of the emission of a spectral lamp it is necessary to know the spectral line shifts and broadenings, determined by collisions with the noble gas atoms. When detailed studies of the spectral light sources first began there was no reliable information available concerning pressure effects on the spectral characteristics of lamps and absorbing cells. In this connection the method of magnetic scanning was investigated and developed. This method yielded reliable information on the factors pertaining to shifts and broadening of lines for a number of elements. In spite of some pioneering work on the application of the magnetic scanning principle for investigating the spectra of colliding atoms, since 1965 systematic developments of the magnetic scanning method and its applications to study of the collision broadenings and shifts of spectral lines of atoms have been carried out independently by two groups, in Paris (Department of Physics, Université Pierre et Marie Curie) and Leningrad (Coherent Optics Laboratory, Leningrad State University).

The work of the French physicists was mainly devoted to research on the collision characteristics and features of the spectral line contours of mercury. At Leningrad State University collision broadenings and shifts of Rb and Cs lines were first determined. Different variants of the Zeeman spectrometer (which are known today) were developed on the basis of this research work.

10.1.2. Principles of Zeeman Spectrometers

The simplest variant of Zeeman spectroscopy consists in recording the intensity of resonance radiation of the investigated element, passing through an absorbing vapor of the same element, as a function of the external magnetic field strength, i.e., magnetic scanning.

The integrated intensity of light, passing through an absorbing cell is expressed by

$$I(B) = \int_0^\infty I_0(\nu)\exp[-k(\nu)l]d\nu \tag{10.1}$$

where $I_0(\nu)$ is the spectral shape of the radiation line while $k(\nu)$ and l are the absorption coefficient of the cell and its length, respectively.

There are two possible variants of the magnetic scanning method. In both cases the intensity of the transmitted light is recorded as a function of the magnetic field strength. In the first, an external magnetic field is imposed on the radiation source. In this case the line contour is described by $I_0(\nu) = I_0(\nu - \nu_0(B))$ and changes its shape when changing B. Usually, one Zeeman component of the radiation line is extracted by polarizing devices. The frequency of this line will be scanned in the vicinity of the absorption line of vapor in the cell. In the second variant, an absorption line of the researched element is scanned by the magnetic field: $k(\nu) = k(\nu - \nu_0(B))$. Here, the frequency position of the radiation line contour is constant. A change in transmission is caused by Zeeman splitting of the absorption line. The transmission function of the vapor cell is rather complicated and depends on the characteristics of the studied atoms and on the experimental conditions. Among other things, this function is determined by the spectral characteristics of the radiation and the absorption line, which depends on the natural width of the upper level of the transitions, on the atomic temperature in the source, and on the absorption line contour. An important factor is the cell's optical length kl, related to the density of the absorbing particles.

Owing to the sensitivity of the transmission function to the parameters of the line contour (shape, width, position on the frequency scale) the Zeeman spectrometer allows one to study fine features of the spectral distribution of the line intensity. It is a typical problem of high resolution spectroscopy.

In the general case, direct extraction of information on each of the above-mentioned characteristics from the transmission function is impossible. The situation becomes even more complicated if the spectral line has a complicated structure (cf. the hyperfine structure of the Rb D-lines, Figure 4.5). Therefore, the use of a Zeeman spectrometer for measuring spectral line characteristics requires computer simulation and processing of the experimental data. The final results may be obtained via comparison of the data with computation in the framework of an accepted model, by varying the parameters of the line contour and adjusting the computed curve to the experimental characteristics.

A significant feature of the transmission function is the sensitivity of this function to the presence of foreign atoms or molecules in the absorbing cell. Pressure-induced variations of the absorption line profile results in smoothing of the transmission function profile. The pressure effect on the absorption line, which in the low-pressure region results in the broadening and shift of the center of gravity of the Lorentzian contour of this line, is at low pressure proportional to the concentration of the admixed particles. The high sensitivity of the transmission curve to small concentrations of admixed particles should be especially mentioned. In some experiments the transmission curve showed its sensitivity to pressure changes in the region of some mbar. Such a high sensitivity allows the Zeeman spectrometer to be employed in the study of admixed particle effects under conditions corresponding to the collisional approximation. The change of the spectral line shape by such small admixtures of atoms is significantly smaller than the apparatus width of a classical high-resolution device. In the instrumental Zeeman spectrometer the contour of the radiation line takes the place of the instrumental function usually involved in classical absorption spectroscopy.

10.2. Scanning of the Irradiation Line

Let us consider the first variant of the Zeeman spectrometer on the basis of magnetic scanning of the irradiation line. This variant was widely applied in earlier works. The lamp in which the vapor of the studied atoms is excited in the plasma of the hf discharge was located between the poles of a magnet. The magnetic field strength was variable up to nearly one Tesla. Radiation emitted in the magnetic field direction passed through a quarter-wave plate and a polarizer. Thus one of the σ-components was extracted and its position in the frequency scale depended linearly on the magnetic field strength:

$$I = I_0(\nu - \nu_0(B)) \tag{10.2}$$

The radiation that passed through an absorbing cell was recorded.

In the first works of the French group [211] the broadening and shift of the mercury 253.7 nm resonance line affected by collisions with He, Ne, and H_2 was measured. The even Hg 198 isotope (nuclear spin quantum number $I = 0$) was placed in the spectral lamp to avoid hyperfine splitting of the scanned contour, which would have complicated the interpretation of the experimental results. The shape of the line was Gaussian (FWHM 4.2×10^{-2} cm^{-1} = 1.7 GHz). The magnetic field varied over a range of ± 970 mT, corresponding to a scanning range of 1300 cm^{-1}. In this way a light source, tunable in its emission frequency, was created.

The transmission function of the cell with mercury vapor can be represented as

$$I(\nu) = I_0(\nu - \nu_0(B)) \exp[-k(\nu)l] \tag{10.3}$$

The absorption coefficient $k(\nu)$ has a Doppler profile whose width is known in the absence of a foreign gas. When a foreign gas is inserted into the absorption cell, the factor $k(\nu)$ is turning out to be a convolution of Doppler and Lorentzian contours. The Lorentzian contour, its width and its position on the frequency scale are functions of the buffer gas pressure. Broadening and shift of the mercury resonance line for foreign gas pressures up to 1000 mbar were measured experimentally by applying this method.

Later the magnetic scanning method was used when other isotopes or a natural mixture of mercury isotopes were acting as cell vapor [216]. Collisions in a mixture of ^{198}Hg–^{199}Hg isotopes and in pure isotopes ^{196}Hg, ^{198}Hg, ^{199}Hg, and ^{200}Hg, as well as collisions within the natural Hg isotopic mixture have been studied. Absorption cells of various types have been manufactured for this purpose. A cell of optical thickness 1 was filled with natural mercury and the absorption of ^{196,198,199,200}Hg isotope line radiation was studied. Atoms of the natural mixture of Hg isotopes were used as perturbing particles. Cells of a second type (optical thickness 10) were filled with a ^{198}Hg (98.6%), ^{199}Hg (1.3%), ^{200}Hg (0.07%) mixture and 0.001% of the other isotopes. The absorption of a line of ^{199}Hg was studied and ^{198}Hg atoms were used as perturbing particles. The total gas pressure in the cell was more than 10–100 times the partial pressure of the isotope under research.

During these experiments the collisional broadening of the 253.7 nm line was found to be only a few times larger than the natural width. However, differences in the spectral contours, caused by collisions of mercury atoms with atoms of other elements, were found and account for the high sensitivity of the method. The broadening cross sections in Hg–Hg collisions were closely related and of the order of $(35 \pm 3) \times 10^3$ nm^2 and therefore 10–20 times larger than the measured cross sections for Hg–Ne, He, H_2 collisions. The cross sections between different isotope colliding pairs (Hg–Hg$'$) were found to be 10% lower.

The method described in [217] was also applied to determine the broadening and shift of the 253.7 nm line of ^{198}Hg atoms caused by collisions with molecules of hydrogen, deuterium, and tritium at pressures below 1000 mbar. A Voigt contour was observed; its shift $\Delta\nu_L$ and broadening $\Delta\nu_D$ were linearly dependent on pressure. The shift $\Delta\nu_L$ is connected with the spectral line broadening cross section σ^2 by the relation

$$\Delta \nu_L = \frac{2N}{Tc}\sigma^2\sqrt{2\pi RT\left(\frac{1}{M_1}+\frac{1}{M_2}\right)} \tag{10.4}$$

where N is the concentration of admixed particles, T is the absolute temperature; while M_1 and M_2 are molecular weights of the mixture components. Despite the change in the mass of the hydrogen isotopes, $\Delta \nu_L$ was constant within the experimental limits.

The most important results of the measurements were the anomalously large values of the ratio $\Delta \nu_D/\Delta \nu_L$ for all researched kinds of perturbing particles in comparison with the theoretical value of 2.7 for the Van der Waals particle interaction. This anomaly is explained in [217] by introducing a more complicated potential of the atomic interaction.

Published results [141, 218] were used in [217] to determine the C_6 and C_{12} factors in the framework of the collision theory of Lindholm–Foley using a Lennard-Jones potential:

$$V(r) = -\frac{C_6}{r^6} + \frac{C_{12}}{r^{12}} \tag{10.5}$$

It was found that the C_6 factor was the same for all researched perturbing gases while the C_{12} factor varied. The close lying C_6 values indicated the similarity of the electronic structures of the admixed molecular particles, and hence their polarizability.

Knowledge of the instrumental function of an optical device is important for any spectroscopic method. Indefiniteness of the apparatus function essentially reduces the accuracy and reliability of measured characteristics. As mentioned above, the emission (absorption) spectral line contour takes the place of the apparatus function in the magnetic scanning method. Therefore, when using a computer procedure it is necessary to know with high accuracy the profile of the emission (absorption) line. As the emission (absorption) line shape is formed in the discharge plasma (in the absorbing cell), placed in the magnetic field, various physical processes in the plasma and the external magnetic field effect can result in a deviation of the contour from the theoretically expected Doppler profile.

For example, when investigating the isotopic structure of the second mercury resonance line $\lambda = 184.9$ nm, which lies in the vacuum ultraviolet region, the authors of [219] had to take into account the effect of self-absorption in the light source, which is essentially important for this line, resulting in a significant deviation of its contour from the theoretical form.

The absorption profiles of every component of the line were assumed to be Voigt profiles and the function $\exp[-k(\nu)l]$ depended on various parameters like positions of the component centers of the absorption line, thickness of the absorption cell, etc.; $\Delta\nu_D$ and k_0, depending on the temperature of the cell, were parameters characterizing the Voigt function of a one-component absorption contour.

The magnetic contour of the transmitted light in various modes of lamp operation, including the lamp self-absorption mode, were recorded in order to obtain the dependence of the cell transmission curve on the parameters of the $I_0(\nu-\nu_0(B))$ function. The emission line contour was found to be formed under the effect of mainly three physical processes: natural broadening, connected with the finite lifetime of the excited state; the Doppler effect, dependent on the effective temperature radiating part of the lamp; and the self-absorption effect, dependent on the source configuration, its effective temperature, and the magnetic field. Taking these factors into consideration the radiation line contour can be expressed symbolically as

$$I_0(\nu-\nu_0(B)) = I_0P(\nu-\nu_0(B))\cdot F(p,n,\nu-\nu_0(B)) \tag{10.6}$$

where P is the Voigt function depending on the natural and Doppler linewidth, F a function describing the self-absorption in the light source, p the self-absorption parameter

$$p = \frac{h\nu}{c}B_{12}P_\alpha(0)N_\alpha \tag{10.7}$$

where B_{12} is the Einstein coefficient, $P_\alpha(0)$ the maximum of the absorption shape, and $N_\alpha = \overline{N}nr_0$; $\overline{N}$ is the average atomic density in the lamp, r_0 the radius of the lamp, and n is a parameter characterizing the degree of homogeneity of the plasma in the lamp.

The cell transmission should be given by

$$f(\nu_0(B)) = \frac{1}{\tau}\frac{\int_0^\infty \tau I_0(\nu-\nu_0(B))\exp[-k(\nu)l]d\nu}{\int_0^\infty I_0(\nu-\nu_0(B))d\nu} \tag{10.8}$$

where τ is the transmission of the cell windows. The optical length of the cell was chosen to provide sufficient signal amplitude.

The method of Zeeman spectroscopy was used in [220] to investigate the effect of collisions on the spectral line shape of the 404.7 nm mercury line ($6\,^3P_0 - 7\,^3S_1$). The population of the lower level of the transition, $6\,^3P_0$, was provided in two steps. In the first step a cell, filled with mercury vapor with a little nitrogen addition, was irradiated by the 253.7 nm resonance line, which excited atoms to the $6\,^3P_1$ state. In the second step the optically forbidden transition $6\,^3P_1 - 6\,^3P_0$ took place under the effect of inelastic collisions with nitrogen molecules. The light source with the ^{198}Hg isotope was located in a magnetic field, allowing Zeeman scanning of the radiation line.

The transmission of a cell containing the natural mixture of mercury isotopes is characterized by the convolution and the absorption coefficients of every hyperfine structure component j of the isotope (assuming all parameters to be constant over the cross section of the light beam):

$$k_{ij}(\nu) = \frac{h\nu}{c} B_{ij} \alpha_i N F_{ij}(\nu) \tag{10.9}$$

where

$$B_{ij} = \frac{c^3}{8\pi L \nu^3} \left(\frac{g_j}{g_i}\right) \frac{1}{\tau(7\,^3S_1)} \tag{10.10}$$

while α_i is the relative abundance of the i-isotope, N the density of the metastable atoms in the 6 $^3P_0$ state, and F_{ij} the normalized line contour. This contour is represented as a convolution of the Doppler and Lorentzian contours, the latter being determined by the natural lifetime and collisional relaxation. Thus

$$k(\nu) = \frac{\lambda^2}{2\pi} \frac{1}{\tau(7\,^3S_1)} \sum_{ij} \left(\frac{g_j}{g_i}\right) F_{ij}(\nu) \overline{\alpha_i N} \tag{10.11}$$

where

$$\overline{\alpha_i N} = \frac{1}{l} \int_0^\infty \alpha_i(z) N(z) dz \tag{10.12}$$

An experimental determination of the parameters of the instrumental contour $I_0(\nu - \nu_0(B))$ was performed preliminarily. It was assumed that its shape (Doppler profile with half-width $\Delta\nu_D$) did not depend on the magnetic field. The transmittance of the absorbing cells was studied for this purpose in the absence of inert gas. The function F_{ij} can be expressed symbolically as

$$F_{ij} = G_i(\nu) \otimes L_N(\nu) \tag{10.13}$$

where $L_N(\nu)$ accounts for natural damping and the Doppler profile G_i can be calculated. Since there is no information on the average population $\overline{\alpha_i N}$, it is difficult to determine $\exp[-k(\nu)l]$. A family of curves of the cell transmittivity as a function of the magnetic field without inert gas was calculated for this purpose. Every curve in this group corresponded to different parameters $\Delta\nu_D$ and $\alpha_i N$. Comparison with experiment then allowed one to choose suitable values of these parameters.

The broadening and shift of the absorption line was then determined by comparison of the calculated dependence $I_0 \exp(-kl)$ with the experimental one. It was assumed that broadenings and shifts of all the components of a line were equal. The transmission profile of a cell at a constant buffer gas pressure, under different excitation modes by the 253.7 nm resonant light, has been investigated for an accurate choice of parameters. This experiment allowed the broadening and shift of the 404.7 nm mercury line to be found by collisions with N_2, Xe, Ne, and He at pressures below 150 mbar.

10.3. Scanning of the Absorption Line: Measurement of Line Shift and Broadening

10.3.1. Features of Absorption Line Scanning

Despite the apparent simplicity, the first variant of the Zeeman spectrometer has a number of disadvantages. The main disadvantage is the dependence of the shape of the spectral lines radiated by the light source on the magnetic field strength. This fact additionally complicates the mathematical modeling of the experiment. Therefore, the variant in which an absorption line of vapor in a cell is scanned by a magnetic field is more widely used in practice.

This variant was used, for example, in measuring of broadenings and shifts of the 326.1 nm cadmium resonance line by inert gases [221]. A longitudinal magnetic field was applied to an absorbing cell and the radiation intensity was measured as a function of the field strength. For easier interpretation of the results, the even isotope ^{114}Cd having no hyperfine structure was used. Taking into account that on the 326.1 nm cadmium line a normal Zeeman triplet is observed and circularly polarized components ($\sigma^{\pm}$) were absorbed, the absorption coefficient was expressed as

$$k(B) = \frac{k_0}{2}\left\{\exp\left[-4\ln 2\left(\frac{\nu-\nu_0+\Delta\nu(B)}{\Delta\nu_D(\text{with gas})}\right)^2\right] + \exp\left[-4\ln 2\left(\frac{\nu-\nu_0-\Delta\nu(B)}{\Delta\nu_D(\text{with gas})}\right)^2\right]\right\} \tag{10.14}$$

where k_0 is the factor of absorption in the line center at frequency ν_0, $\Delta\nu_D$(with gas) is the Doppler under the influence of the disturbing foreign gas, while

$$\Delta\nu(B) = \frac{\mu_B B}{h} \tag{10.15}$$

is the frequency shift of a σ-component of the normal Zeeman triplet with $\mu_B = e\hbar/2m$ (Bohr's magneton, 9.274×10^{-24} J/T).

To simplify calculations, the Voigt shape of an absorption line broadened through collisions with additive gas was substituted by the Doppler profile given above. This substitution did not introduce errors into the determination of the line broadenings at low pressures of additional particles for low optical density of the absorbing vapor ($k_0 l \ll 1$). The radiation line contour $I_0(\omega)$ was also assumed to be Doppler.

By expanding Eq. (10.1) in series, using (10.14), and subsequently integrating all the terms, we obtain

$$I = I_{max} \sum_{n=0}^{\infty} \sum_{m=0}^{n} \frac{(-1)^n \left(\frac{k_0 l}{2}\right)^n C_n^m}{n! \sqrt{1+n\alpha^2/\gamma^2}} \times \exp\left\{-4\ln 2 \left(\frac{\Delta\nu(B)}{\Delta\nu_D(\text{with gas})}\right)^2 \left[\frac{(n-2m)^2}{n+\gamma^2/\alpha^2} - n\right]\right\} \quad (10.16)$$

where $\alpha = (I_{max}/I_0)^2/\pi$, $\gamma = \Delta\nu_D(\text{with gas})/\Delta\nu_D$, $C_n^m = (n(n-1)\cdots(n-m+1)/m!)$ is the well-known binomial factor; I_0 and $\Delta\nu_D$ refer to values determined in the absence of buffer gas in an absorption cell ($kl \approx 1$), while the parameters $k_0 l$ and $\Delta\nu_D$ were determined from this function. After determining α and γ in measurements with the foreign gas, we find the collisional broadening of an absorption line to be

$$\Delta\nu_{col} = (\gamma - 1)\Delta\nu_D \quad (10.17)$$

A modification of the spectrometer scheme allowed an increase in the accuracy of the collisional broadening measurements. A certain spectral interval was selected and the transmission contour was measured within this frequency band. Spectrometers with high resolving power, such as grating monochromators and Fabry–Perot interferometers, were used for this purpose. This modification allowed measurements to be conducted at the center of the absorption line, increasing the sensitivity of the method threefold.

The collisional shift of the line was determined by measuring the cell transmittivity for one σ-component. Cadmium lamp radiation passed through a polarizer and a $\lambda/4$ plate, and the transmission of the cell in the magnetic field was studied. If the maximum of the absorption factor is shifted by

$$s = \nu_{max} - \nu_0 \quad (10.18)$$

because of collisional interaction with atoms of the foreign gas, the absorption factor for the $\sigma^{\pm}$-component can be expressed as

$$k_\nu(\sigma^{\pm}) = \frac{k_0}{2} \exp\left\{-4\ln 2 \left(\frac{\nu_0 - \nu + \Delta\nu(B) + s}{\Delta\nu_D(\text{with gas})}\right)^2\right\} \quad (10.19)$$

The shift in the maximum of the transmission function by foreign gases was measured at the central frequency (ν_0). The ratio $I(\sigma^+)/I(\sigma^-)$ was equal to 1 in

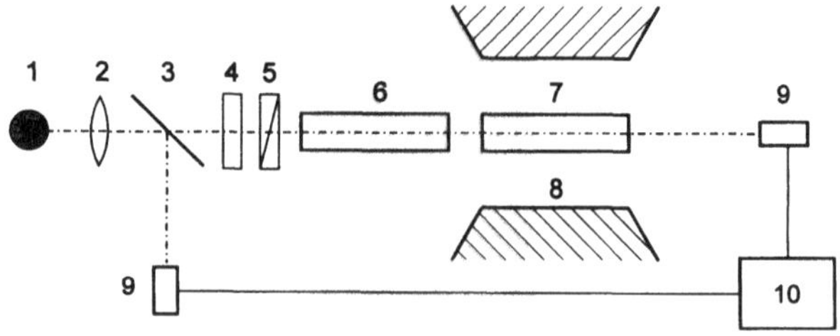

FIGURE 10.1. Experimental scheme of a magnetic scanning spectrometer for investigation of pressure shift of Rb resonance lines: 1, light source (hf EL); 2, lens; 3, beam splitter; 4, interference filter (separation of one of the D-lines); 5, polarizer; 6, hyperfine filter (principle see Figure 4.5); 7, working cell; 8, magnet; 9, photodetector; 10, signal comparator.

the absence of the foreign gas. He, Ar, and Xe addition resulted in a decrease or increase of this ratio, which allowed one to determine the sign of the shift of the absorption line during collisions.

An estimate of the accuracy of this method for measuring the shift and broadening of an absorption line gave systematic errors no larger than 5%. The statistical error of the line broadening was about 25%, and about 7% for the line shift.

The method of absorption line scanning was realized in [191] for investigations of the collisional shifts of the Rb D_1-line ($\lambda = 794.7$ nm, $5\,^2S_{1/2}$–$5\,^2P_{1/2}$).

In this experiment the favorable positions of the hyperfine components of the D_1-line of ^{85}Rb and ^{87}Rb isotopes relative to each other were used (cf. Figure 4.5). The transmission function of the cell showed a minimum while the absorption line of the vapor was scanned by the magnetic field. This minimum was due to the coincidence of one Zeeman π-component of the ^{87}Rb line in a field of 430 mT with the short-wave component (group a) of the hyperfine structure of the ^{85}Rb D_1-line. The Zeeman π-component of ^{87}Rb line corresponds to the transition $5\,^2S_{1/2}, F = 2, m_F = 2$–$5\,^2P_{1/2}, F = 2, m_F = 2$.

As mentioned earlier, when the optical thickness of a cell is rather small, the shape of the transmission signal can be expressed as a convolution of the radiation and absorption line contours. If foreign particles are absent, the shape of an absorption line is determined by the Doppler effect and the natural linewidth. Upon addition of foreign particles, the absorption line will broaden and shift. It is possible to measure the collision shift of a line by determining the change in the position of the minimum of the cell transmission curve as a function of the pressure of the admixed gas in the cell. However, the D_1-line of ^{85}Rb radiation consists of four hyperfine structure components. Hence, the coincidence of a Zeeman component of an absorption line with another hyperfine component of the emission line will result in another minimum of the magnetic curve of vapor transmittivity. This minimum will be observed in fields of the order of 150 mT.

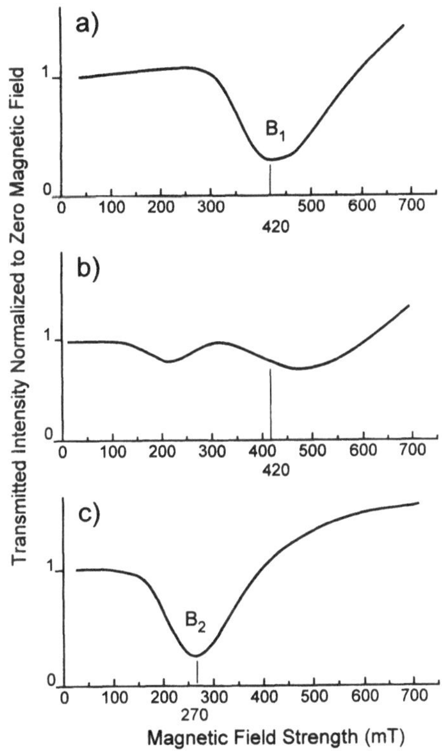

FIGURE 10.2. Intensity of light passing through the working cell: (a) π-polarized light, cell: ^{87}Rb without buffer gas; (b) π-polarized light, cell: ^{87}Rb and Ar at 61 mbar; (c) σ-polarized light, cell: ^{87}Rb without buffer gas.

Under admixture of gases with pressures of the order of 60 mbar, both minima will broaden and the error in the line-shift measurement will increase.

The ^{85}Rb–^{87}Rb hyperfine filtering (see Figure 7.2) was used in the scheme of the Zeeman spectrometer to avoid the second minimum effect. By placing a cell with vapor of ^{87}Rb across part of the ^{85}Rb radiation, it is possible to achieve a significant decrease in the long-wave ^{85}Rb hyperfine group (group b) intensity. The complete removal of one hyperfine component is possible when buffer gas, shifting and broadening the components of the ^{87}Rb absorption contour, is added to the cell (see Figure 7.2). As a result a light beam, the spectrum of which is modified by the filtration, enters the working cell and the absorption line contour is favorably positioned in the spectrum relative to the component of the ^{87}Rb absorption line contour.

The experiment was carried out by the following scheme (see Figure 10.1). Radiation of the light source with ^{85}Rb vapor passes through an interference filter (to separate the D_1- and D_2-lines), a polarizer, a hyperfine filter (for absorption of the long-wave component group), and is passed through a cell with ^{87}Rb vapor placed in a magnetic field. The intensity of the transmitted radiation was detected by means of a bridge scheme, using two photo-detectors.

Figure 10.2a shows the measured transmission (for π-polarized light) of a cell with pure rubidium as a function of the magnetic field strength. The minimum of the signal is in line with mutual superposition of the component maxima of the radiation and absorption lines. Figure 10.2b shows transmission in the presence of admixed gas. The minimum is shifted to higher field strengths; another, weakly pronounced minimum appeared at lower field strengths, while the modulation depth is decreasing. The accuracy of the shifts values obtained from this curve was limited and measurement of broadening was practically impossible.

The σ-transitions of the radiation line have also been investigated. Their other dependencies of the intensity on the magnetic field allow some effects of interfering lines to be removed. Also, comparison of measurements in π- and σ-polarizations has provoked interest. Figure 10.2c shows the shape of the transmission function for σ-components. It is seen to differ from the previous case of π-components by a sharper dependence of the transmission of the scanned light frequency on the magnetic field. This contrast manifests itself as a shift of the maximum to smaller fields ($B_1 = 420$ mT, $B_2 = 270$ mT), and as a decrease in sensitivity to the measured shift.

10.3.2. A Zeeman Spectrometer for the Study of Lines Having Hyperfine Structure

Reliable information on the spectral collisional effect was obtained by the Zeeman scanning technique for those atoms with a hyperfine structure which had favorable positions of the hyperfine components in the emission spectrum of different isotopes. The creation of the general scheme of a Zeeman spectrometer for atoms with hyperfine structure is therefore an important problem.

The solution of this problem for ^{87}Rb is described elsewhere [222, 223]. A spectral lamp was not used directly now as a light source for the Zeeman spectrometer. In order to have a radiation source whose spectral intensity distribution can be considered as known with rather high accuracy, another cell was used that emitted fluorescence radiation when excited by the light of a hf EL. The line contour of this fluorescence light source is the instrumental function of the Zeeman spectrometer. Various isotopes of a single element were used in the light source and absorption cell.

The line contour of fluorescence light depends on the radiation direction relative to the irradiation direction. If these directions coincide, the fluorescence line shape will be a product of the shape of the irradiating line of the source and the shape of the absorption factor. If these directions are opposite, the radiation spectrum will be the same product with distortion, caused by a double natural width of the line. If the directions are perpendicular to each other, the contour of an individual component is a Doppler profile or, taking into account the natural width of the line, it is a Voigt profile. In this case the relative intensities of the spectral line components depend on the irradiation spectrum. Inevitable decrease in the light flow upon substitution of a lamp by the resonance cell is not a serious handicap in view of modern methods to record weak light flow. It must be added that the apparatus function can be rigidly fixed by thermostabilization of the fluorescent cell.

Let us consider the experiment [222] in which the line of ^{87}Rb vapor fluorescence was investigated. The Doppler width of the line was determined by the temperature of the cell [Eq. (3.13)].

The Rb D_1-line, $\lambda = 794.7$ nm, has four hyperfine components. Their relative intensities depend on the excitation conditions. For isotrope excitation they can be expressed through the populations of the excited state N_F and the transition probabilities A_{ki} or the relative oscillator strengths f:

$$\begin{aligned} I_1 &= N_{F=1} f_{F=1\to F'=1}, \qquad & I_2 &= N_{F=1} f_{F=1\to F'=2} \\ I_3 &= N_{F=2} f_{F=2\to F'=1}, \qquad & I_4 &= N_{F=2} f_{F=2\to F'=2} \end{aligned} \tag{10.20}$$

As the transitions probabilities for the considered transition are constant, there are two unknown values: $N_{F=1}$ and $N_{F=2}$. As $f_{F=1\to F'=2} = f_{F=2\to F'=1} = f_{F=2\to F'=2}$, the relative component intensities can be expressed by one unknown parameter:

$$\beta = \frac{N_{F=1}}{N_{F=2}} = \frac{I_2}{I_3} \tag{10.21}$$

However, as the excitation is carried out by directed light, the excited state can be characterized not only by populations, but also a longitudinal alignment must be taken into account.

This alignment leads to a variation in the radiation intensity in that particular direction compared to the case of isotopic excitation of the investigated atoms. An alignment can disturb the relationship between the intensities of the hyperfine components of the radiation line of the fluorescence cell. Quantitative estimations of this effect concerning the excited states of ^{87}Rb, and subsequently the relationship between hyperfine intensities, were carried out for standard experimental conditions. In this way it could be shown that corrections connected with the alignment were negligible. We will also neglect the influence of the second isotope, which makes some changes in the fluorescence line contour.

Thus, in the case being considered, the fluorescence spectrum or apparatus function of the magnetic spectrometer is represented as

$$I_0(\nu) = \sum_{j=1}^{4} I_{j,0} \exp\left\{-4\ln 2\left(\frac{(\nu - \nu_{j,0})}{\Delta\nu_D}\right)^2\right\} \tag{10.22}$$

where I_{0j} is the relative intensity of the hyperfine component j of the fluorescence line. As stated above, the relative intensities can be described by only one parameter, β, determining the apparatus function. The precise value of this parameter can be obtained by an additional experiment. The shape of each hyperfine component is determined by the fluorescence cell temperature.

Light from the source is directed through the cell with the vapor of the studied element. Light was partially absorbed in the cell and the transmitted light was recorded by a photoreceiver. When applying the magnetic field to the cell, the absorption line contour, as noted, is deformed and the integral absorption varies. The intensity of the passed light is recorded as a function of the magnetic field. This dependency could be used to determine the shape of the original contour of the absorption line (when $B = 0$).

It was assumed in this model that an absorption factor represents individual Zeeman components identical in shape. The positions of these components on the frequency scale are determined by the Zeeman splitting and the shift under the effect of collisions. Maximal observable relative intensities of these components are determined by the transition probabilities. The contour of each Zeeman component is described by a Voigt function. In so doing the Gaussian part of this function is determined by the temperature of the abscrption cell, and the Lorentzian part either by the natural width for pure vapor or by the sum of the natural width and the collisional broadening by the additional gas in the cell. We neglect the resonant interaction of atoms of the metal, as the pressure of vapor in a cell was less than 10^{-6} mbar. Besides, we assume that under the effect of the additional gas all components of the absorption line are subjected to the same shift Δ and the same broadenings $\Delta\nu_{col}$.

The absorption factor connected with one Zeeman component can be expressed in the following form:

$$k_i = k_{i,0}(\nu_{i,0})\phi \tag{10.23}$$

where ϕ is the Voigt function.

In describing the absorption of the cell with a gas, we "shift" (in the frequency scale) all components by adding the value s to every ν_{0i}. We "broaden" the components by adding $\Delta\nu_{col}$ to the value $\Delta\nu_{nat}$. By defining

$$s = \nu_{i,0}(\text{with gas}) - \nu_{i,0}, \qquad a = \frac{\sqrt{\ln 2}(\Delta\nu_{nat} + \Delta\nu_{col})}{\Delta\nu'_D}$$

$$y = \frac{2\sqrt{\ln 2}(\nu' - \nu_{i,0} + s)}{\Delta\nu'_D} \tag{10.24}$$

we will obtain the general form of an absorption contour:

$$k(\nu) = \sum_i k_{i,0} \frac{a}{\pi} \int_{-\infty}^{\infty} \frac{e^{-y^2} dy}{a^2 + \left[\frac{2\sqrt{\ln 2}(\nu' - \nu_{i,0} + s)}{\Delta\nu'_D} - y \right]^2} \tag{10.25}$$

Summation over index i will account for the contribution of all Zeeman components of the transition to the absorption factor.

The positions of the components on the frequency scale, $\nu_{i,0}$, and the absorption factor of each component in the maximum, $k_{i,0}$, proportional to the transition probability, represent the spectral distribution when the magnetic field is absent.

10.3.3. Computer Simulations of Characteristic Features of the Observed Signals

We will consider only investigations of the ^{87}Rb D_1-line absorption. Taking into account the above-mentioned assumptions, the cell transmission curve in a magnetic field is expressed by

$$I(B) = I_0 \int_0^{\infty} \sum_{j=1}^{4} \beta e^{Y} d\nu \tag{10.26}$$

with

$$Y = \left(\frac{2\sqrt{\ln 2}(\nu - \nu_{j,0})}{\Delta\nu_D} \right)^2 - \ell \sum_i k_{i,0} \frac{a}{\pi} \int_{-\infty}^{\infty} \frac{e^{-y^2} dy}{a^2 + \left[\frac{2(\nu' - \nu_{i,0} + s)\sqrt{\ln 2}}{\Delta\nu_D} - y \right]^2}$$

Extensive calculations of the signal contour were performed on a computer. For the shape of one signal, determined by Eq. (10.26), the Voigt integral calculation had to be repeated 25,000 times for π-polarized light and 45,000 times for σ-polarized light. The Voigt integral [151] was necessary to determine the absorption factor of each Zeeman component at the specified frequency.

In order to decrease the CPU time the Voigt integral calculation was carried out before beginning the general calculations for points of the contour situated at distances up to 20 Doppler widths from the center of the line with a step of 0.1 of the Doppler width. Further, these values were used in the following manner. The value $\Delta\nu = (\nu_{i,0} - \nu)/\Delta\nu'_D$ (which required the determination of an absorption factor) was rounded off to an integer value. It was used as an index in the determination of the appropriate term in the initial file of numbers. Thus the calculation of the Voigt integral was replaced by the application of definite values of the absorption factor at specified frequencies.

Limits and the integration step were chosen such that the time of count was minimal for the integral calculation. However, this factor did not significantly influence the accuracy of the final calculations. Selected integration limits (4.5 and 6.5 GHz relative to the center of gravity of a level) were restricted to 3 Doppler widths from the center of the outward-lying components of the hyperfine structure. The step (0.15 GHz) approximately corresponded to 1/3 of the Doppler width. These limits were sufficient, as a double extension of the integration limits and a 10-fold increase in the number of steps changed the result only in the fourth significant figure.

All the described operations represent the calculation of only one point of the signal. In order to determine the general form of the magnetic scanning signal, the expression was calculated for values of the magnetic field from 0 to 810 mT, at 30-mT intervals (which was the range of fields covered experimentally). The complete form of the π-polarization light signal was found by repeating this process 27 times. The signal for the σ-polarized light entered an absorption in the field of 550 mT. Approximately 20 points in the fields from 0 to 550 mT were sufficient for its description.

Information about the following values in Eq. (10.26) is necessary for these calculations:

(1) $\Delta\nu_D, \Delta\nu'_D$ are the values of the Doppler width specified by the temperature of the fluorescence cell ($T = 25$ °C, $\Delta\nu_D = 500$ MHz) and by the temperature of the absorbing cell ($T = 50$ °C, $\Delta\nu'_D = 520$ MHz), respectively.

(2) β characterizes the relative intensities of hyperfine components in the irradiation contour.

(3) ℓ denotes the effective optical thickness of the absorption cell.

(4) a takes into account the collisional broadening of a line [Eq. (10.24)].

(5) s is the shift parameter, characterizing the change in position of the maxima of the Zeeman absorption components via the interaction of Rb with buffer gases.

Preliminary information on the values of a and s is absent; these parameters should be determined during the work. Other parameters may be determined by evaluating the dependence $I(B)$ during the passage of radiation through the cell containing pure Rb vapor.

In a field of approximately 500 mT, the intensity of the transmitted light is determined mainly by the behavior of the Zeeman component No. 26. This component is the only one which changes from one initial component group to the other, in this way scanning a large frequency range (see Figure 10.3). Other

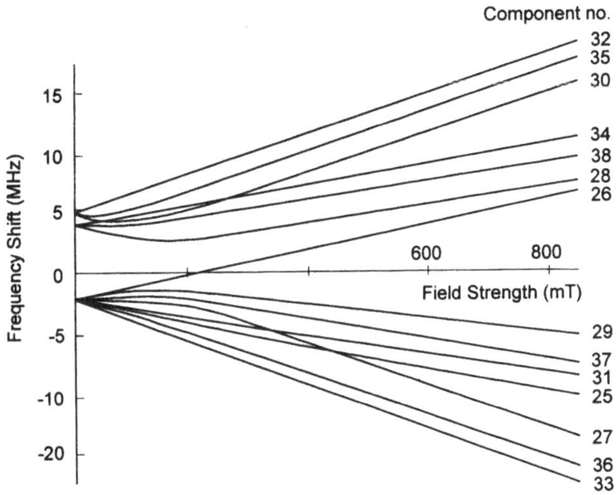

FIGURE 10.3. Frequency of the Zeeman components of the ^{87}Rb D_1-line (π-polarization).

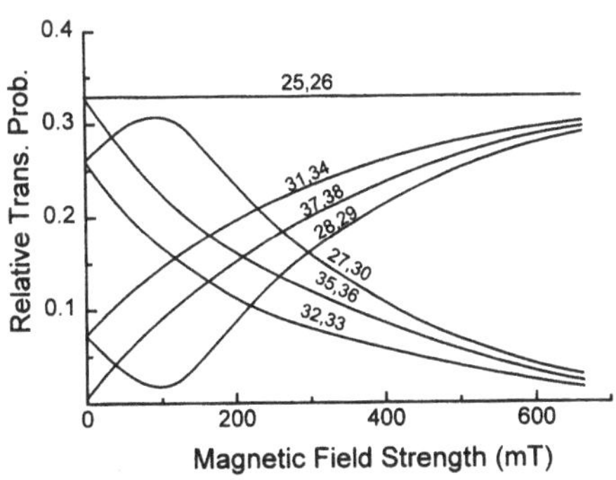

FIGURE 10.4. Transition probabilities for the Zeeman components of Figure 10.4 as a function of the magnetic field.

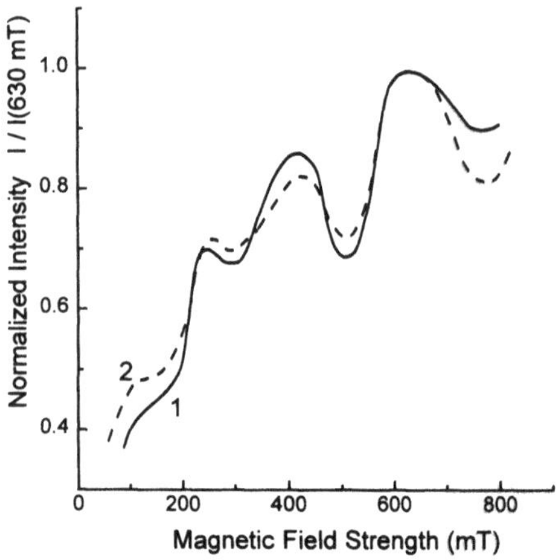

FIGURE 10.5. Sensitivity of the computed transmission curve to parameter (population of the hyperfine sublevels of 5 $^2P_{1/2}$): 1, $\beta = 2.5$; 2, $\beta = 0.6$.

components practically do not influence the intensity of the passing light. The transition probabilities of some components are shown in Figure 10.4.

Let us consider the effect of each parameter on the form of the magnetic scanning curve. Curves were calculated, and distinguished by the value of one parameter. For reliability of comparison, all calculated (and later experimental) curves were normalized to the value in the field of 630 mT. By analyzing this family of curves it is possible to distinguish the characteristics of the signal which are most sensitive and least sensitive to each parameter.

The variation range of parameter β is determined by the radiating lamp characteristics. If the spectrum of the lamp is white (equal population of the two upper hyperfine levels, $N_{F=1} = N_{F=2}$), β will be 1. When the upper levels are populated by absorption from equally populated lower levels, then the parameter β will be 0.6 (for ^{87}Rb).

Figure 10.5 shows the dependence of the calculated curve on parameter β. The effect of β on the form of the curve enables us to resolve four regions: from 90 up to 330 mT, where the decrease in the passing light intensity corresponds to an increase in the parameter; from 330 up to 450 mT, where the dependence becomes opposite; from 450 to 570 mT, where the dependence adopts the initial character; and the region from 690 up to 780 mT, which is most sensitive to this parameter. In this region the lower intensity of the passing light corresponds to a lower value of the parameter. Boundaries of these regions correspond to the points independent of parameter β. The use of these points significantly simplifies the comparison of calculated with experimental curves.

Using σ-polarized light it is also possible to specify some regions with a different dependence of the form of curves on the given parameter. The boundaries of these regions also correspond to points independent of the parameter β.

The parameter ℓ characterizes the optical thickness of the working cell. Besides, it contains the proportionality factors connected with transition probabilities for the Zeeman components of the absorption line. The value of ℓ was estimated on the basis of the following quantities: the geometrical length l of the cell was 3 cm; the density of the saturated vapor was of the order of 10^{-6} mbar (at a cell temperature of $T = 50$ °C); the Doppler width was estimated to be $\Delta\nu_D = 520$ MHz; the oscillator strength was 0.4. According to [151] the absorption factor of a single line at the center is

$$k_0 = \frac{2}{\Delta\nu_D}\sqrt{\frac{\ln 2}{\pi}}\frac{\pi l^2}{mc}Nf \tag{10.27}$$

In our case the line consists of two resolved components, so this formula must be transformed. As a consequence of the high sensitivity of the initial points of the curve to parameter β, changes connected with parameter ℓ were controlled at points which corresponded to fields of 300–330 and 450 mT (for π-polarized light) and 150–170 and 360 mT (for σ-polarized light). These points were independent of β (see Figure 10.5).

The parameter $\Delta\nu'_D$ characterizes the Doppler width of the absorption line. This parameter practically does not influence the form of the contour (for a field of the order of 500 mT; see Figure 10.6). In the region from 90 up to 500 mT, lower values of this parameter correspond to a greater intensity of the passing light. The most characteristic region is that from 120 up to 180 mT. In this region the parameter decrease can result in the appearance of an additional minimum on the curve. In the case of σ-polarization the character of the dependence of the curve on parameter $\Delta\nu'_D$ remains as before. For π-polarization, the effect of the parameter is limited to a magnetic field up to 330 mT. This is caused by a stronger dependence of the σ-polarization light intensity on the magnetic field strength.

The parameter $\Delta\nu_D$ takes into account the Doppler width of the irradiation line. The sensitivity of the curve to this parameter is significantly lower compared to parameter $\Delta\nu'_D$. Generally, it leads to a smoothing of extrema and to a marginal change in the passing light intensity.

The parameter a is one of the values to be determined by means of the method. Its effect on the character of the calculated curve results in a smoothing of the signal trend. All the aforesaid relates also to the parameter ℓ, but a change of the parameter ℓ causes a variation in inclination angle of the curves while parameter a practically does not change the angle. The most sensitive points correspond to fields of 480, 510, and 750 mT (π-polarization); the least sensitive points correspond to fields of 240 and 420 mT. From this general analysis we can see that it is possible to distinguish curves which differ in parameter a by 40 MHz when measuring the intensity of the passing light with an error of 2%.

The parameter s possesses a significant effect on the shape of the curve right up to the change of minima to maxima. In π-polarized light (see Figure 10.7a) a

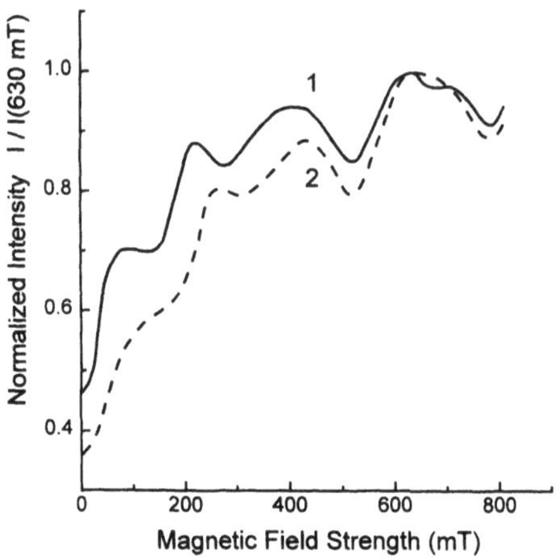

FIGURE 10.6. Sensitivity of the computed transmission curve to parameter $\Delta\nu'_D$ (Doppler width of transmission): 1, $\Delta\nu'_D =$ 460 MHz; 2, $\Delta\nu'_D = 600$ MHz.

"red" shift of the absorption line (curves 1) causes a maximum on the abrupt part of the curve in the region of 140 mT, and in large fields the minima are shifted to larger values of the field strength. In σ-polarized light (see Figure 10.7b) the minimum at $B = 390$ mT is shifted to larger fields and a characteristic bend appears in the region of the lower fields (around 200 mT). Under the "blue" shift (curves 2) maxima at larger fields are shifted to lower values. In π-polarized light (in fields of 390 and 240 mT) we see the convergence of neighboring maxima while the minimum between them disappears. Evaluation of the sensitivity of the calculated curve to parameter s allows one to distinguish between s-values differing by only 30 MHz.

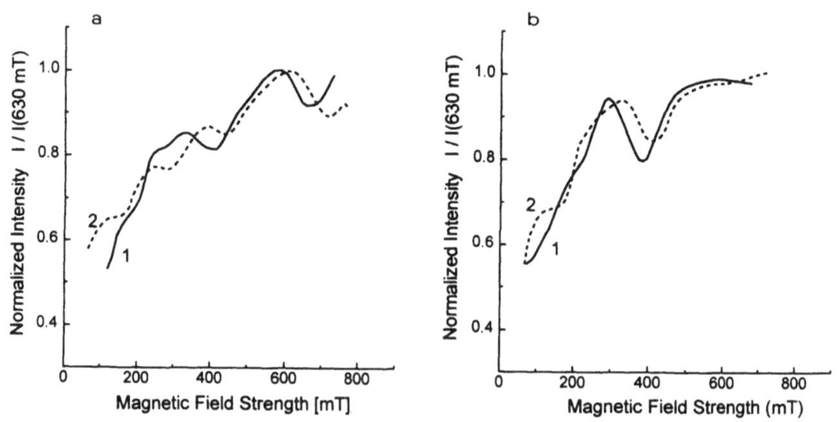

FIGURE 10.7. Sensitivity of the computed transmission curve to parameter s (line shift): (a) π-polarization, (b) σ-polarization: 1, $s = -200$ MHz; 2, $s = 200$ MHz.

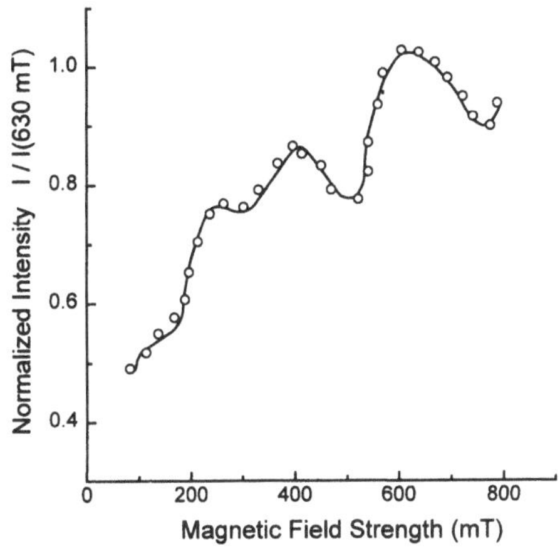

FIGURE 10.8. Comparison of the computed curve with simulated experimental points for an error of 10% (normally distributed).

The estimated inaccuracy of the magnetic field determination was taken into account by introducing a parameter α in the theoretical expression for the field strength. The characteristic behavior of the calculated signal change was studied by varying α to find the most sensitive regions of the curve. The sensitivity of the curve to parameter α (for π-polarized light) increased with growing magnetic field strength. In the case of σ-polarization, the intensity of the passing light, starting from 450 mT, varied slightly and the effect of α was insignificant. Thus, the 480 and 510 mT points for the signal in π-polarized light and the 390 and 410 mT points for the signal in σ-polarized light were extracted. The ordinates at these points showed high sensitivity to changes of α. A sensitivity of the same order was determined at a number of points located in stronger fields. However, the variation in optical thickness parameter ℓ and parameter β exerted a similar effect at these points. The numerical value of the correction was determined by comparing the calculated curves with the experimental signal. Repeating these comparisons many times showed that α was determined to $0.972 \pm 0.5\%$ in a number of experiments.

We have analyzed the contour of the transmittivity of a cell with vapor of the investigated element in the magnetic field and its dependence on the phenomenological parameters in the Zeeman spectrometer model. Let us now consider the real experimental situation.

The possibility of obtaining information from the experimental curve by comparing it with a calculated curve is limited by the errors of the experiment. In the case of significant errors, comparison with simulations can result in a coincidence of the experimental curve with a few calculated curves which differ by their parametric sets.

The influence of experimental measurement error on the unambiguity of the chosen parametric set was determined by mathematical modeling of the measurement process. Using a random number generator with known distribution, the ordinate of every point of the calculated curve was varied over the range conditioned by a dispersion of the random number distribution. The value Σ, which is equal to the calculated sum of the squares of the differences of the signal ordinates, was entered in the computer and used as a measure of the coincidence of "experimental" and calculated curves. It was found out that for an increase in artificial error of spread right up to 8%, there was only one initial set of parameters which caused the best or the same coincidence with the initial curve (following the rule of the normal distribution of errors), i.e., equal to or less than the control value Σ. If the error was more than 8%, it became possible to improve the coincidence by varying the initial parameters. Figure 10.8 shows an example of a curve obtained with an error of 10% in the normal distribution. The results testify that unambiguity of broadening and shift determination takes place under the assumption of a normal error distribution, even for an error which is significantly higher than that in the (real) experiment.

From the viewpoint of mathematical statistics, a rigorous evaluation of the informative opportunities of the magnetic spectrometer technique, carried out in [224], confirms the above practical conclusions. Moreover, the estimation shows the possibility of employing this method to solve a number of problems connected with the determination of the spectral contour of a line source.

10.3.4. Experimental Investigations of the Broadening and Shift of Spectral Lines

10.3.4.1. A Magnetic Spectrometer for Determination of the Collision Broadening and Shift of the ^{87}Rb D_I-Line

Let us consider in detail the setup of a magnetic spectrometer used in [222] to determine the broadening and shift of the ^{87}Rb D_1-line under collisions with atoms of inert gases.

Figure 10.9a shows a block scheme of the equipment. The light source (1) is a hf EL with ^{87}Rb vapor, placed between the turns of the hf generator coil. Its light is used to excite ^{87}Rb atoms in the fluorescence cell (2) acting as a real light source in the investigations. The emitted light of cell (2) passes through the working cell (4), placed between the poles of magnet (8) and is fed via a polarizer (5) to the entrance slit of a monochromator (6), selecting the investigated line. The intensity is detected by a photomultiplier (7) as a function of the magnetic field strength. The signal was recorded in the photon counting mode, because of the very small light intensity. Fluorescence excitation was performed by lamp bulbs of spherical form (13 mm in diameter), filled with ^{87}Rb and Kr vapor at a pressure of 2 mbar. The intensity of the radiation and its spectral composition

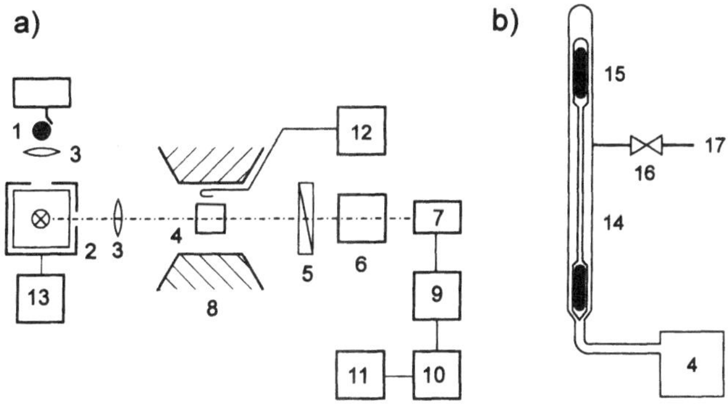

FIGURE 10.9. Zeeman spectrometer for measuring pressure effects on the Rb D_1-line: 1, hf EL as exciting light source for fluorescence cell; 2, 3, lens; 4, working cell; 5, polarizer; 6, monochromator; 7, photomultiplier; 8, magnet; 9, photon counting system; 10 frequency meter; 11, stabilized power supply for magnet and magnetometer; 12, radio-frequency generator and antenna; 13, thermostabilizer for fluorescence cell. (b) Connection of the working cell with the vacuum system, avoiding losses of Rb: 14, glass tube; 15, glass rod movable by magnets, tight-fitting at its conical end; 16, valve; 17, vacuum system, allowing one to pump out and fill the cell with noble gases.

depended on the operating mode (current of generator, temperature of the lamp, etc.). The fluorescence light contour of cell (2), observed here perpendicularly to the excitation direction, is independent of the characteristics of the exciting light if one neglects the hyperfine structure. In the case of the presence of a hyperfine structure (as for ^{87}Rb and ^{85}Rb) the relative population of the hyperfine sublevels depends on the spectrum of the irradiating light. This fact was taken into account by parameter β in our calculations. Obviously, β must be held constant during the measurements.

Besides the intensity redistribution by the hyperfine components, the presence of ^{85}Rb admixtures in the resonant cell and in the lamp affects the signal form. The components of the lines of this isotope were disregarded in the calculations (their effect was considered to be negligible). During the experiment, conditions were selected, in which this effect is minimal; anyway, ^{85}Rb lines are excited by the wings of the main isotope (^{87}Rb) lines. For an increase in discharge current, the intensity is increasing but also the line contour is significantly broadening. In the investigations, a generator mode was used (anode current, 29 mA) for which intensive excitation was ensured, but for which no deviation between the calculated and measured signals was obtained.

The resonance cell (2) was a bulb of cylindric form ($d = 40$ mm, $l = 60$ mm), filled with ^{87}Rb vapor. In order to maintain constant Rb vapor density, the cell was placed in a special oven with double glass windows. One window was a lens of focal length 7 cm. This distance corresponded to half the internal oven volume length. The cell was placed in such a way that the exciting radiation entered the cell near its end directed to the outlet window. That permitted minimization of all distortions of the fluorescence contour connected with self-absorption. Special diaphragms were installed in the oven to exclude lamp radiation, scattered by glass surfaces.

The working cell was formed as a cylinder ($d = 32$ mm, $l = 30$ mm) with parallel windows, connected to vacuum equipment through a small capillary tube. A special shutter valve system was used to avoid Rb diffusion from the working cavity. Figure 10.9b shows this system. The glass tube (14) was connected to the vacuum equipment. The lower part of the tube was narrowed at the site of connection with cell (4). A rod (15), placed in the tube (14), was used as a shutter valve. This rod was made of a long glass tube with metallic cylinders, fused into the ends. The lower cylinder was made from a nonmagnetic material, used to lower the center of gravity of the rod. The other cylinder was made from soft iron and enabled displacement of the rod in a vertical direction with the help of magnets. Hence this cylinder provided a connection between the absorption cell and the vacuum equipment by which it was possible to pump out the cell and fill it with buffer gases.

The working cell was placed in an oven made of nonmagnetic material with a bifilarly winded coil and asbestos thermoisolation, serving to achieve sufficient density of Rb vapor. The temperature T of the cell, controlled by a thermocouple, was used to calculate the Doppler width of the absorption line; T was held constant during the experiment. The scanning magnetic field around the cell was created by a solenoid, made from copper tubes ($d = 5$ mm) cooled with flowing water. The coil was fed with direct current from a special power supply.

The relative inhomogeneity of the field, $\Delta B/B$, was less then 0.1% at a radius of 3 cm (these measurements were carried out at a field strength of 700 mT). Along the axis of the magnet the nonuniformity was a few times smaller. Taking into account that the working cell had a radius of 15 mm, it was possible to assume the field strength B in the cell region to be constant with an accuracy of 0.1%. The magnetic field strength was measured by means of a Hall gauge, which was calibrated with the aid of a nuclear magnetic resonance magnetometer.

The monochromator (6) equipped with a blazed diffraction grating with 2600 lines/mm (70×80 mm area) and with mirror objectives of focus length 70 cm was used to select the Rb D_1-line from the fluorescence light. The dispersion of this monochromator allowed one to use an entrance slit of 7 mm width, observing the spectrum in second order.

A polarizing foil was installed in front of the monochromator to determine the working polarization, as well as a red glass filter to suppress light of higher orders of diffraction.

A photomultiplier (7) was used as a detector. This photomultiplier is characterized by a low dark current and high sensitivity. Although the photomultiplier was placed a significant distance (about 2 m) from the electromagnet a strong

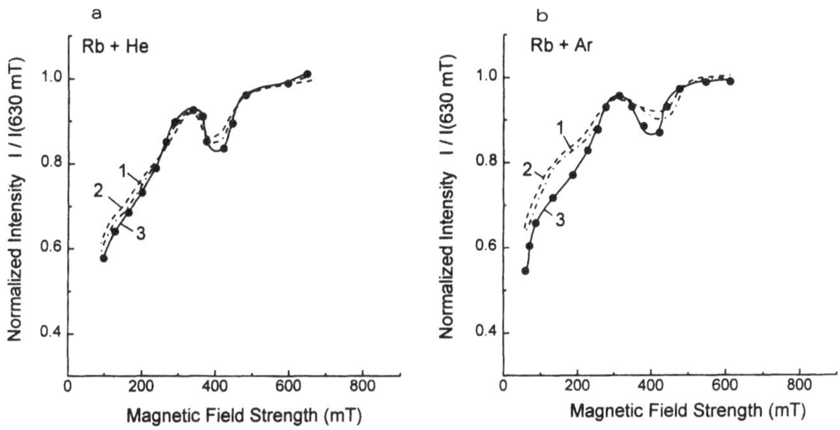

FIGURE 10.10. Variation in the transmission curve for σ-polarized light as a function of foreign gas pressure, (a) He, (b) Ar: 1, 13 mbar; 2, 26 mbar; 3, 40 mbar. Points denote experiment (shown only for curves 3), lines denote fit.

dependence of the photomultiplier sensitivity on the magnetic field intensity was noted during control experiments. Therefore, two screens of permalloy foil were applied to provide independent operation. One screen was installed directly on the photomultiplier, the other was located on the light tight enclosure so as to protect the end section of the device. The signal was recorded by the photon counting method. The electronic circuit used was characterized by a large dynamic range and interference-suppressing features. The latter was achieved by the use of two ensembles included in a circuit of anticoincidence — the channel of interferences and the channel of the main signal.

In the experiments the radiation intensity transmitted through the cell with the Rb vapor and buffer gas was recorded as a function of the strength of the magnetic field applied to the cell.

The experiments were divided in two parts:

(1) Auxiliary measurements, characterizing the resonance absorption by the ^{87}Rb vapor without addition of buffer gas, were carried out before the principle measurements and after their completion to estimate the stability of all elements of the circuit.

(2) Principle measurements. In these measurements buffer gas (1–7 mbar) as well as Rb vapor was present in the working cell. For an increase of the reliability, the signals were recorded for light polarized parallel to the magnetic field strength vector (π-polarization), as well as for light polarized perpendicular to it (σ-polarization).

In addition to controlling the system stability at the beginning and end of a set of measurements (by comparison of curves without gas), the stability during each measurement was also tested. Stability control was realized by comparing signals in field of 120 mT. This point was chosen because in this region the absorption is only slightly dependent on the field strength. Control of field stability was also carried out by recording the intensity of the transmitted light in a region (≈ 200 mT) of the $I(B)$ curve, characterized by the maximal slope.

Experimental curves were compared with calculations. For comparison, the signals were normalized with respect to the maximum intensity of the passing light, corresponding to a field of 630 mT.

Figures 10.10 and 10.11 illustrated an increase in buffer gas pressure yielded a smoothing of the experimental curves. This is natural, as broadening of the Zeeman components leads to an increased absorption factor at the line wings and to a decrease of the absorption factor at the central frequency. The degree of smoothing depends on the buffer gas pressure and its individual properties.

It is possible to estimate the value and direction of the line shift by the general shape of the signal. It was mentioned above that, for σ-polarized light, some minima placed at field strengths of 390 and 420 mT and, for π-polarized light, at 720, 750, and 780 mT had maximal sensitivity. A shift of the minima into regions of lower fields corresponds to a line shift to the short-wave region of the spectrum and, conversely, a shift of the minima in the opposite direction corresponds to a line shift to the long-wave region of the spectrum. Figure 10.10a,b show the signal for σ-polarized light for two mixtures: Rb–helium and Rb–argon. By comparing these mixtures we can see the differences in the direction of the shifts caused by Ar and He. In Figure 10.11 one can see that, for identical pressures, Ar has a stronger smoothing effect on the curve than Kr. This fact testifies to the higher broadening factor of argon.

10.3.4.2. Accuracy of Spectroscopic Measurements of Atomic Constants

The analysis of the $I(B)$ dependencies was performed by a computer-simulation technique comparing simulated and experimental curves (see Section 10.3.3). In the first step a set of auxiliary parameters must be determined by means of the curves, obtained in the absence of the interaction of the Rb atoms with the buffer gases. In the second step, these parameters are used to compute the unknown values of the broadening a and shift s of the resonance line from the coincidence of the curves modified by the effect of the buffer gas pressure.

The most critical part of this comparison process was the calculation of the auxiliary parameters β, ℓ, $\Delta\nu_D$, $\Delta\nu'_D$, and α. The values of the Doppler widths were calculated using the temperatures of the resonance and absorption cells. At a resonant cell temperature $T = 30$ °C, typical for the experiments, the Doppler width of the radiation line was $\Delta\nu_D = 500$ MHz. The average temperature of the absorption cell was 50 °C, corresponding to $\Delta\nu'_D = 520$ MHz. The other

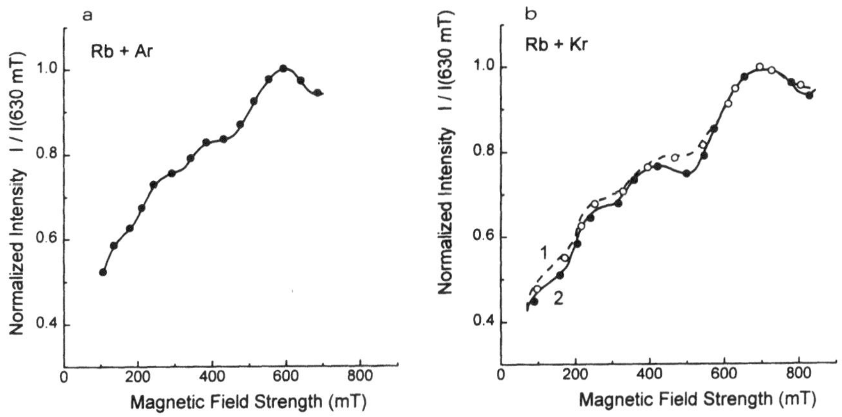

FIGURE 10.11. Variation in the transmission curve for π-polarized light as a function of foreign gas pressure. (a) Ar, 37 mbar, (b) Kr: 1, 26 mbar; 2, 38 mbar. Points denote experiment, lines denote fit.

three parameters were determined from the form of the experimental curve by the method of the successive approximations.

The relative value of the transmitted light intensity $I(B)$ corresponding to a field of 300 mT was used to estimate the value of the optical density parameter, ℓ. The sensitivity of this point to the parameters β and α was negligibly small and all variations in the ordinate are caused by a change of optical thickness. In order to obtain information on the value of parameter α, which characterizes the shift in the magnetometer gauge data, the region between 480 and 510 mT of the experimental curve (for π-polarized light) was used. The relative difference between the ordinates of these points, characterized by a small sensitivity to other parameters, allowed the value of the parameter α to be obtained with sufficient accuracy. The value $\beta = 0.9$ was selected as the zero approximation to characterize the relationship between hyperfine component intensities. This value characterized the typical conditions of the experiment in the best way.

The curve, calculated using these zero-approximation parameters, was then compared with the experimental one. The result of the comparison was characterized by the sum of the squares of the ordinate differences between the experimental and computed contours.

The minimum of this sum was calculated with an accuracy determined by the experimental errors. The best fit of the computed curve to the experimental data was sought by adjusting the parameters until the mean-square deviation Σ of the experimental points from the calculated curve, determined by

$$\Sigma = \left[\frac{1}{n} \sum_{j=1}^{n} (I_{j,e} - I_{j,c})^2 \right]^{1/2} / I(630mT) \tag{10.28}$$

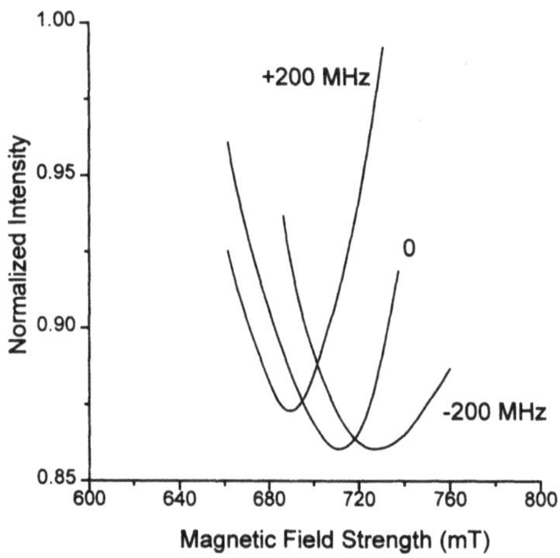

FIGURE 10.12. Variation in the position of the minimum of the computed curves with the shift parameter s.

(where I_e and I_c are the values of the measured and calculated intensities, while n was the number of points of a curve), became less than the error of the experimental measurements. This error was determined by the same formula, but instead of $I_e - I_c$, the difference I_e of the ordinates of two experimental curves at the beginning and end of the experimental series was used.

The next step in determining the broadening and shift values was to compare the experimental and calculated curves, modified by the admixture of a foreign gas. In this case the parameters, calculated in the first step, were used as fixed parameters, and only parameters a and s were varied.

An approximate value of the shift parameter s can be found from the position of the minimum in the 680–740 mT region. Figure 10.12 shows a magnified part of the curve $I(B)$, in this region of the magnetic field. This minimum stems from the overlap of a component, which corresponds to the transition $5\,^2S_{1/2}$, $m_I = -\frac{3}{2}$, $m_j = -\frac{1}{2} \leftrightarrow 5\,^2P_{1/2}$, $m_I = -\frac{3}{2}$, $m_j = -\frac{1}{2}$ with the hyperfine component $5\,^2P_{1/2}$, $F = 2 \leftrightarrow 5\,^2S_{1/2}$, $F = 1$ of the irradiation line located distance 4575 MHz from the center of gravity of the transition (see Figure 4.5). The contribution of other components of the absorption line in this region is negligibly small. Therefore, the position of this minimum depends on the line shift (and the parameter s) practically linearly. The dependence of the signal ordinate with σ-polarized light in the vicinity of 270 mT on the parameter a was used to determine this parameter in the zero approximation. The zero approximations, calculated from records in σ-polarization light, were employed to analyze the experimental curves obtained with π-polarized light.

Figures 10.13–10.15 illustrate the comparison of calculated and experimen-

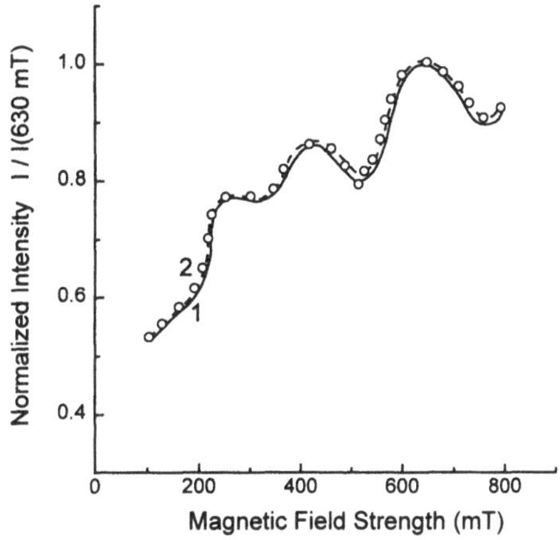

FIGURE 10.13. Selection of parameter ℓ describing the optical length (π-polarization). Points denote experiment; 1, $\ell = 2.34$; 2, $\ell = 2.38$.

tal signals as well as the course in obtaining initial values of the broadening and shift. Figure 10.13 shows the calculation of the auxiliary parameters. Initially the ℓ and α values were determined by the methods mentioned above; ℓ was determined to be 2.34 and α to be 0.978, resulting in sufficient coincidence of the curves. The following plots describe the method of choice of parameters a and s. As known, the line shift caused by He (see Figure 10.14) is directed to the short-wave region, as illustrated by the plots. Changing the shift direction (solid curve) brought about the best fit of the experimental points with the theoretical curve. In this case an absence of full coincidence was caused by an insufficiently large value of parameter a (0.7 instead of 0.95, see Figure 10.15). The calculated curve (Figure 10.15a) presents more relief than the experimental one and is located lower than the experimental one, testifying to a necessary increase in the

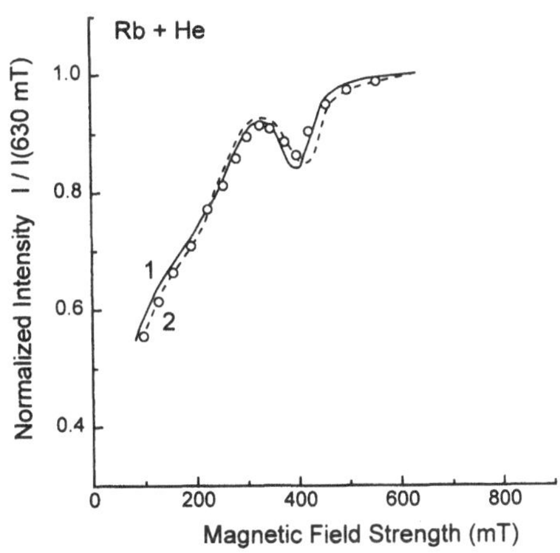

FIGURE 10.14. Selection of shift parameter s (π-polarization), assuming a broadening parameter $a = 0.7$. Points denote experiment: 1, $s = 190$ MHz; 2, $s = -105$ MHz.

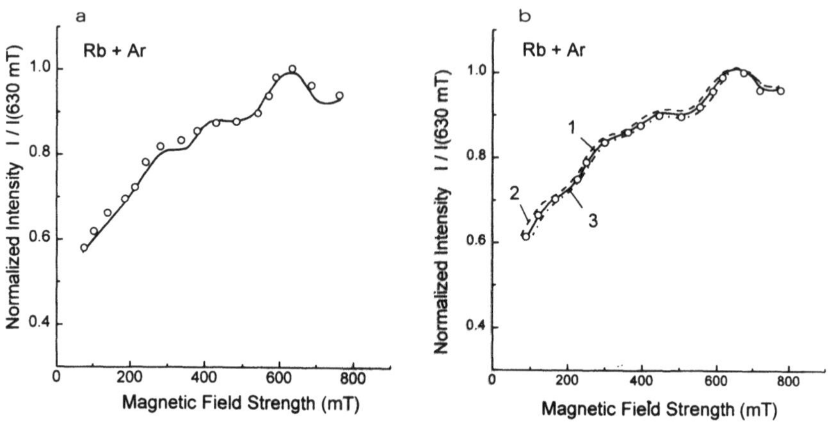

FIGURE 10.15. Adjustment of the computed curve to experimental data (π-polarization). Argon pressure 40 mbar. Points denote experiment: (a) $a = 0.7$. (b) 1, $a = 0.95$, $\Sigma = 5.34 \times 10^{-4}$; 2, $a = 0.89$, $\Sigma = 9.13 \times 10^{-4}$; 3, $a = 0.97$, $\Sigma = 6.71 \times 10^{-4}$.

value a. The difference between the calculated and experimental curves becomes smaller in Figure 10.15b. In this case, besides visual agreement, the coincidence is evaluated by the control value Σ (the curve with $\Sigma = 0.00057$ corresponding to $\alpha = 0.97$ is preferred to the curve with $\Sigma = 0.00091$, where $a = 0.88$).

The accuracy of the parameters under study depends significantly on the error in the experimental measurements. Fluctuations of transmitted light intensity are determined by the experimental conditions: by the change of the Rb vapor density in the resonant and absorption cells, by the change of intensity of the spectral lamp radiation. Experimental determination of the magnetic field strength introduces an error in the abscissa-axis scale.

Instabilities related to the vapor density in the resonant cell are caused by environmental temperature changes. In order to maintain constant temperature the cell was placed in a thermostat, which controlled the specified temperature with an accuracy of 0.1 °C. A temperature change of 1 °C for typical experimental conditions corresponds to a 6% density change in the Rb vapor of. Therefore, the variation in the resonance radiation intensity should fluctuate no more than 0.6%.

The absorption cell, contrary to the resonant cell, was connected to the vacuum equipment. Therefore, the density of vapor in it was a complex function of the temperature of the cell walls, and the chamber with Rb. *A priori* estimates were not carried out. Another source of error is caused by the magnetic field, for two reasons: possible oscillation of the magnetic field, and an error connected with the measurement process. A special circuit of stabilization was used in the experiment to reduce the first kind of these errors. This circuit allowed to reduce the field strength oscillations to be reduced to less than 0.3%. The second error can be divided into random and systematic components. Random errors,

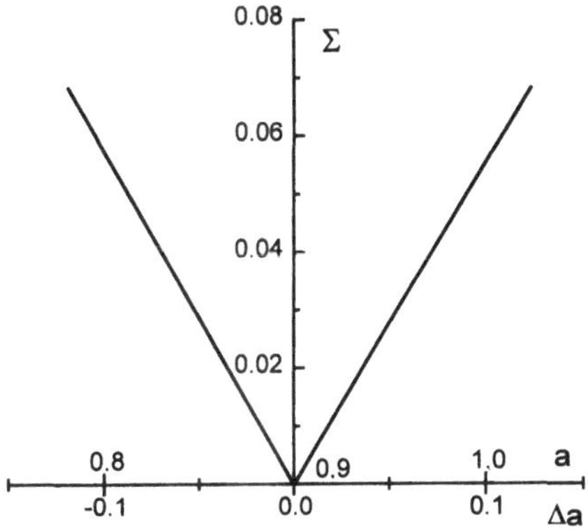

FIGURE 10.16. Sensitivity of the computed curve to the broadening parameter a (π-polarization).

determined by reading the applied Hall gauge magnetometer, were about 0.7%. The systematic error was related to the temperature effect on the Hall gauge operation. A significant part of this error was compensated by introducing the parameter α. Nonuniformity of the magnetic field had the same effect as an error in measuring the field strength. The field nonuniformity created an error of less than 0.1%, as mentioned earlier, and is negligible for a and s determination.

The radiation stability of the spectral lamp is very important in the measurements. Without stabilizing of the discharge current and temperature of the oven in which the lamp is located, intensity variations are so high that a determination of the broadening becomes impossible. Stabilization of current and temperature provides an intensity stability of 10% for 10 h. Experimentally, the effect of all factors can be estimated simultaneously by comparing two curves obtained at significantly different time intervals (a few hours, which suffices to run a full set of experiments). Fluctuations in the experimental conditions were no more than 1%.

The deviation of the experimental points from the calculated curve was characterized by the control value Σ. This deviation was determined by the spread of the integral intensity measurements as well as by errors in the approximations of the varied parameters.

The dependence of Σ on the parameter values is characterized by the calculated sensitivity of the curve to a given parameter. Figure 10.16 shows the dependence of Σ on the variation Δa of the broadening parameter a.

The error in measuring the ordinate of every point of the curve was estimated to be 0.5%. This spread results in a value of $\Sigma = 0.0019$ for a given number of measured points. The value of Δa, corresponding to this value of Σ, is assumed to be the error of the parameter determination. Figure 10.16 shows, that $\Delta a/a = 6\%$. In the same way we estimated $\Delta s/s = 10\%$. The real error is possibly somewhat smaller than in estimates based on coincidence of the calculated and experimental curves.

A possible parameter interchangeability was checked in the following manner. The coincidence with the experimental curve was worsened by varying one parameter. Then an attempt was made to achieve some improvement of the coincidence by varying the other parameters. This check showed that an interchangeability effect was observed for the parameters of the optical density ℓ and for the parameters of pressure broadening: a 1% error in the determination of ℓ led to a 2% error in the determination of a. It should be mentioned that ℓ was determined with an error no worse than 0.3% (in most cases, 0.1%) from curves recorded without gas addition. The error in the determination of the broadening, related to the inaccuracy of ℓ, was estimated to be 0.6%.

10.3.4.3. Results

The final values of collisional broadening and shift of the Rb D_1-line are listed in Tables 10.1 and 10.2 as well as other known data for this line. A comparison of different experimental data is impossible without clear allowance for the experimental conditions and especially for the range of pressures, since lot of available shift and broadening coefficients have been determined at much higher pressures than those determined by the method described here. In [225–227] the values of broadenings and shifts were measured at 20–50 bar pressure, in [228, 229] at 0.5–3 bar pressure, and in [119, 230, 231] the lowest limit of the measured pressure was several hundred mbar. An extrapolation of these results, obtained in regions of greater pressure, to low pressures of foreign gas is necessary when comparing these data with our values. This procedure is frequently complicated by nonlinearity in the dependence of the measured values of the pressure. The nonlinearity of line broadening by foreign gases at low pressures was explained elsewhere [230]. Non-Lorentzian wings of the spectral lines and their change with pressure are most important. Therefore the determination of the studied values of broadening and shift depends on how one interprets the experimental data. The values in Tables 10.1 and 10.2, taken from the cited works, were obtained with allowance for the difference in density of the perturbing gas, and the corresponding corrections. In [225, 227, 229] the effect of He, Ar, and Xe at different pressures on the Rb D_1-line shape was investigated, and in [228, 229] the intensity distribution in the line wings was specially measured at different pressures of Xe in the range 0.5–3.0 bar at a temperature of 250 °C. A linear variation in the width and shift of the central part of the spectral line was observed. The slope of these linear dependences determines the values presented in Table 10.1 ($s = 4.3 \times 10^{-4}$ MHz/mbar). Measurements for argon and helium were carried out at a pressure of 20–50 bar and at an average temperature of 250 °C in [225]. A linear change in the width with p was observed here as well. The dependence of the shift on

TABLE 10.1. Shift [MHz/mbar] of the Rb D_1-Line during Collisions with Noble Gas Atoms

	Foreign gas					
	He	Ne	Ar	Kr	Xe	Reference
Exp.	5.25				−4.27	[225, 226, 228]
		0.3	−6.26	−4.5		[227]
	6.0 ± 1.5	1.5 ± 3	−4.5 ± 1.5	−3 ± 1.5	−5.25 ± 1.52	[234]
		−0.28 ± 0.22	−5.25 ± 0.15	−5.40 ± 0.37	−5.63 ± 0.52	[229]
				3.75 ± 0.7		[237]
	4.65 ± 0.37	0.45 ± 0.22	−4.95 ± 0.3	−4.12 ± 0.37	−6.0 ± 1.1	[236]
Theory	6.7					[239]
		−4.88	−4.88	−3.53	−3.56	[241]
		−4.05	−4.05	−4.12	−4.73	[240]

the helium pressure is approximated consistently by a linear function, while in the case of argon the dependence was approximated by a $p^{3/2}$-law.

Our interpretation of data from [225] leads to $a = 15.7 \times 10^{-4}$ MHz/mbar for the Rb D_1-line ($\lambda = 794.8$ nm) broadened by argon. Similar analysis of the same data in [230] gave $a = 21.3 \times 10^{-4}$ MHz/mbar. From our point of view these corrections are not necessary, because the interpreted plot (Figure 2 in [225]) illustrates the linear dependence of broadening on pressure. It should be mentioned that in [227] the expediency of the red satellite substraction is pointed out only for shifts and a similar correction is considered to be insufficient in the measurement of broadenings.

The nonlinear dependence of the D_1-line shift by argon, observed in [225], did not allow this shift value to be estimated. Presented in that work are some plots characterizing the behavior of the investigated line at pressures in the order of some ten bars. Therefore an extrapolation to the low-pressure region is not reliable.

TABLE 10.2. Broadening [MHz/mbar] of the Rb D_1-Line during Collisions with Noble Gas Atoms

	Foreign gas					
	He	Ne	Ar	Kr	Xe	Reference
Exp.	14.85		15.68		13.28	[225, 226, 228]
		3.75	12.5	7.5		[227]
	13.8 ± 0.22	7.05 ± 0.97	13.5 ± 1.72	13.5 ± 1.28	15.52 ± 1.72	[229]
				10.5 ± 1.5		[237]
	13.73 ± 0.67	630 ± 1.5	14.55 ± 0.75	12.98 ± 0.68	15.0 ± 1.5	[236]
Theory	18					[239]
			12.53	16.8	15.75	[241]
			11.02	11.40	13.05	[240]

The values of the shift by helium (5.25×10^{-4} MHz/mbar), listed in Table 10.1, were obtained from the linear dependence of the shift on the pressure, presented in [225]. The measurements were carried out at a temperature of 520 °C. The temperature dependence of the shift parameter by helium is not understood sufficiently, therefore it is difficult to estimate the correction of the shift parameter for a temperature of 323 °C.

Broadenings and shifts were measured in the region below 40 bar at various temperatures [226, 227]. In the case of argon a linear change in width was observed in the 3–20 bar pressure region at temperatures of 100 and 300 °C. The red satellite subtraction was carried out when determining the shift. The 12.5×10^{-4} MHz/mbar value, listed in Table 10.1, represents the slope of this linear dependence at 100 °C. Measurements with neon were carried out in the same pressure range, but in a wider temperature range. Approximately linear dependence was observed in the 3–15 bar region of pressure at temperature 142 °C, leading to 3.75×10^{-4} MHz/mbar.

In [230] the data of [227] were interpreted in another way. The broadening of the D_1-line was assumed to be linearly dependent in the whole pressure region, leading to a value 7.8×10^{-4} MHz/mbar for the broadening. In the 10–35 bar region at a temperature of 240 °C the dependence of the D_1-line shift on pressure was linear. A significant deviation from linearity was observed during a temperature increase to 500 °C.

When investigating the Rb D_1-line shift by krypton in the 15–150 bar pressure region at 152 °C, the red satellite substraction was carried out in [227], leading to 4.5×10^{-4} MHz/mbar, listed in Table 10.1. A value of 4.5×10^{-2} MHz/Pa, obtained at the same temperature, is listed in Table 10.1 for broadening by krypton. Unfortunately, the origin of this value, given in [227] is unknown, as the dependence of the Rb D_1-line width on the krypton pressure and the range of pressure is not presented there.

Detailed investigations of the interaction of Rb atoms with buffer gases were carried out in a number of papers [230–232], published by the U.S. National Institute of Standards and Technology (NIST). Measurement of the relative spectral intensity distribution was carried out in these works. Let us consider the general idea of these experiments. The spectrum of the Rb vapor fluorescence in a cell, filled with noble gases at 100–1000 mbar pressure at a temperature of 47 °C, was investigated. To exclude self-absorption, the density of Rb was maintained in the order of 10^{-6} mbar. The exciting light contained only the Rb D_1- and D_2-lines for a selective population of the $^2P_{1/2}$ and $^2P_{3/2}$ states of Rb. The fluorescence spectrum was monitored with the help of a double monochromator with diffraction grating and detected by a photomultiplier. The intensity profile of the lines was investigated for the wavelength range ± 5.0 nm from their center. A second multiplier recorded the integral fluorescence intensity.

In order to extract the Lorentzian core of a spectral line, it is necessary to take into account the instrumental function of the spectroscopic device and the presence of the hyperfine structure of the Rb lines. It was assumed that the fluorescence spectrum represented a sum of four Voigt contours, corresponding to the four hyperfine components of the Rb D_1-line. The instrumental function was determined by recording the narrow krypton lines $\lambda = 780.1$ and 785.3 nm, radiated by the lamp at a temperature insignificantly different from the temperature of the cell. Thus, the recorded profile represented a convolution of the Doppler contour of the source with the apparatus function. The Doppler width for Rb and Kr may be considered to be equal because of nearly equal temperature and closely related atomic masses. The convolution of the Lorentzian contours, corresponding to the four hyperfine components of the Rb D_1-line, leads to the desired calculable functions:

$$s(\nu) = \int_0^{\infty} I(\nu) F_\nu \left(\nu - \nu'\right) d\nu' \qquad I(\nu) = \sum_{j=1}^{4} \frac{A I_j \Gamma^2}{\Gamma^2 + (\nu_j - \nu)^2} \tag{10.29}$$

where ν_j is the central transition frequency of a hyperfine component of the line and I_j is the intensity of component j. Data on broadening and shift of a spectral line were obtained by comparing the theoretically calculated dependence and the experimental contours.

The smallest obtained value of the width of the apparatus function was 0.012 nm (approximately 6 GHz). Thus for measurements at a pressure around 100 mbar the ratio of the recorded contour width to the instrumental function width was found to be 3 : 1 for argon and 7 : 1 for neon. The method of statistical processing was not described in [230], so we cannot estimate the accuracy of the measurements carried out near the low-pressure limit (100 mbar). All results of this work, listed in Table 10.1, originate from experiments over pressure and temperature ranges close to our experimental conditions [233–236].

The papers [237, 238] are devoted to measurement of the shift and broadening of the Rb resonance absorption line by high resolution interference spectroscopy. The main device of the equipment is a complex Fabry–Perot interferometer, which consists of two interferometers in series with a 1 : 8 ratio of their free spectral ranges. The overall free spectral range of this device was 18,750 MHz. The resolution of the double interferometer was about 3×10^6 for $\lambda = 652.8$ nm, corresponding to an apparatus function width of 140 MHz (cf. Chapter 4). The light source was a lamp filled with the ^{87}Rb isotope. The experimental equipment allowed one to investigate three spectra: the spectrum of the source; the spectrum of radiation, passed through a cell with Rb; the spectrum of the radiation, passed through a cell with Rb and Kr at a pressure of 10 mbar. Numerical processing of the obtained spectra allowed the absorption factors to be determined. It was assumed that the absorption line contour was shaped by the Doppler profile of the radiating atoms, broadened by the pressure of the added gas, natural broadening, and the apparatus distortions. Mathematical processing of the contour was carried out assuming a Voigt profile for every hyperfine component. The Lorentzian part of the Voigt profile was determined using a computer program, according to the method of minimum squares.

Theorétical calculations of broadenings and shifts of the Rb D-lines are described in [239–242]. In [239] calculations were carried out in the collision approximations for a density of the perturbing atoms less than 10^{16} cm^{-3}. Only two-body collisions were considered. The interatomic exchange Rouff potential [243], describing well the interaction of the alkali metals with light inert gases, was used as interaction potential for Rb–He collisions. It was assumed that the motion can be approximated by rectilinear trajectories. The scattering matrix was determined neglecting any connection between the $P_{1/2}$–$P_{3/2}$ states. The obtained values of broadening (1.8×10^{-3} MHz/mbar) and shift (5×10^{-4} MHz/mbar) are listed in Table 10.1, too. Collisional broadening and shift of the resonance doublets of Rb and Cs have been computed [241]. Calculations were carried out in the collisional approximation (for pressures less than 3×10^3 mbar) using an interatomic potential of the Baylis and Pascal type [244, 245], which well describes the processes of collisions with heavy inert gases. The values of parameters characterizing the Baylis potential were calculated theoretically in this work; they were not entered as varied phenomenological values. The obtained parameters of the Rb D_1-line broadenings and shifts by argon, krypton, and xenon are listed in Table 10.1.

Calculations of the Rb resonance line broadenings and shifts by heavy inert gases in the collision approximation using a Van der Waals potential were presented in [240]. A model taking into account a break in the hyperfine connection at the moment of collision was used in the calculations. We can see from Tables 10.1 and 10.2 that the values given in this work (broadening and shift of the Rb D_1-line under collisions with atoms of He, Ne, Ar, Kr, and Xe) agree with the data of other authors.

The data [233, 236] agree with results of [229] concerning the value and direction of the shift. A law corresponding to the results of [225, 227] can be found for the measured values for argon and krypton. Our value for the shift caused by krypton (4.12×10^{-4} MHz/mbar) fits all results, except the value obtained in [229]. This value differs significantly from the other values.

The line broadening data are in better agreement with each other than the values of the shifts. The 3.75×10^{-4} MHz/mbar broadening by neon was obtained in [227] from the interpretation of a plot which illustrated a small nonlinear dependence of linewidth on neon pressure. Interpretation of this dependence as linear by the authors of [229] allowed them to obtain 7.5×10^{-4} MHz/mbar. This value is in better agreement with our value.

It should be mentioned that the experimental values of the Rb line broadenings by argon and krypton differ from the theoretically calculated ones given in [241]. The theory indicates that the D_1-line broadening by krypton is larger than broadening by argon. However, our experimental observations give an opposite result. We note the same tendency in analyzing the experimental results of [230]. A similar law can be noted for the resonance D_1-line of cesium: the 7.5×10^{-4} MHz/mbar broadening by argon [246] is slightly larger than the 7.05×10^{-4} MHz/mbar by krypton [246, 247].

Theoretical calculations of Rb line broadenings and shifts have not been carried out for all inert gases (for example, there are no calculations available for neon). Published experimental results indicates that neon has the least effect on the spectral characteristics of Rb.

As calculated in [239], the value of the shift by helium coincides sufficiently with the results of the Zeeman scanning technique, but there is significant disagreement (25%) between the broadening values. The pressure effects, caused by heavy gases (such as argon, krypton, and xenon), correspond on average to approximately 25% of calculations performed in [241] and [240].

The dependence of broadening and shift values on the atomic weight of a noble gas atom should be stressed. Following predictions in [241], during an increase in the atomic weight of the gas, the shift should decrease and a maximum of the broadening parameter should be observed for krypton, but in [240] an increase of shift is predicted. The Zeeman scanning technique shows nonmonotonic behavior with the shift caused by krypton being minimal within the row of the three heavy inert gases. A similar anomaly is observed for the broadening parameters. Anomalously large values of the broadening parameters for argon raised doubts about the purity of the Ar gas used. However, repeated measurements with certainly pure Ar gave the same result.

In conclusion, the development of the Zeeman scanning method to investigate spectra of colliding atoms has allowed different problems to be solved. The modification of the Zeeman spectrometer, where the same vapors are used in the radiation source and in the absorbing resonance cell, provided a precise shape for the irradiation contour and allowed the absorption control to be described with a high degree of accuracy.

Extensive work on the mathematical modeling of the experiment, allows one to argue that the given variant of the Zeeman spectrometer is expedient in also solving problems connected with very small distortions of spectral lines contours, not only of a collisional nature.

11

Polarization Spectroscopy of High-Frequency Discharges

11.1. Physical Principles of Polarization Spectroscopy

The present chapter is devoted to the study and exploitation of the polarization of the emission spectrum of a high-frequency discharge for sensing its parameters. Contrary to the emission or absorption intensity, the polarization of the optical emission in the linear spectrum was not used so widely as a tool for quantitative diagnostics until recently.

The study of spectropolarimetric effects in ionized gases indicates that the polarization of optical radiation in the line spectrum in the majority of cases results from the quadrupole orientation of momenta of the excited atomic ensemble, induced by excitation in a plasma or by self-alignment [128]. Different physical reasons for the excited particle self-alignment in gas discharge plasmas are known. Most promising for diagnostics is the electron impact self-alignment, which arises from macroscopic ordering of the velocities of the exciting electrons.

Quantitative spectropolarimetric diagnostics is based on the general relation between the self-alignment tensor of excited atomic particles under impact excitation and the quadrupole moment of the velocity distribution function or the momentum flow tensor of fast electrons [248]. By these means spectropolarimetry enables one to perform remote sensing of energy delivery, internal fields, and boundary processes, for which ordinary spectroscopic intensity measurement methods are principally inadequate.

11.1.1. Stokes Parameters of the Detected Light Beam

Spectropolarimetric plasma diagnostics is based on the study of the optical radiation propagating along the line of sight in the solid angle τ which is determined by the aperture of the spectropolarimeter. The emission of the line spectrum of an ionized gas results from dipole transitions of atomic particles (atoms, molecules, ions) between upper and lower states with momenta J and J_1. If the solid angle

τ is small enough, the optical radiation may be approximated by a unidirectional photon flux.

Polarization properties of this flux may be easily described in the detector frame of reference with the OZ axis oriented along the direction of propagation. Because of the transversality of the electromagnetic waves the state of photons in the beam may be represented as a superposition of pure states with right and left circular polarization:

$$|\varepsilon\rangle = \alpha_{+1}\,|+1\rangle + \alpha_{-1}\,|-1\rangle \tag{11.1}$$

where $\alpha_{\pm 1}$ are the probability amplitudes to find the photon in the states $|\pm 1\rangle$.

If the analyzer transmits photons with polarization vector $\vec{e}$, the probability to detect the photon in the state $|\varepsilon\rangle$ is proportional to $\langle\vec{e}|\varepsilon\rangle^2$. Restricting ourselves to the case of even isotopes of noble gases with well resolved fine structure, the detected intensity is expressed by

$$I(\vec{e}) = \overline{|\langle\vec{e}\mid\varepsilon\rangle|^2} = \sum_{\mu\mu'=\pm 1} \langle\vec{e}\mid\mu'\rangle\ \overline{\alpha_\mu\alpha^*_{\mu}}\ \langle\mu\mid\vec{e}\rangle = \sum_{\mu\mu'} e^*_{\mu'} e_\mu \overline{\alpha_\mu\alpha^*_{\mu}} \tag{11.2}$$

where e_μ are cyclic components of the vector $\vec{e}$, $e_{\pm 1} = \mp\frac{1}{\sqrt{2}}(e_x \pm e_y)$ and the overline means time averaging.

On the other hand, the amplitudes α_μ may be expressed in terms of the matrix elements of the corresponding cyclic components of the electric dipole moment vector operator $\widehat{\vec{d}}_\mu$ [249]:

$$\alpha_\mu \sim \sum_{mm_1} f_m \left\langle J_1 m_1 \middle| \widehat{\vec{d}}_\mu \middle| Jm \right\rangle \tag{11.3}$$

Here $|Jm\rangle$ and $|J_1 m_1\rangle$ indicate the wave functions of the magnetic substates of the upper and lower states of the optical transition (m and m_1 depict magnetic quantum numbers), while f_m is the probability amplitude to find atomic particles in the state $|Jm\rangle$.

The intensity of radiation with polarization $\vec{e}$ is then given by

$$I(\vec{e}) = B(\omega)\tau N \sum_{\mu\mu'} e^*_{\mu'} e_\mu \sum_{m_1 mm'} \left\langle Jm' \middle| \widehat{\vec{d}}^{\,+}_\mu \middle| J_1 m_1 \right\rangle \rho_{mm'} \left\langle J_1 m_1 \middle| \widehat{\vec{d}}_{\mu'} \middle| Jm \right\rangle \tag{11.4}$$

where N is the number density of the atoms;

$$B(\omega) = \omega^3/2\pi c^3\hbar \tag{11.5}$$

where $\omega = 2\pi\nu$ is the angular frequency of the spectral transitions, c is the light velocity, and $\hbar$ is Planck's constant divided by 2π;

$$\rho_{mm'} = \overline{f_m f^*_{m'}} \tag{11.6}$$

is the density matrix of the upper level in the representation of magnetic quantum numbers (temporal or ensemble averaging is assumed here)

Making use of the Wigner–Eckart theorem and the symmetry conditions for $3j$ symbols [250], we write

$$\left\langle Jm' \left| \hat{\vec{d}}^{+}_{\mu} \right| J_1 m_1 \right\rangle = (-1)^{J_1 - m_1} \begin{pmatrix} J_1 & 1 & J \\ -m_1 & \mu & m' \end{pmatrix} \left\langle J_1 \left\| \hat{\vec{d}} \right\| J \right\rangle^* \tag{11.7}$$

$$= (-1)^{J_1 - J + \mu} \frac{\left\langle J_1 \left\| \hat{\vec{d}} \right\| J \right\rangle^*}{\left\langle J_1 \left\| \hat{\vec{d}} \right\| J \right\rangle} \left\langle Jm' \left| \hat{\vec{d}}^{+}_{-\mu} \right| J_1 m_1 \right\rangle$$

The expression (11.4) may be transformed as follows keeping in mind that $(e_\mu)^* = (-1)^\mu (e^*)_{-\mu}$ and $\mu_1 \mu' = \pm 1$:

$$I(\vec{e}) = (-1)^{J - J_1} B(\omega) \tau N \frac{\left\langle J_1 \left\| \hat{\vec{d}} \right\| J \right\rangle^*}{\left\langle J \left\| \hat{\vec{d}} \right\| J_1 \right\rangle} \tag{11.8}$$

$$\times \sum_{m_1 m m'} \rho_{mm'} \left\langle Jm' \mid \left(\vec{e}^* \hat{\vec{d}}\right) \mid J_1 m_1 \right\rangle \left\langle J_1 m_1 \left| \left(\vec{e}\, \hat{\vec{d}}\right) \right| Jm \right\rangle$$

Introducing the scalar projection operator $\hat{P}$ on the $|J_1\rangle$ state which is invariant under rotation of the coordinate system, namely,

$$\hat{P}_{J_1} = \sum_{m_1} |J_1 m_1\rangle \langle J_1 m_1| \tag{11.9}$$

we may write

$$I(\vec{e}) = (-1)^{J - J_1} B(\omega) \tau N \frac{\left\langle J_1 \left\| \hat{\vec{d}} \right\| J \right\rangle^*}{\left\langle J \left\| \hat{\vec{d}} \right\| J_1 \right\rangle} S_{P_J} \left[\hat{\rho}_J \left(\vec{e}^* \hat{\vec{d}}\right) \right] P_{J_1} \left(\vec{e}\, \hat{\vec{d}}\right) \tag{11.10}$$

where S_{P_J} stands for the spur or the summation over the magnetic substates of the upper state.

If we employ the same approach as that developed by Fano and Macek [251] and again use transversality of the electromagnetic wave, we introduce the following parameterization of the Cartesian components of the vector $\vec{e}$ in the detector frame of reference:

$$\vec{e} = (e_x, e_y, e_z) = \left(\cos\beta, e^{i\varphi} \sin\beta, 0\right) \tag{11.11}$$

where β and φ are real parameters.

It is easy to see that such a parameterization describes all possible states of polarization. The values $\varphi = 0, \pm 2\pi, \pm 4\pi, \ldots$ and $\varphi = \pm\pi, \pm 3\pi, \pm 5\pi, \ldots$ correspond to linear polarizations with the polarization plane oriented at angles β or $-\beta$ with respect to the OX axis. Other φ values correspond to different types of elliptic polarization. In particular, the values $\beta = \pm\pi/4, \varphi = \pi/2$ give the right and left circular polarizations.

The product of operators $\left(\vec{e}^* \widehat{\vec{d}}\right)\left(\vec{e}\,\widehat{\vec{d}}\right)$ may be rewritten as follows:

$$\left(\vec{e}^* \widehat{\vec{d}}\right)\left(\vec{e}\,\widehat{\vec{d}}\right) = \frac{1}{3}\left|\widehat{\vec{d}}\right|^2 - \frac{1}{6}\left(3\widehat{d}_z^2 - \left|\widehat{\vec{d}}\right|^2\right) + \frac{1}{2}\left(\widehat{d}_x^2 + \widehat{d}_y^2\right)\cos 2\beta \tag{11.12}$$
$$+\frac{1}{2}\sin 2\beta\left[\left(\widehat{d}_x\widehat{d}_y + \widehat{d}_y\widehat{d}_x\right)\cos\varphi - i\left(\widehat{d}_x\widehat{d}_y - \widehat{d}_y\widehat{d}_x\right)\sin\varphi\right]$$

For $\varphi = \pi/2$, Eq. (11.12) is easily reduced to that which was carried out in [251]. The reason for such transformations is that the operators describing the polarization of optical radiation may be represented in terms of the superposition of irreducible tensor operators of different ranks composed of the components of the vector operator $\widehat{\vec{d}}$ using the irreducible tensor product operation [250]:

$$\left\{\widehat{\vec{d}} \otimes \widehat{\vec{d}}\right\}_q^{(\kappa)} = \sqrt{2\kappa+1}\sum_{\mu\mu'}(-1)^q \begin{pmatrix} 1 & 1 & \kappa \\ \mu & \mu' & -q \end{pmatrix} \widehat{\vec{d}}_\mu \widehat{\vec{d}}_{\mu'} \tag{11.13}$$

($\kappa = 0, 1, 2;\ -\kappa \le q \le \kappa$).

By these means the intensity of radiation is expressed as

$$I(\vec{e}) = (-1)^{J-J_1} B(\omega)\tau N \frac{\left\langle J_1 \left\| \widehat{\vec{d}} \right\| J \right\rangle^*}{\left\langle J \left\| \widehat{\vec{d}} \right\| J_1 \right\rangle} - \frac{1}{\sqrt{3}}\widehat{\Pi}_0^{(0)} \tag{11.14}$$
$$-\frac{1}{\sqrt{6}}\widehat{\Pi}_0^{(2)} + \frac{1}{2}\widehat{\Pi}_{2+}^{(2)}\cos 2\beta + \frac{1}{2i}\widehat{\Pi}_{2-}^{(2)}\sin 2\beta\cos\varphi$$

where $\widehat{\Pi}_q^{(\kappa)} = \left\{\widehat{\vec{d}} \otimes \hat{P}_{J_1}\widehat{\vec{d}}\right\}^{(\kappa)}$ and $\widehat{\Pi}_{q\pm}^{(\kappa)} = \widehat{\Pi}_q^{(\kappa)} \pm \widehat{\Pi}_{-q}^{(\kappa)}$. The angle brackets stand for the mean value of an operator over the state $|J\rangle$ defined as $\left\langle \widehat{A} \right\rangle_J = \mathrm{Tr}_J \left\langle \widehat{\rho}\widehat{A} \right\rangle$.

This expression for the intensity (11.14) enables us to represent the intensity of radiation in terms of the polarization moments of the atomic density matrix ρ_J for the state $|J\rangle$. These polarization moments $\rho_q^{(\kappa)}$ may be regarded as the mean values of the irreducible tensor operators $\widehat{T}_q^{(\kappa)}$ built up from the wave function $|Jm\rangle$:

$$\widehat{T}_q^{(\kappa)} = \sqrt{2\kappa+1}\sum_{mm'}(-1)^{J-m+q}\begin{pmatrix} J & J & \kappa \\ m' & -m & -q \end{pmatrix}|Jm'\rangle\langle Jm| \tag{11.15}$$

Therefore $\rho_q^{(\kappa)} = (-1)^q \left\langle \widehat{T}_q^{(\kappa)} \right\rangle_J$ and the reduced matrix elements of operator $\widehat{T}_q^{(\kappa)}$ are expressed as $\left\langle J \left\| \widehat{T}^{(\kappa)} \right\| J \right\rangle = \sqrt{2\kappa+1}$. Using the Wigner–Eckart theorem, the mean values of the irreducible tensor operator may be expressed in terms of the polarization moments of atomic density matrix. In particular, we have

$$\left\langle \widehat{\Pi}_q^{(\kappa)} \right\rangle_J = \frac{\left\langle J \left\| \widehat{\Pi}^{(\kappa)} \right\| J \right\rangle}{\left\langle J \left\| \widehat{T}^{(\kappa)} \right\| J \right\rangle} \left\langle \widehat{T}_q^{(\kappa)} \right\rangle_J = (-1)^q \frac{\left\langle J \left\| \widehat{\Pi}^{(\kappa)} \right\| J \right\rangle}{\sqrt{2\kappa+1}} \rho_{-q}^{(\kappa)} \tag{11.16}$$

The reduced matrix elements of the $\widehat{\Pi}_q^{(\kappa)}$ operators may be calculated making use of the following relation [250]:

$$\left\langle J \left\| \left\{ \widehat{\vec{A}} \otimes \widehat{\vec{B}} \right\}^{(\kappa)} \right\| J \right\rangle = (-1)^{2J-\kappa}\sqrt{2\kappa+1}\sum_i \begin{Bmatrix} 1 & 1 & \kappa \\ J & J & J_i \end{Bmatrix} \left\langle J \left\| \widehat{\vec{A}} \right\| J_i \right\rangle \left\langle J_i \left\| \widehat{\vec{B}} \right\| J \right\rangle \tag{11.17}$$

where $\widehat{\vec{A}}$ and $\widehat{\vec{B}}$ are vector operators; $\begin{Bmatrix} 1 & 1 & \kappa \\ J & J & J_i \end{Bmatrix}$ is the $6j$-symbol and the summation is performed over all atomic states. The reduced matrix element of the operator $\widehat{\Pi}^{(\kappa)}$ is expressed as

$$\left\langle J \left\| \widehat{\Pi}^{(\kappa)} \right\| J \right\rangle = (-1)^{2J-\kappa}\sqrt{2\kappa+1} \begin{Bmatrix} 1 & 1 & \kappa \\ J & J & J_i \end{Bmatrix} \times \left\langle J \left\| \widehat{\vec{d}} \right\| J_1 \right\rangle \left\langle J_1 \left\| \widehat{\vec{d}} \right\| J \right\rangle \tag{11.18}$$

and with the help of relations (11.8), (11.17), and (11.18) we get for the intensity

$$\begin{aligned} I(\vec{e}) = I(\beta,\varphi) = {} & (-1)^{J+J_1} B(\omega)\tau N \left| \left\langle J \left\| \widehat{\vec{d}} \right\| J_1 \right\rangle \right|^2 \\ & \times \left[-\frac{1}{\sqrt{3}} \begin{Bmatrix} 1 & 1 & 0 \\ J & J & J_i \end{Bmatrix} - \frac{1}{\sqrt{6}} \begin{Bmatrix} 1 & 1 & 2 \\ J & J & J_1 \end{Bmatrix} \rho_0^{(2)} \right. \\ & + \frac{1}{2} \begin{Bmatrix} 1 & 1 & 2 \\ J & J & J_i \end{Bmatrix} \rho_{2+}^{(2)} \cos 2\beta - \frac{1}{2i} \begin{Bmatrix} 1 & 1 & 2 \\ J & J & J_i \end{Bmatrix} \rho_{2-}^{(2)} \sin 2\beta \cos\varphi \\ & \left. - \frac{1}{\sqrt{2}} \begin{Bmatrix} 1 & 1 & 1 \\ J & J & J_i \end{Bmatrix} \rho_0^{(1)} \sin 2\beta \sin\varphi \right] \end{aligned} \tag{11.19}$$

where

$$\rho_{q\pm}^{(\kappa)} = \rho_q^{(\kappa)} \pm \rho_{-q}^{(\kappa)} \tag{11.20}$$

The polarization characteristics of the detected light beam may be described completely by the set of Stokes parameters I, P, Q, and C [249, 252]. Expression (11.20) allows the Stokes parameters of the detected beam to be expressed in terms of the polarization moments of the atomic density matrix:

$$\begin{aligned} I &= \frac{2}{3}\frac{B(\omega)\tau N}{(2J+1)^{1/2}}\left|\left\langle J \left\| \hat{d} \right\| J_1 \right\rangle\right|^2 I_0; \\ P &= \frac{3}{2}(-1)^{J+J_1}(2J+1)^{1/2}\left\{\begin{array}{ccc} 1 & 1 & 2 \\ J & J & J_1 \end{array}\right\} I_0^{-1}\rho_{2+}^{(2)} \\ Q &= -i\frac{3}{2}(-1)^{J+J_1}(2J+1)^{1/2}\left\{\begin{array}{ccc} 1 & 1 & 2 \\ J & J & J_1 \end{array}\right\} I_0^{-1}\rho_{2-}^{(2)} \\ C &= \frac{3}{\sqrt{2}}(2J+1)^{1/2}(-1)^{J+J_1}\left\{\begin{array}{ccc} 1 & 1 & 1 \\ J & J & J_i \end{array}\right\} I_0^{-1}\rho_0^{(1)} \end{aligned} \tag{11.21}$$

where

$$I_0 = \rho_0^{(0)} + \frac{3}{\sqrt{6}}(-1)^{J+J_1}(2J+1)^{1/2}\left\{\begin{array}{ccc} 1 & 1 & 2 \\ J & J & J_1 \end{array}\right\}\rho_0^{(2)} \tag{11.22}$$

These expressions show that the zero-order polarization moment $\rho_0^{(0)}$ determines the total light intensity of the beam; the polarization moments of the first $\rho_q^{(1)}$ and second $\rho_q^{(2)}$ ranks are responsible for the circular and linear polarizations, respectively.

This interpretation corresponds to the physical meaning of polarization moments which characterize the angular symmetry of an excited atomic ensemble as the macroscopic ordering of the spatial distribution of momenta. The zero-order polarization moment

$$\rho_0^{(0)} = (2J+1)^{1/2}\,\mathrm{Tr}(\rho_J) \tag{11.23}$$

is a scalar proportional to the number density of the excited atoms in a certain state. The three elements $\rho_{0,\pm 1}^{(1)}$ may be regarded as cyclic components of a vector known as the orientation vector. It represents the dipole ordering of momenta and hence the macroscopic momentum of the ensemble of the excited atoms. The five components $\rho_{0,\pm 1,\pm 2}^{(2)}$ form the alignment tensor describing the quadrupole ordering of the momenta of the excited atoms. The $q = 0$ component of the alignment tensor, $\rho_0^{(2)}$, is called the longitudinal alignment. It represents the noncoherent superposition of the magnetic sublevels $|m\rangle$ of the excited atomic state:

$$\rho_0^{(2)} = \sqrt{5}\left[(2J+3)(2J+1)(J+1)J(2J-1)\right]^{-1/2} \sum_m \left[3m^2 - J(J+1)\right] \rho_{mm} \tag{11.24}$$

The other components of the alignment tensor with $q \neq 0$ describe the coherent superposition of magnetic sublevels with $\Delta m = q$.

Transformation of polarization moments under the rotation of the reference frame $(X,Y,Z) \longrightarrow (X',Y',Z')$ defined by Euler angles is given by the Wigner D-matrix [250]:

$$^{Z'}\rho_{q'}^{(\kappa)} = \sum_q D_{qq'}^{(\kappa)*}(Z,Z')\ ^{Z}\rho_q^{(\kappa)} \tag{11.25}$$

where $D_{qq'}^{(\kappa)}(Z,Z')$ are the matrix elements. In particular, for q or q' equal to zero, $D_{0q}^{(\kappa)}(\vec{n},Z) = C_q^{(\kappa)}(\theta,\varphi)$.

$C_q^{(\kappa)}(\theta,\varphi)$ are spherical harmonics of the polar angles θ,φ, which determine the direction of the axis $\vec{n}$ in the coordinate system (X,Y,Z) [250].

11.1.2. Polarization under the Electron Impact Excitation

Let the elementary excitation process in a low-temperature plasma (electron impact, unpolarized light reabsorption) be characterized by the axial symmetry with respect to the axis $\vec{n}$. If the initial state is spherically symmetric, only longitudinal components of the polarization moments ($q = 0$) will appear in the frame of reference with the quantization axis directed along $\vec{n}$. The remaining components not being invariant under rotation of the coordinate system around the quantization axis will not appear. Therefore

$$^{\vec{n}}\rho_q^{(\kappa)} = {}^{\vec{n}}\rho_0^{(\kappa)}\delta_{0q} \tag{11.26}$$

(δ_{0q} are the Kronig symbols). In the laboratory frame of reference (X,Y,Z) we have

$$^{Z}\rho_q^{(\kappa)}(\vec{n}) = D_{0q}^{(\kappa)}(\vec{n},Z)\ ^{\vec{n}}\rho_0^{(\kappa)} = C_q^{(\kappa)}(\vec{n}) \tag{11.27}$$

Analyzing the electron impact self-alignment in a plasma, we should point out that the kinetics of electrons in an ionized gas is described by the velocity distribution function $f(\vec{v})$.

For symmetry considerations in the velocity space of electrons the multipole representation of the velocity distribution function is usually applied [250]:

$$f(\vec{v}) = \sum_{\kappa=0}^{\infty} \sum_{q=-\kappa}^{\kappa} C_q^{(\kappa)}(\vec{n}) f_q^{(\kappa)}(\vec{v}) \tag{11.28}$$

where

$$f_q^{(\kappa)} = \frac{2\kappa+1}{4\pi} \int d\vec{n} C_q^{(\kappa)}(\vec{n}) f(\vec{v}) \tag{11.29}$$

The zero-order component $f_0^{(0)}(v)$ accounts for the isotropic part of the distribution function. It is connected to the electron concentration N_e in the plasma:

$$\int_0^{\infty} f_0^{(0)}(v) v^2 dv = N_e/4\pi \tag{11.30}$$

Anisotropy of the electron motion is taken into account by multipole moments of higher orders. The vector combined from the cyclic components $f_q^{(1)}(\vec{v})$ characterizes the local electron flux. The tensor $f_q^{(2)}(\vec{v})$ or the quadrupole moment of the velocity distribution function reflects the quadrupole ordering of the electron velocities. It is proportional to the momentum flux tensor of the electrons [154] and therefore describes the structural characteristics of an ionized gas object.

To examine the impact self-alignment of atomic particles in an ionized gas let us divide the exciting electrons in a macroscopic small volume of the plasma into a number of elementary beams with number density of electrons $f(\vec{v})d\vec{v}$.

Such an elementary excitation will be axially symmetric with respect to the $\vec{n}$ vector. Hence in the collision frame of reference with quantization axis directed along $\vec{n}$ only longitudinal components of the polarization moments will appear and may be expressed in terms of the excitation cross section as follows:

$$^{\vec{n}}\rho_0^{(\kappa)}(\vec{v}) = \Gamma_\kappa v \sigma^{(\kappa)}(v) f(\vec{v}) d\vec{v} \tag{11.31}$$

Here Γ_κ are relaxation constants for the polarization moments,

$$\sigma^{(\kappa)}(v) = (2\kappa+1)^{1/2} \sum_m (-1)^{J-m} \begin{pmatrix} J & J & \kappa \\ m & -m & 0 \end{pmatrix} \sigma_m(v) \tag{11.32}$$

where $\sigma_m(v)$ are electron impact cross sections of the magnetic sublevels of the studied state as a function of the electron velocity. For $\kappa = 0$ we get

$$\sigma^{(0)} = (2J+1)^{1/2} \sum_m \sigma_m(v) = (2J+1)^{1/2} \sigma(v) \tag{11.33}$$

So the cross section of the excitation of the polarization moment with $\kappa = 0$ is proportional to the total excitation cross section of the state $\sigma(v)$. The alignment excitation cross section ($\kappa = 2$) will be given by

$$\sigma^{(2)}(v) = \sqrt{5}\left[(2J+3)(2J+1)(J+1)J(2J+1)\right]^{1/2} \cdot \sum_m \left[3m^2 - J(J+1)\right] \sigma_m(v) \tag{11.34}$$

The polarization moments induced due to elementary excitation by the electron beam in the laboratory frame of reference (X, Y, Z) take the form

$$^Z\rho_q^{(\kappa)}(\vec{v}) = \Gamma_\kappa^{-1} C_q^{(\kappa)}(\vec{n}) v \sigma^{(\kappa)}(v) f(\vec{v}) d\vec{v} \tag{11.35}$$

The total polarization moments of the macroscopic small volume of ionized gas may be calculated by integrating in the velocity space between the limits of the threshold excitation velocity $v_t = (2\varepsilon_t/m)^{1/2}$ (ε_t is the threshold excitation energy of the state under study) and infinity:

$$^Z\rho_q^{(\kappa)} = \frac{4\pi}{2\kappa+1} \int_{v_t}^{\infty} dv v^3 \sigma_{(v)}^{\kappa} f_q^{(\kappa)}(v) \tag{11.36}$$

This formula shows that there are straightforward connections between the symmetry of the electron velocity distribution and the type of the ordering of momenta of the excited atoms. In other words, the polarization moments $\rho_q^{(\kappa)}$ are determined by the multipole momenta of the velocity distribution function of the same rank. Generally, this result is the consequence of the orthogonality condition of the irreducible representations of the rotation group.

Making use of these relations we may write expressions for the Stokes parameters of the detected light beam for the given spectral line in terms of the multipole moments of the velocity distribution function:

$$I = \frac{2}{3}(2J+1)^{-1/2} B(\omega) N \tau \left| \left\langle J \middle\| \hat{\vec{d}} \middle\| J_1 \right\rangle \right|^2 I_0 \tag{11.37}$$

$$P = \frac{4\pi}{5I_0\Gamma_2} \frac{3}{2} (-1)^{J+J_1} (2J+1)^{1/2} \begin{Bmatrix} 1 & 1 & 2 \\ J & J & J_1 \end{Bmatrix} \int_{v_t}^{\infty} dv v^3 \sigma^{(2)}(v) f_{2+}^{(2)}(v)$$

$$Q = -\frac{i4\pi}{5I_0\Gamma_2} \frac{3}{2} (-1)^{J+J_1} (2J+1)^{1/2} \begin{Bmatrix} 1 & 1 & 2 \\ J & J & J_1 \end{Bmatrix} \int_{v_t}^{\infty} dv v^3 \sigma^{(2)}(v) f_{2-}^{(2)}(v)$$

$$I_0 = \frac{4\pi}{\Gamma_0} \int_{v_t}^{\infty} dv v^3 \sigma^{(0)}(v) f_0^{(0)}(v) + \frac{8\pi}{5\sqrt{6}\Gamma_2} \frac{3}{2} (-1)^{J+J_1} (2J+1)^{1/2} \begin{Bmatrix} 1 & 1 & 2 \\ J & J & J_1 \end{Bmatrix} \int_{v_t}^{\infty} dv v^3 \sigma^2(v) f_0^{(2)}(v)$$

where $f_{2+}^{(2)}(v) = f_2^{(2)}(v) + f_{-2}^{(2)}(v)$ and the multipole moments of the velocity distribution function are calculated in the detector frame of reference.

If the studied plasma is axially symmetric with respect to the axis $\vec{n}$ and the OZ-axis of the laboratory frame of reference is parallel to $\vec{n}$, we have

$$P = \left[R(JJ_1)^{-1} + \cos^2\theta - 1/3\right]^{-1} \sin^2\theta \cos 2\varphi \tag{11.38}$$

$$Q = \left[R(JJ_1)^{-1} + \cos^2\theta - 1/3\right]^{-1} \sin^2\theta \sin 2\varphi$$

$$R(JJ_1) = \frac{3}{10}\sqrt{\frac{3}{2}}(-1)^{J+J_1}(2J+1)^{1/2} \begin{Bmatrix} 1 & 1 & 2 \\ J & J & J_1 \end{Bmatrix} \frac{\Gamma_0}{\Gamma_2}$$

$$\times \int_{v_t}^{\infty} dv v^3 \sigma^{(2)}(v)^{\vec{n}} f_0^{(2)}(v) \bigg/ \int_0^{\infty} dv v^3 \sigma^{(0)}(v) f_0^{(0)}(v)$$

where θ and φ are polar angles which determine the direction of the $\vec{n}$ axis in the detector frame of reference.

For a low degree of polarization $\left(\rho_0^{(2)} \ll \rho^{(0)}\right)$ the quantity $R(JJ_1)$ is the degree of polarization of the optical radiation of atoms detected in the frame of reference with the axis collinearly oriented to $\vec{n}$.

These expressions show that the quadrupole moment of the velocity distribution function is totally responsible for the Stokes parameters P and Q of the detected light. Therefore measuring the Stokes parameters for different spectral lines under electron impact excitation and making use of the spectroscopic constants (cross sections of excitation and alignment of atomic states, relaxation constants) it is possible to determine the quadrupole moment of the velocity distribution function for energies exceeding the excitation potentials. This quantity is principally inaccessible for traditional optical diagnostic methods. Study of the spatial distribution of polarization spectroscopic characteristics across the image of an ionized gas entity enables one to estimate the transport of energy in different parts of the plasma and hence to find out the relative role of those parts in the general energy balance of a plasma object. By these means spectropolarimetric diagnostics allows one to perform remote sensing of structural energy characteristics.

11.1.3. Solution of the Inverse Problems of Spectropolarimetric Diagnostics

A quantitative solution of the inverse problem of spectropolarimetry is possible, if the energy dependencies of the cross sections $\sigma^{(0)}(\varepsilon)$ and $\sigma^{(2)}(\varepsilon)$ are known. The cross section of excitation of an atomic state $\sigma^{(0)}(\varepsilon)$ can be determined by means of many spectroscopic techniques.

The determination of the electron impact alignment cross section became necessary when the quantitative spectropolarimetry was developed [248]. The electron impact alignment cross section, by its nature, reflects the formation of the quadrupole ordering of momenta of excited atoms under electron–atom collisions, and it is expressed as the combination of the magnetic sublevel excitation cross sections $\sigma_m(\varepsilon)$. If only one sublevel $m = 0$ is populated (for $J = 0 - J = 1$ transitions), we have

$$\sigma^{(2)}/\sigma = -\sqrt{5}\,[J(J+1)/(2J+3)(2J+1)(J-1)]^{1/2} < 0 \tag{11.39}$$

For the opposite case (only $m = \pm 1$ sublevels are populated) we have

$$\sigma^{(2)}/\sigma = -\sqrt{5}\,[J(2J-1)/(2J+3)(2J+1)(J+1)]^{1/2} > 0 \tag{11.40}$$

The simplest experimental way to determine the electron impact cross section is based on measurement of the degree of the linear polarization of an atomic ensemble in the direction perpendicular to the beam P_0 under stationary electron beam excitation. The polarization moments in the collision frame of reference are given by (11.26). Using expressions (11.31) and (11.37) for $P_0(v)$ we obtain

$$P_0(v) = \frac{1}{\frac{1}{R(v)} - \frac{1}{3}} \tag{11.41}$$

where

$$R(v) = \left(\frac{3}{2}\right)^{3/2} (-1)^{J+J_1}(2J+1) \begin{Bmatrix} 1 & 1 & 2 \\ J & J & J_i \end{Bmatrix} \frac{\Gamma_0}{\Gamma_2} \frac{\sigma^{(2)}(v)}{\sigma(v)} \tag{11.42}$$

Therefore the electron impact alignment cross section may be expressed in terms of the experimentally measured quantities $P_0(v)$ and $\sigma(v)$ by the following relation:

$$\sigma^{(2)}(v) = \left(\frac{2}{3}\right)^{3/2} \frac{\Gamma_2}{\Gamma_0} \frac{(-1)^{J+J_1}}{(2J+1)} \begin{Bmatrix} 1 & 1 & 2 \\ J & J & J_i \end{Bmatrix} \frac{P_0(v)\sigma(v)}{1+P_0(v)/3} \tag{11.43}$$

In spite of the in principle simple spectropolarimetric measurements of $P_0(v)$ under electron beam excitation, such measurements have been realized only for a limited number of relatively strong optical transitions in helium. The main problem is that the spectropolarimetric experiments for the threshold energies are very difficult because of the low light intensities.

Electron impact alignment for a large number of atomic states may be obtained using experimental and theoretical data of the electron–photon angular correlation parameters [249]. This type of experiment is characterized by the lack of axial symmetry with respect to the excitation electron beam. When describing the angular correlation experiments, Stokes parameters are replaced by the following characteristics:

$$P_l = \left(P^2 + Q^2\right)^{1/2} \tag{11.44}$$

which is the maximal degree of polarization when the analyzer is fixed at the angle $\gamma_0 = (1/2)\arctan(Q/p)$, and the mean value of the Z-projection of the total momentum J_1. The set of A_q and Q_q parameters [249] also may be used. These parameters are determined as the mean values of different combinations of the momentum projection:

$$A_0 = \left\langle 3J_x^2 - J^2 \right\rangle / J(J+1) \tag{11.45}$$

The following parameters are usually used to describe an atomic state polarization with $J = 1$:

$$\lambda = \sigma_0/\sigma \tag{11.46}$$

where

$$\sigma = \sigma_0 + \sigma_1 + \sigma_{-1} \tag{11.47}$$

is the total cross section of the impact excitation. Making use of eqs. (11.41)–(11.44) and averaging over all scattering directions, one obtains the following relations for the alignment cross section, expressed by different parameters:

$$\sigma^{(2)}(v) = \frac{2\pi}{\sqrt{6}} \int_0^\pi d\theta \sin\theta\ \sigma(\theta, v)\left[1 - 3\lambda(\theta, v)\right] \tag{11.48}$$

$$\sigma^{(2)}(v) = \frac{\pi}{\sqrt{6}} \int_0^\pi d\theta \sin\theta\ \sigma(\theta, v)\left[-1 - 3P(\theta, v)\right]$$

$$\sigma^{(2)}(v) = \frac{\pi}{\sqrt{6}} \int_0^\pi d\theta \sin\theta\ \sigma(\theta, v)\left\{-1 - 3P_l(\theta, v)\cos^2\left[2\gamma_0(\theta, v)\right]\right\}$$

For $J \neq 1$ we have

$$\sigma^{(2)}(v) = 2\pi\sqrt{5}\left[\frac{J(J+1)}{(2J+3)(2J+1)(2J-1)}\right]^{1/2} \int_0^\pi d\theta \sin\theta\ \sigma(\theta, v) A_0(\theta, v) \tag{11.49}$$

where $\sigma(\theta, v)$ is the differential cross section of the electron scattering.

If the exciting electron energy is close to the threshold of an atomic state, the alignment cross section may be calculated using the Percieval–Seatton approach [249]. In this case, using Eq. (11.34) we have

$$\sigma_t^{(2)} = \alpha(s)\sigma_t$$

$$\alpha(s) = \sqrt{5}\left[\frac{J(J+1)}{(2J+3)(2J+1)(2J-1)}\right]^{1/2}\left[\frac{3S}{J(J+1)} - 1\right]$$

where $S = 0, \pm 1$ is the spin quantum number of the studied state while σ_t and $\sigma_t^{(2)}$ are threshold values of the impact excitation and alignment cross sections.

The alignment cross section for high excitation energies may be calculated in the Born approximation using symmetry considerations [251]. Making use of the rotational transformation of the polarization moments and averaging over all directions of the scattered electrons, the alignment cross section may be expressed according to [252]:

$$\sigma^{(2)}(v) = \pi\alpha(s) \int_0^\pi d\theta \sin\theta \, \sigma(\theta, v) \left(3 \frac{1 + \delta\cos^2\theta - 2\sqrt{\delta}\cos\theta}{1 + \delta - 2\sqrt{\delta}\cos\theta} \right) \tag{11.50}$$

where $\delta = \varepsilon_1/\varepsilon$, ε is the incident electron energy, $\varepsilon_1 = \varepsilon - \varepsilon_t$, ε_t is the threshold energy of the studied level.

11.2. *Realization of Spectropolarimetric Sensing*

11.2.1. *Polarimetric Spectral Measurements of Spatially Inhomogeneous Plasmas*

The integral intensity of the polarized light detected by a monochromator for a Doppler broadened line will be expressed as

$$I_\omega(\vec{e}) = 2S \int_{L_1}^{L_2} dz I(\vec{e}, z) S'(\xi_z) \tag{11.51}$$

where $L_{1,2}$ are coordinates of the boundaries of the plasma along the line of sight, S is the cross section of the entrance slit, $I(\vec{e}, z)$ is the intensity with the polarization $\vec{e}$ irradiated by the differentially small volume with coordinate z,

$$S'(\xi_z) = \frac{1}{\sqrt{\pi}} \int_{-\infty}^{\infty} d\beta \exp\left(-\beta^2 - \xi e^{-\beta^2}\right) \tag{11.52}$$

and

$$\xi_z = \int_{L_1}^{z} dz' \chi_0(z') \tag{11.53}$$

Here, $\chi_0(z')$ is the absorption coefficient at the central frequency ω_0,

$$\beta = c\left[(\omega - \omega_0)/\omega_0\right](M/2k_B T)^{1/2}$$

c is the velocity of light, M the atomic mass, k_B the Boltzmann constant, and T is the atomic ensemble's temperature.

Using these equations, we can express the Stokes parameters of a spectral line in terms of z-distributions of the polarization moments:

$$I = 2/3(2J+1)^{-1/2}B(\omega)N_a\tau S\left|\left\langle J\left\|\widehat{\vec{d}}\right\|J\right\rangle\right|^2 I_0 \tag{11.54}$$

$$P = J_0^{-1}\int_{L_1}^{L_2} dzP(z)j(z)S'(\xi_z)$$

$$Q = J_0^{-1}\int_{L_1}^{L_2} dzQ(z)j(z)S'(\xi_z)$$

where

$$I_0 = \int_{L_1}^{L_2} dzj(z)S'(\xi_z) \tag{11.55}$$

$$j(z) = \rho_0^{(0)}(z) + \frac{3}{\sqrt{6}}(-1)^{J+J_1}(2J+1)^{1/2}\begin{Bmatrix} 1 & 1 & 2 \\ J & J & J_1 \end{Bmatrix}\rho_0^{(2)}(z)$$

$$P(z) = \frac{3}{2}j(z)^{-1}(-1)^{J+J_1}(2J+1)^{1/2}\begin{Bmatrix} 1 & 1 & 2 \\ J & J & J_1 \end{Bmatrix}\rho_{2+}^{(2)}(z)$$

$$Q(z) = \frac{3}{2}j(z)^{-1}(-1)^{J+J_1}(2J+1)^{1/2}\begin{Bmatrix} 1 & 1 & 2 \\ J & J & J_1 \end{Bmatrix}(-i)\rho_{2-}^{(2)}(z)$$

Most plasmas in practical applications are characterized by cylindrical symmetry with respect to a certain axis. In the X',Y,Z' frame of reference, the OZ' axis is parallel to the $\vec{r}$ vector. All spectroscopic parameters of the plasma are r-dependent:

$$^{z'}\rho_q^{(\kappa)} = {}^{z'}\rho_q^{(\kappa)}(r), \qquad j = j(r), \qquad \chi_0 = \chi_0(r) \tag{11.56}$$

The transformation to the detector frame (parallel to the symmetry axis) is expressed by Eq. (11.27), so finally we obtain

$$^{z}\rho_{2+}^{(2)} = \frac{1+\cos^2\varphi^{z'}}{2}\rho_{2+}^{(2)} + \sqrt{\frac{3}{2}}\sin^2\varphi^{z'}\rho_0^{(2)} + \frac{1}{2}\sin 2\varphi^{z'}\rho_{1-}^{(2)} \tag{11.57}$$

$$^{z}\rho_{2-}^{(2)} = \cos\varphi^{z'}\rho_{2-}^{(2)} + \sin\varphi^{z'}\rho_{1\mp}^{(2)} \tag{11.58}$$

where

$$\varphi = \left(\widehat{0x, 0x'}\right), \qquad \sin\varphi = x/r \tag{11.59}$$

Due to the axial symmetry $^{z'}\rho_{1+}^{(2)} = {}^{z'}\rho_{1-}^{(2)} = 0$.

Substituting the z-variable instead of $r = \sqrt{x^2 + z^2}$ for a low degree of alignment we obtain

$$P(x) = \frac{3}{2J_0(x)}(-1)^{J+J_1}(2J+1)^{1/2}\begin{Bmatrix} 1 & 1 & 2 \\ J & J & J_1 \end{Bmatrix} \tag{11.60}$$
$$\times \int_x^R \frac{drr}{\sqrt{r^2 - x^2}}$$
$$\times \left[\frac{2 - x^2/r^2}{2}\,{}^{z'}\rho_{2+}^{(2)}(r) + \sqrt{\frac{3}{2}\frac{x^2}{r^2}}\,{}^{z'}\rho_0^{(2)}(r)\right]$$
$$\times \left[S'(\xi_{rR}) + S'(\xi_{xr} + \xi_{xR})\right]$$
$$Q(x) = \frac{3}{i2J_0(x)}(-1)^{J+J_1}(2J+1)^{1/2}\begin{Bmatrix} 1 & 1 & 2 \\ J & J & J_1 \end{Bmatrix}$$
$$\times \int_x^R \frac{drr}{\sqrt{r^2 - x^2}}\rho_{1+}^{(2)}(r)\left[S'(\xi_{rR}) + S'(\xi_{xr} + \xi_{xR})\right]$$
$$I_0(x) = \int_x^R \frac{drr}{\sqrt{r^2 - x^2}}\rho_0^{(0)}(r)\left[S'(\xi_{rR}) + S'(\xi_{xr} + \xi_{xR})\right]$$

where

$$\xi_{r_1 r_2} = \int_{r_1}^{r_2} \frac{dr' r' \chi_0(r')}{\sqrt{r'^2 - x^2}} \tag{11.61}$$

and R is the radius of the cross section.

When the expression (11.52) for function $S'(\xi)$ is introduced we get for $I_0(x)$

$$I_0(x) = \frac{2}{\sqrt{\pi}}\int_{-\infty}^{\infty} d\beta e^{-\beta^2}\exp\left[-e^{-\beta^2}\int_x^R \frac{drr\chi_0(r)}{\sqrt{r^2 - x^2}}\right] \tag{11.62}$$
$$\times \int_x^R \frac{drr\rho_0^{(0)}(r)}{\sqrt{r^2 - x^2}}\cosh\left[-e^{-\beta^2}\int_x^r \frac{dr'\chi_0(r')r'}{\sqrt{r'^2 - x^2}}\right]$$

If the transparency for the central frequency is high enough,

$$\Pi_0(x) = \exp\left[-2\int_x^R \frac{r\,\chi_0(r)}{\sqrt{r^2 - x^2}}dr\right] \tag{11.63}$$
$$= \exp(-2\xi_{xR}) > 0.5$$

then the system of equations may be simplified to

$$I_0(x) = 2S'(\xi_{xR})J'(x), \qquad I'(x) = \int_x^R \frac{drr\rho_0^{(0)}}{\sqrt{r^2-x^2}} \tag{11.64}$$

$$P(x) = \frac{3}{2J'(x)}(-1)^{J+J_1}(2J+1)^{1/2}\begin{Bmatrix} 1 & 1 & 2 \\ J & J & J_1 \end{Bmatrix}$$

$$\times \int_x^R \frac{drr}{\sqrt{r^2-x^2}}\left[\frac{2-x^2/r^2}{2}\,{}^{z'}\rho_{2+}^{(2)}(r) + \sqrt{\frac{3}{2}\frac{x^2}{r^2}}\,{}^{z'}\rho_0^{(2)}(r)\right]$$

$$Q(x) = \frac{3}{2J'(x)}(-1)^{J+J_1}(2J+1)^{1/2}\begin{Bmatrix} 1 & 1 & 2 \\ J & J & J_1 \end{Bmatrix}\int_x^R \frac{drr}{\sqrt{r^2-x^2}}\,{}^{z'}\rho_{1+}^{(2)}(r)$$

11.2.2. Anisotropy of Electron Motion and Spectropolarimetric Effects in Different Parts of High-Frequency Capacitive Discharges

Experimental measurements of the distribution of the Stokes parameters for different spectral lines of noble gases emitted by a cylindrical capacitive low-pressure discharge with variable distance between the outer electrodes ($f_\sim = 100$ MHz, $L = 0$–30 cm) confirmed the complex structural nature of this discharge [133, 134, 253]. Experiments have shown a strong polarization maximum in the vicinity of the electrode dark space on all spectral lines with different excitation potentials. The shape of this polarization maximum reflects the process of the relaxation of momenta of fast electrons formed at the electrode sheath boundary and propagating to the center of the discharge. In the central part of the capacitive discharge a constant value of the degree of linear polarization has been observed for all spectral lines. This central region blows up when the distance between the electrodes rises. Some energy-selective spectropolarimetric effects have been observed near the side glass wall of the discharge cell. For lines with low-lying upper levels a monotonic fall in the degree of polarization was observed while moving to the wall. But for spectral lines originating from highly excited levels a strong polarization maximum has been detected near the side wall of the discharge cell [254]. The degree of linear polarization in this side wall maximum was 2–3 times higher as compared to the central part.

The electrons in the central part of the discharge may be regarded as moving in the potential wall formed by the stationary radial potential profile near the side glass walls $\varphi(r) = \varphi(z,r) - \varphi(z,0)$ (r is the coordinate of the point under study and z is the coordinate along the discharge axis), with the potential drop $\Delta\varphi$ within the thin Debye sheath near the side wall. The total potential drop between the wall and the axis of the plasma distinguishes two groups of electrons: imprisoned electrons with total energy less than this potential drop, and nontrapped electrons with kinetic energy exceeding the total potential drop. Electrons belonging to the last group can leave the plasma and impinge on the dielectric side wall.

The most effective way to describe the imprisoned electrons is to use the following set of variables [254]: ψ is the azimuthal angle in the plane perpendicular to the axis, $\varepsilon_0 = m_e v^2/2 + e\varphi(r)$ is the total energy of an electron for the given cross section of the cell, $\mu = m_e v_\psi r$ is the momentum of electron; and $\alpha = v_y\sqrt{m_e/2\varepsilon_0}$ (v_y is the axial electron velocity projection). The kinetic equation for the velocity distribution function $f = f(y, r, \psi, \varepsilon_0, y, \alpha)$ will take the form

$$\frac{\partial f}{\partial t} + \frac{\partial f}{\partial z}\dot{z} + \frac{\partial f}{\partial r}\dot{r} + \frac{\partial f}{\partial \psi}\dot{\psi} + \frac{\partial f}{\partial \varepsilon_0}\dot{\varepsilon}_0 + \frac{\partial f}{\partial \mu}\dot{\mu} + \frac{\partial f}{\partial \alpha}\dot{\alpha} = \Xi(f) \qquad (11.65)$$

where $\Xi(f)$ is the collision integral accounting for the electron–atom collisions.

For a cylindrical discharge cell, the distribution function is axially and azimuthally symmetric so $\partial f/\partial z = \partial f/\partial \psi = \dot{\mu} = 0$. Expressing the time derivatives by $\dot{\varepsilon}_0 = ev_z E_z$ and $\dot{\alpha} = \left(eE_z/\sqrt{2m_e\varepsilon_0}\right)\left(1-\alpha^2\right)$ we have

$$\frac{\partial f}{\partial t} + v_r\frac{\partial f}{\partial r} + \frac{\partial f}{\partial \varepsilon_0}ev_z E_z + \frac{\partial f}{\partial \alpha}\frac{eE_z}{\sqrt{2m_e\varepsilon_0}}\left(1-\alpha^2\right) = \Xi(f) \qquad (11.66)$$

Here $E_z = E_\sim \cos\omega_\sim t$, $\omega_\sim = 2\pi f_\sim$ is the angular frequency of the electric field, and v_r is the radial component of the velocity vector:

$$v_r = \pm\sqrt{\frac{2}{m_e}\left[\varepsilon_0\left(1-\alpha^2\right) - e\varphi(r) - \mu^2/2m_e r^2\right]} \qquad (11.67)$$

where the "+" sign corresponds to the motion of an electron from the center to the wall of the cell and the "−" sign determines the opposite motion.

On the discharge axis it is possible to considerably simplify this equation. In this case $\partial f/\partial r = 0$, $\alpha = \cos(\vec{v}, OY)$, ε_0 becomes the kinetic energy of electron, and so we have

$$\frac{\partial f}{\partial t} + \alpha a\frac{\partial f}{\partial v} + \frac{a}{v}\frac{\partial f}{\partial \alpha}\left(1-\alpha^2\right) = \Xi(f) \qquad (11.68)$$

where we used the velocity $v = \sqrt{\frac{2m_e}{\varepsilon_0}}$ and acceleration $a = \frac{eE_z}{m_e} = a_\sim\cos\omega_\sim t$. The kinetic equation in this form depends directly on the angular variable α, so it allows us to study the angular anisotropy of the velocity distribution function.

The multipole representation for the velocity distribution function, taking into consideration the axial symmetry, is expressed as

$$f(v,\alpha) = \sum_{\kappa=0}^{\infty} f^{(\kappa)}(v)P_\kappa(\alpha) \qquad (11.69)$$

where $P_\kappa(\alpha)$ are the Legendre polynomials. The kinetic equation may be written in the form

$$\sum_{\kappa=0}^{\infty}\left(\frac{\partial}{\partial t}+a\alpha\frac{\partial}{\partial v}+a\frac{(1-\alpha^2)}{v}\frac{\partial}{\partial \alpha}\right)f^{(\kappa)}(v)P_\kappa(\alpha)=\sum_{\kappa=0}^{\infty}\Xi^{(\kappa)}(f)P_\kappa(\alpha) \quad (11.70)$$

where

$$\Xi^{(\kappa)}(f)=\frac{2\kappa+1}{4\pi}\int d\vec{n}C_0^{(\kappa)*}(\vec{n})\Xi(f) \quad (11.71)$$

The collision integral for the low temperature plasma incorporates only electron–atom collisions and may be expressed as

$$\Xi[f(\varepsilon,\vec{n})]=\phi[f(\varepsilon,\vec{n})]+\nu(\varepsilon)\left[\overline{f(\varepsilon,\vec{n})}-f(\varepsilon,\vec{n})\right]+\Xi^*[f(\varepsilon,\vec{n})] \quad (11.72)$$

where $\phi[f(\varepsilon,\vec{n})]$ is the Fokker–Planck term describing the energy drift of electrons due to elastic electron–atom collisions, $\nu(\varepsilon)$ is the frequency of the elastic electron–atom collisions, $\vec{n}=\vec{v}/v$ is the velocity unit vector, $\overline{f(\varepsilon,\vec{n})}$ is the local distribution function averaged over all directions:

$$\overline{f(\varepsilon,\vec{n})}=\int d\vec{n}'f(\varepsilon,\vec{n}')\sigma_{el}(\varepsilon,\vec{n}',\vec{n})\Big/\int d\vec{n}'\sigma_{el}(\varepsilon,\vec{n}',\vec{n}) \quad (11.73)$$

Here $\sigma_{el}(\varepsilon,\vec{n}',\vec{n})$ is the differential cross section of the inelastic electron–atom scattering $(\vec{n}',\vec{n})$. If σ_{el} does not depend on the scattering angle, $\overline{f(\varepsilon,\vec{n})}=f_0^0(\varepsilon)$ [$f_0^0(\varepsilon)$ is the isotropic part of the velocity distribution function]; $\Xi^*f(\varepsilon,\vec{n})$ describes the inelastic electron–atom interaction. Inelastic electron–atom collisions described by the factor $\nu^*(\varepsilon)f(\varepsilon,\vec{n})$ [$\nu^*(\varepsilon)$ is the inelastic electron–atom frequency] contribute to the loss of the fast electrons and, on the other hand, they enhance the quantity of slow electrons in the velocity distribution function. We restrict ourselves only to the first mechanism because fast electrons are responsible for the impact excitation of optical emission of the plasma. Taking into account that the electron temperature is much higher than the temperature of neutral atoms and ions, we may express the Fokker–Planck terms as follows:

$$\phi[f(v)]=\frac{\gamma}{v^2}\frac{\partial}{\partial v}v^3\nu_{tr}(v)f_{(v)}^{(0)} \quad (11.74)$$

where $\gamma=m_e/M$ is the ratio of the masses of the electron and atom while $\nu_{tr}(v)$ is the transport collision frequency of electrons.

As the isotropic Fokker–Planck term affects only the isotropic part of the velocity distribution function, the zero-order moment of the collision integral is given by

$$\Xi^{(0)}(f)=\frac{\gamma}{v^2}\frac{\partial}{\partial v}v^3\nu_{tr}(v)f_{(v)}^{(0)}-\nu^*f_{(v)}^{(0)} \quad (11.75)$$

The contribution of the second term of this equation to $\Xi^{(0)}(f)$ describes the change of the velocity vector direction when the elastic collision is zero. For higher-order moments (11.71), this second term becomes predominant and the influence of the Fokker–Planck term vanishes. Inelastic collisions may be neglected because the frequency of such collisions is much lower than for the elastic interactions. Using the axial symmetry of the elastic cross section with respect to the direction of the incident electron $\vec{n}'$, $(\sigma_{el}(\vec{n}',\vec{n}) = \sigma_{el}(\beta), \beta = \cos(\vec{n}',\vec{n}))$ and the transformation of the spherical functions under rotations of the frame of reference, we can obtain for the moments of the collision integral ($\kappa > 1$) the relation

$$\Xi^{(\kappa)}[f(\vec{v})] = -\nu_\kappa(v) f^{(\kappa)}(v) \tag{11.76}$$

where the coefficients $\nu_\kappa(v)$ are the effective collision frequencies of rank κ:

$$\nu_\kappa(v) = N_a v \int_{-1}^{1} d\beta [1 - P_\kappa(\beta)]\, \sigma_{el}(\beta) \tag{11.77}$$

and N_a is the concentration of neutral atoms.

Making use of the recursive relations for Legendre polynomials

$$\begin{aligned} \alpha P_\kappa(\alpha) &= \frac{\kappa}{\kappa+1} P_{\kappa-1}(\alpha) + \frac{\kappa+1}{2\kappa+1} P_{\kappa+1}(\alpha) \\ \left(1-\alpha^2\right) \frac{\partial}{\partial\alpha} P_\kappa(\alpha) &= \frac{\kappa(\kappa+1)}{2\kappa+1} \left[P_{\kappa-1}(\alpha) - P_{\kappa+1}(\alpha)\right] \end{aligned} \tag{11.78}$$

it is possible to transform equation (11.70) into the infinite system of equations

$$\begin{aligned} \frac{\partial f^{(0)}(v)}{\partial t} + \frac{a}{3}\left(\frac{\partial}{\partial v} + \frac{2}{v}\right) f^{(1)} v = \frac{\gamma}{v^2} \frac{\partial}{\partial v} v^3 \nu_1(v) f^{(0)} v - \nu^* f^{(0)}(v) \frac{\partial f^{(\kappa)}}{\partial t} \\ + \frac{\kappa}{2\kappa-1} a \left(\frac{\partial}{\partial v} + \frac{\kappa-1}{v}\right) f^{(\kappa-1)} \\ + \frac{\kappa+1}{2\kappa+3} a \left(\frac{\partial}{\partial v} + \frac{\kappa+2}{v}\right) f^{(\kappa+1)} \\ = -\nu_\kappa f^{(\kappa)} \qquad (\kappa \geq 1) \end{aligned} \tag{11.79}$$

This system is very useful for low anisotropy of the velocity distribution function, when it is possible to restrict ourselves to the study of small order multipole moments. If $f^{(\kappa+1)} \ll f^{(\kappa-1)}$, the higher-order terms in Eq. (11.79) may be neglected and the multipole moments of rank κ become dependent only on the multipole moments of lower rank. It is valid for a low temperature plasma because $f^{(\kappa+1)}/f^{(\kappa-1)} \sim m/M$. The equations for the first and second multipole moments take the form

$$\frac{\partial f^{(1)}}{\partial t} + a\frac{\partial f^{(0)}}{\partial t} = -\nu_1 f^{(1)} \qquad (11.80)$$
$$\frac{\partial f^{(2)}}{\partial t} + \nu_2 f^{(2)} = \frac{2}{3}v\frac{\partial}{\partial v}\frac{1}{v}af^{(1)}$$

Hence the mean value for the quadrupole moment of the velocity distribution averaged over the high-frequency field period is expressed as

$$\left\langle f^{(2)}(v)\right\rangle = \frac{2}{3v_2}v\frac{\partial}{\partial v}\frac{1}{v}\left\langle af^{(1)}(v)\right\rangle \qquad (11.81)$$

where the angle brackets denote time averaging.

In order to calculate the value $\left\langle af^{(1)}(v)\right\rangle$ let us represent the time dependence of the moments of the velocity distribution function as a Fourier expansion:

$$f^{(\kappa)}(v,t) = \sum_{n=-\infty}^{\infty} f_n^{(\kappa)}e^{in\omega_\sim t}, \quad \text{where} \quad a(t) = \frac{1}{2}a_\sim\left(e^{i\omega_\sim t} + e^{-i\omega_\sim t}\right) \qquad (11.82)$$

For a low-temperature plasma in the weak harmonic electric field it is possible to restrict ourselves only to temporal harmonics with small n.

Using Eq. (11.82) we have

$$\left\langle af^{(1)}(v)\right\rangle = \frac{1}{2}a_\sim\left(f_1^{(1)} + f_{-1}^{(1)}\right) \qquad (11.83)$$

and the equations for the harmonics $f_{\pm 1}^{(1)}$ take the form

$$i\omega_\sim f_1^{(1)} + \nu_1 f_1^{(1)} = -\frac{1}{2}\frac{\partial}{\partial v}a_\sim\left(f_0^{(0)} + f_2^{(0)}\right) - i\omega_\sim f_{-1}^{(1)} + \nu_1 f_{-1}^{(1)}$$
$$= -\frac{1}{2}\frac{\partial}{\partial v}a_\sim\left(f_0^{(0)} + f_{-2}^{(2)}\right) \qquad (11.84)$$

Neglecting $f_{\pm 2}^{(2)}$ in comparison with $f_0^{(0)}$ we have

$$\left\langle af^{(1)}(v)\right\rangle = -\frac{1}{2}\frac{\nu_1(v)a_\sim^2}{\omega_\sim^2 + \nu_1^2(v)}\frac{\partial}{\partial v}f_0^{(0)}(v) \qquad (11.85)$$

and

$$\left\langle f^{(2)}(v)\right\rangle = \frac{a_\sim^2}{3\nu_e(v)}v\frac{\partial}{\partial v}\frac{1}{v}\frac{v_2(v)}{\omega_\sim^2 + \nu_1^2}\frac{\partial}{\partial v}\left\langle f^{(0)}(v)\right\rangle \qquad (11.86)$$

For the low polarization degree P detected in the central parts of the capacitive discharge, $P \sim R(JJ_1)$ is valid. Using the expression for the quadrupole moment (11.86) we have

$$P = -\frac{3a_{\sim}^2}{10\sqrt{6}}(-1)^{J+J_1}(2J+1)^{1/2}\begin{Bmatrix} 1 & 1 & 2 \\ J & J & J_1 \end{Bmatrix} \times \frac{\Gamma_0}{\Gamma_2}\frac{\int_{v_t}^{\infty}\sigma^{(2)}\frac{v^x}{\nu_2}\frac{\partial}{\partial v}\frac{1}{v}\frac{\nu_1}{\omega_{\sim}^2+\nu_1^2}\frac{\partial}{\partial v}\left\langle f^{(0)}(v)\right\rangle dv}{\int_{v_t}^{\infty}\sigma^{(0)}v^3\left\langle f^{(0)}(v)\right\rangle dv} \tag{11.87}$$

So if the low-lying atomic state is excited collisionally by the imprisoned electrons in the central part of the discharge plasma, the degree of polarization of the optical transition from this state is dependent on the amplitude of the variable electric field strength and on the energy dependence of the zeroth-order moment of the velocity distribution function.

11.2.3. Boundary Effects

The main reason for the anisotropy of the fast electrons with energy exceeding the potential drop between plasma and wall is the formation of cones of losses in the velocity distribution [255]. The effect of these cones of losses has been confirmed experimentally by the detection of selective polarization maxima near the side wall of a capacitive discharge [253]. In order to compute the shape of the cone of losses formed near the side dielectric wall of a cylindrical cell let us use the frame of reference related to the radius vector axis. In this coordinate system the electron velocity vector is characterized by the polar and azimuthal angles θ and ψ. The shape of the cone of losses in the studied small volume is determined by the critical electron trajectories with velocity vector $v(\theta_0, \psi)$ ($v_r = 0$ for $r = r_0$, while r_0 is the discharge radius). In the case of a cylindrical discharge tube the cone of losses is not rotationally symmetric with respect to the $\vec{r}$ vector. It is characterized by the function $\theta_0(\psi)$. Electron trajectories (which for the given macroscopically small volumes inside the plasma are characterized by the polar angles $0 < \theta_1 < \theta_0$ and $\pi - \theta_0, \theta < \pi$) terminate on the nearest and opposite walls, respectively. The remaining trajectories describe the infinite spiral motion around the discharge axis. Supposing a stationary radial potential profile $\varphi(r)$ at every cross section of the cell and at every time, the function $\theta_0(\psi)$ may be obtained from the laws of concentration of the energy, momentum, and z-component of the electron velocity:

$$m_e\left(v_r^2 + v_\psi^2\right)/2 + e\varphi(r) = \varepsilon_0 = \text{const} \tag{11.88}$$

$$v_z = \text{const} \; m_e v_\psi r = \mu = \text{const} \quad (11.89)$$

Finally, we have

$$\sin^2\theta_0 = S_0(\varepsilon, r) / \left[\xi^2 + \left(1 - \xi^2\right)\sin^2\psi\right] \quad (11.90)$$

where

$$S_0(\varepsilon, r) = \left[\varepsilon + e\varphi(r) - eV_0\right]/\varepsilon, \qquad \varepsilon = m_e v^2/2 = \varepsilon_0 - e\varphi(r) \quad (11.91)$$
$$\xi = r/r_0$$

For the discharge axis this relation takes the form

$$\sin^2\theta_1 \sin\psi = S_0(\xi, r) \quad (11.92)$$

which corresponds to the axially symmetric cone of losses with respect to OX axis. In the vicinity of the wall $\xi \longrightarrow 1$, $\sin^2\theta_0 = S_0(\xi, r)$ and the axis of the cone of losses becomes oriented along the radius vector $\vec{r}$.

For the low-pressure limit when the mean free path of electrons exceeds the radius of the discharge cell, there are no electrons inside the cone of losses and the velocity distribution function within the cone of losses is zero. The velocity distribution function outside the cone of losses may be regarded as isotropic because the oscillating field amplitude is much smaller than the stationary radial electric field.

If the mean free path is comparable to the discharge cell radius r_0, the cone of losses will be filled by elastically scattered electrons. The collisional filling of cones of losses will be calculated using the Liouville–Vlasov principle [154]. The distribution function of electrons along the electron trajectory in the phase space may be changed only due to collisions. The kinetic equation for the velocity distribution function is dependent only on the total electron energy ε_0 and the length of the electron trajectory ℓ from a given point. It will take the form ($\dot{\varepsilon} = 0$)

$$v(\ell)\left[\partial f(\varepsilon_0, \ell)/\partial \ell\right] = \Xi(f) \quad (11.93)$$

where

$$v(\ell) = \sqrt{2\left[\varepsilon_0 - e\varphi(r)\right]/m_e} \quad (11.94)$$

In this case of strong anisotropy of the fast electrons the Fokker–Planck elastic term in Eq. (11.70) is much smaller in comparison with the strong changes in the velocity vector of electrons described by the second term in Eq. (11.72). Inelastic collisions will not bring significant contributions because of the low inelastic collision frequency in comparison with the frequency for elastic collisions. Therefore we may write

$$\frac{\partial f(\varepsilon_0, \ell)}{\partial \ell} = \left[\overline{f(\varepsilon_0, \ell)} - f(\varepsilon_0, \ell)\right]/\lambda_e(\ell) \quad (11.95)$$

where $\lambda_e(\ell) = v(\ell)/\nu(\ell)$ is the mean free path of the electron.

The physical meaning of the first term on the right-hand side of Eq. (11.95) is the population of the trajectory by electrons from other trajectories as a result of elastic electron–atom collisions. The second term represents the collisional knocking-out of electrons of this trajectory. If the trajectory which originates on the wall is empty $[f(\varepsilon_0,\ell) = 0]$, then at a distance of the order of λ_e along the electron path the distribution increases to the mean value of the plasma. So in order to determine $f(\varepsilon_0,\ell)$ at any point it is necessary to know it for all locations in the plasma volume and for all trajectories. This fact makes the determination of $f(\varepsilon_0,\ell)$ a self-consistent problem. In order to solve it, one may use the isotropic distribution function for the central plasma $f_0(\varepsilon_0)$ as a first approximation for $\overline{f_0(\varepsilon_0,\ell)}$ in the iteration process. The solution of Eq. (11.95) in this case will take the form

$$f(\varepsilon_0,\ell) = f(\varepsilon,0)\exp\left[-\int_0^\ell \frac{d\ell'}{\lambda_e(\ell')}\right] + \int_0^\ell \exp\left[-\int_{\ell'}^\ell \frac{d\ell''}{\lambda_e(\ell'')}\right]\frac{\overline{f_0(\varepsilon_0)d\ell'}}{\lambda_e(\ell')} \tag{11.96}$$

where $f(\varepsilon_0,0)$ is the distribution function for the initial point on the trajectory for $\ell = 0$.

The next approximation is easily obtained by substitution of $\overline{f_0(\varepsilon_0,\ell)}$ from the first approximation in the right-hand side of Eq. (11.96). Using the fact that the mean free path of electrons in noble gases for the energy range under study varies very slowly ($\lambda_e \approx$ const), the solution in the first approximation will take the form

$$f(\varepsilon_0,\ell) = f_0(\varepsilon_0) + [f(\varepsilon_0,\ell) - f_0(\varepsilon_0)]\exp[-\ell/\lambda_e] \tag{11.97}$$

This solution accounts for the effect of collisional filling of the cone of losses moving from the wall to the center of the discharge.

Finally, the distribution function out of cones of losses was approximated as isotropic, $f_0(\varepsilon_0)$ and inside the cones by the quantities $f_1(\varepsilon_0)$ and $f_2(\varepsilon_0)$, which account for the collision filling of the cones formed by the influence of the nearest and opposite walls of the discharge tube with respect to the studied volume:

$$f_{1,2}(\varepsilon_0) = f_0(\varepsilon_0)\left[1 - \exp\left(-\frac{R \mp r}{\lambda_e}\right)\right] \tag{11.98}$$

The distribution of linear polarization across the image of the central part of the cylindrical high-frequency discharge for a low degree of alignment of the plasma using Eqs. (11.60), (11.63), and (11.64) may be represented as

$$P(x) = \frac{3}{10}(-1)^{J+J_1}(2J+1)^{1/2}\begin{Bmatrix} 1 & 1 & 2 \\ J & J & J_1 \end{Bmatrix}\frac{\Gamma_0}{\Gamma_2} \times \int_{\varepsilon_t}^\infty d\varepsilon\varepsilon\sigma^{(2)}(\varepsilon)P_1(\varepsilon,x)I_1(\varepsilon,x)\Big/\int_{\varepsilon_t}^\infty d\varepsilon\varepsilon\sigma^{(0)}(\varepsilon)I_1(\varepsilon,x) \tag{11.99}$$

where

$$P_1(\varepsilon,x) = \frac{1}{I_1(\varepsilon,x)} \int_x^R \frac{drr}{\sqrt{r^2-x^2}} \left[\frac{1}{2}\left(r - \frac{x^2}{r^2}\right) {}^{\vec{r}}f_{2+}^{(2)}(\varepsilon,r) + \sqrt{\frac{3}{2}\frac{x^2}{r^2}}\, {}^{\vec{r}}f_0^{(2)}(\varepsilon,r) \right] \tag{11.100}$$

$$I_1(\varepsilon,x) = \int_x^R \frac{drr}{\sqrt{r^2-x^2}} f_0^{(0)}(\varepsilon,r) \tag{11.101}$$

For the multipole moments of the velocity distribution function of fast electrons, using Eqs. (11.88) and (11.96) we have

$$f_0^{(0)}(\varepsilon,r) = \frac{2}{\pi} f_0(\varepsilon_0) \left[\gamma(r) \int_a^1 \frac{d\tau\sqrt{1-B(\varepsilon,r)}}{\sqrt{1-\tau^2}} + \frac{\pi}{2}(1-\gamma(r)) \right] \tag{11.102}$$

$${}^{\vec{r}}f_0^{(2)}(\varepsilon,r) = -\frac{5}{\pi}\gamma(r) f_0(\varepsilon_0) \int_a^1 \frac{d\tau B(\varepsilon,r)}{\sqrt{1-\tau^2}} \sqrt{1-B(\varepsilon,r)}$$

$${}^{\vec{r}}f_{2+}^{(2)}(\varepsilon,r) = \sqrt{\frac{2}{3}}\frac{5}{\pi}\gamma(r) f_0(\varepsilon_0) \int_a^1 \frac{d\tau\,(1-2\tau^2)}{\sqrt{1-\tau^2}} \left[2 + B(\varepsilon,r)\sqrt{1-B(\varepsilon,r)}\right]$$

where

$$a = \begin{cases} \left[(S_0-\xi)^2/(1-\xi)^2\right]^{1/2} & \text{for } \xi < S_0(\varepsilon,\nu) \\ 0 & \text{for } \xi > S_0(\varepsilon,\nu) \end{cases}$$

$$\tau = \sin\psi, \qquad B(\varepsilon,r) = S_0(\xi,r)/\left[\xi^2 - (1-\xi^2)\,\tau^2\right]$$

$$\gamma(r) = \frac{1}{2}\exp\left[-\frac{R}{\lambda_e}(1+\xi)\right] + \exp\left[-\frac{R}{\lambda_e}(1-\xi)\right]$$

In order to calculate polarization profiles for the given cross section, we require the energy dependence of the distribution function $f_0(\varepsilon)$. This function may be calculated by time averaging Eq. (11.86) over the period of the high-frequency field and substituting expression (11.80). Finally, we obtain the equation

$$\frac{1}{v^2}\frac{\partial}{\partial v} v^2 \left[\frac{1}{6}\frac{\nu_1(v) a_\sim^2}{\omega_\sim^2 + \nu_1(v)} \frac{\partial}{\partial v} + \gamma v \nu_1(v) \right] \left\langle f^{(0)}(v) \right\rangle = \nu^* \left\langle f^{(0)}(v) \right\rangle \tag{11.103}$$

Using the variable $\varepsilon = m_e v^2/2$ this equation takes the form

$$\frac{\partial}{\partial\varepsilon}\left[D(\varepsilon)\frac{\partial}{\partial\varepsilon} + V(\varepsilon) \right] \left\langle f^{(0)}(\varepsilon) \right\rangle = U(\varepsilon) \left\langle f^{(0)}(\varepsilon) \right\rangle \tag{11.104}$$

where

$$D(\varepsilon) = \frac{1}{3} a_{\sim}^{2} m_e v_1(\varepsilon) \varepsilon^{3/2} / \left(\omega_{\sim}^{2} + \nu_1^{2}(\varepsilon)\right) \tag{11.105}$$
$$V(\varepsilon) = 2\gamma \nu_1(\varepsilon)\varepsilon^{3/2}, \qquad U(\varepsilon) = \nu^{*}(\varepsilon)\varepsilon^{3/2}$$

while the boundary condition is $\left\langle f^{(0)}(\varepsilon)\right\rangle \varepsilon \underset{\varepsilon\to\infty}{\to} 0.$

Let us find the solution in the form

$$\left\langle f^{(0)}(\varepsilon)\right\rangle = A \exp\left[\int_{\varepsilon_t}^{\varepsilon} d\varepsilon' \eta\left(\varepsilon'\right)\right] \tag{11.106}$$

where A is a constant; $|\eta(\varepsilon)|$ characterizes the local profile of the distribution function connected to the local temperature of electrons with energy close to ε. For $\eta(\varepsilon)$ we may write the following equation:

$$D\frac{\partial \eta}{\partial \varepsilon} + D\eta^2 + \left(D' + V\right)\eta + V' - U = 0 \tag{11.107}$$

This equation is simpler for the numerical solution because only the first derivative of the function η is involved.

Introducing transformations we get

$$\frac{\partial \eta}{\partial \varepsilon} + [\eta(\varepsilon) - X_1(\varepsilon)][\eta(\varepsilon) - X_2(\varepsilon)] = 0 \tag{11.108}$$

where

$$X_{1,2}(\varepsilon) = -\left[V(\varepsilon) \pm \sqrt{V^2(\varepsilon) + 4D(\varepsilon)U(\varepsilon)}\right] / 2D(\varepsilon) \tag{11.109}$$

The numerical solution of the equation began from energies exceeding the threshold value ε_a. The first approximation $\tilde{\eta}(\varepsilon_a)$ was taken equal to $X_1(\varepsilon_a)$. Then, using the Runge–Kutta procedure a step-by-step solution of the equation (with step width h) to lower energies close to ε_t was realized. The result of this procedure is the approximate solution $\tilde{\eta}_{\varepsilon_a,h}(\varepsilon)$, which is stable with respect to small variations in the initial conditions and the value ε_a. The difference of the calculated function $\eta(\varepsilon)$ from $X_1(\varepsilon)$ was less than 6%. This may be a general feature of Eq. (11.107) for a slow variation in the coefficients. Finally, an approximate expression for the energy distribution function of the electrons was used:

$$\left\langle f^{(0)}(\varepsilon)\right\rangle = A \exp\left[\int_{\varepsilon_t}^{\varepsilon} d\varepsilon' X_1\left(\varepsilon'\right)\right] \tag{11.110}$$

11.3. Experimental Application of the Spectropolarimetric Technique to High-Frequency Discharges

11.3.1. Spectropolarimetric Determination of Energy Input into the Near-Electrode Region of a Capacitive Discharge

One of the principle problems of construction and optimization of the hf ELs is the increase of energy input into the plasma from the oscillator. The distribution of the plasma in the small discharge vessels is extremely nonhomogeneous, because the characteristic kinetic lengths of particles are comparable to the linear discharge dimensions and the boundary plasma-surface effects play a significant role in maintaining the discharge. Different electric and optical methods of plasma diagnostics indicated that the electrode sheaths of the E-discharge play an important role in its energy balance and optical efficiency [256–261]. Reliable remote sensing of such a plasma with the well-expressed structural features such as effective local energy input and inhomogeneous distribution of internal fields could not be implemented on the basis of intensity spectroscopy. Only the polarization of atomic particles could be utilized for sensing the structural parameters of the discharge for the purposes of optimization of spectral source regimes and constructions.

The spectropolarimetric method has been used to determine the energy transfer from an external generator through the electrode sheath into the gas-discharge plasma of a low-pressure hf electrodeless capacitive discharge, in a frequency range $\nu_i < \nu < \nu_e$ ($\nu_{i,e}$ being the ion and electron plasma frequencies). In this range the energy input from the external generator in the near-electrode region of the plasma is provided by Fermi acceleration of plasma electrons [252, 254, 262]. The magnetic spectropolarimeter used is described in detail elsewhere [263].

The discharge was excited by means of capacitive electrodes in a cylindrical glass cell 60 mm long and 40 mm in diameter, using generator voltages of 100–200 V at a frequency of 100 MHz. Helium and argon at pressures of 0.02–0.04 mbar showed for all spectral lines, independently of the excitation potential of the upper level of the transition, a maximum of the degree of polarization in the vicinity of the boundary of the near-electrode layer, induced by electron-impact excitation. The velocity distribution function of electrons in this region is composed of two parts: the velocity distribution $f_1(\vec{v})$ of electrons incident on the layer and having a negative z-component of the velocity, and the distribution $f_2(\vec{v})$ of reflected electrons for which $v_z > 0$ (with the z axis perpendicular to the plane of the capacitor plates). Under the collisionless conditions, $f_2(\vec{v})$ can be calculated with the aid of Liouville's theorem: $f_2(v_x, v_{z_2}, v_y, t_2) = f_1(v_x, v_{z_1}, v_y, t_1)$ where t_1 and t_2 are the moments of falling and flying out of an electron from the near-electrode sheath, while v_{z_1} and v_{z_2} are the velocity components of the electron at these

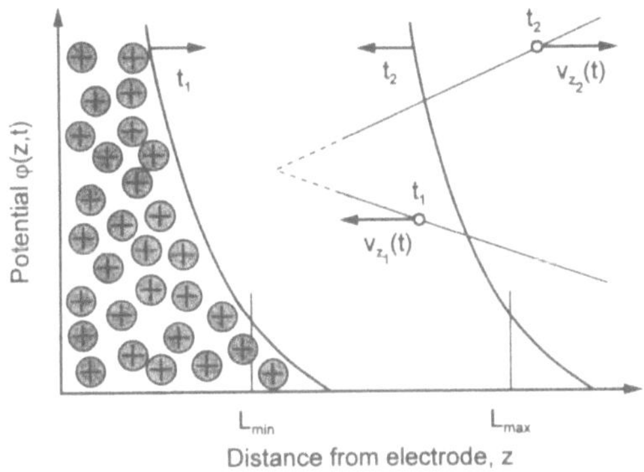

FIGURE 11.1. Model of the plasma–electron interactions with the oscillating potential profile of the electrode sheath of an hf capacitive discharge.

moments (Figure 11.1). Assuming that the distribution of the electrons incident on the boundary of the electrode sheath is stationary, one can write the expression for the distribution function of the reflected electrons, averaged over the field period $T = 2\pi/\omega$, in the form

$$f_2(v_x, v_{z_2}, v_y) = \frac{1}{T}\int_0^T dt_2 f_1\left[v_x, v_{z_1}(v_{z_2}, t_2), v_y\right] \tag{11.111}$$

The distribution function of the incident electrons in this formulation is represented as a complex function of v_{z_2} and its determination is reduced to the calculation of the type of electron trajectory during reflection from the oscillating near-electrode sheath.

The similar-approach computation of the distribution function at the nonstationary boundary of a near-electrode layer was performed for an argon discharge at a pressure of 0.036 mbar and generator voltage 200 V [253]. Polarization measurements on the 801.4, 763.5, and 706.7 nm lines have confirmed the results of these calculations. The degree of polarization of the indicated argon lines was determined in the near-electrode polarization maximum. The simplest model of a rigid layer boundary from which the incident electron is directly reflected was used under the assumption of a sawtooth time dependence of its coordinate. For groups of fast electrons, which are responsible for direct excitation of the atomic states studied and whose velocities satisfy the condition $v > 2u$ (u being the velocity of the layer boundary), multipole moments of the distribution function were obtained in [265]. The degree of polarization was calculated as a function of the parameter u/v_1, using the Born cross section for alignment by electron impact. The parameter u/v_1 was determined by comparing the calculated and experimental values. The proof of the adequacy of the approach was the agreement of the value of u based on data of three spectral lines, and good agreement between the velocity of beam electrons and the result of its direct measurement in the plasma by the magnetic deflection method [254, 265].

For calculating $\overline{f_2(\vec{v})}$ in a more realistic model, distributions of the potential $\varphi(y,t)$ and of the electric field strength $E(z,t)$ in the near-electrode region should be taken into account [254, 265, 266]. By considering the uniform distribution of ions in the discharge gap, the potential and the electric field strength are described by the expressions

$$\begin{aligned} &\varphi(z,t) = 2\pi e N_0 [z - L(t)]^2 \qquad (z < (L(t)) \\ &E(z,t) = -\partial\varphi/\partial z = z\pi e N_0 x [L(t) - z] \qquad (z < L(t)) \\ &\varphi(z,t) = E(z,t) = 0 \qquad (z \geq (L(t)) \end{aligned} \tag{11.112}$$

where $L(t)$ is the instantaneous coordinate of the layer boundary, N_0 the concentration of ions in the layer, and the potential is counted from the plasma potential outside the electrode sheath. The instantaneous value of the potential at the electrode is $V(t) = \varphi(0,t) = 2\pi e N_0 L^2(t)$.

If the impedance of the discharge gap exceeds the output resistance of the low-frequency generator, then, neglecting a small stationary electrode–plasma potential drop, the time dependence of the potential is: $V(t) = V_\sim(1 + \sin\omega_\sim t)$, where

$$L(t) = \sqrt{V_\sim(1 + \sin\omega_\sim t)/2\pi e N_0} = L_0\sqrt{(1 + \sin\omega_\sim t)/2} \tag{11.113}$$

Here $L_0 = \sqrt{V_\sim/\pi e N_0}$ is the lower boundary of the electrode positive space charge sheath (see Figure 11.1)

Such a regime may be called a potential regime. If the output resistance of the generator is appreciably higher than the impedance of the discharge, a current regime is created instead in which the current across the discharge gap and, correspondingly, the charge on the electrodes are sinusoidal. The thickness of the near-electrode layer in the current regime is determined from the condition of compensation of the positive space charge of the sheath by the charge on the electrode: $L(t) = L_0(1 + \sin\omega_\sim t)/2$ (Figure 11.1).

The electron trajectories in the electrode sheath $z(t)$ at different v_{z_2}, t_2, and hence different dependence $v_{z_1}(v_{z_2}, t_2)$, was computed by solving the equation of motion

$$m\frac{d^2z}{dt^2} = -eE(z,t) \tag{11.114}$$

in which $E(z,t)$ was determined as a function of the discharge regime. An assumption of a fairly high oscillation frequency of the nonrigid parabolic boundary of the electrode sheath, corresponds to the conditions of our experiments. An electron, during one oscillation period of the external field, may fail to receive sufficient momentum to escape from the electrode sheath and may remain there during a time longer than the field oscillation period, repeatedly interacting with the boundary. Phases of free flight outside the zone of positive space charge and of the accelerated motion under the action of the repulsive electric field are alternating until the moment t_2. The motion of an electron in the first phase is described by a linear time dependence $z(t) = L_0 - v_{z_1}(t - t_1)$, and the moment of interaction of the electron with the layer boundary is determined from $L_0 - v_{z_1}(t' - t) = L(t')$. Within the framework of the algorithm these equations were solved by successive approximations. The electron trajectory during the second phase of motion in the layer was calculated from the equation of motion with a time-dependent right-hand part. Within the current regime of the discharge, the following analytic solution of this equation exists:

$$z(t) = D_1 \sin(\omega_e t + \phi) + D_2 \sin \omega_\sim t \tag{11.115}$$

where $D_2 = L_0 \omega_e / 2(\omega_e^2 - \omega^2)$ and $\omega_e = (4\pi e^2 N_0 / m_e)^{1/2} = 2V^{1/2}/L_0 m_e^{1/2}$.

The constants D_1 and ϕ are determined by the initial conditions at time t', namely, $dz(t')/dt = v_y$, $z(t') = L(t) D_1 \omega_e \cos(\omega_e t' + \phi) + D_2 \omega_\sim \cos \omega_\sim t' = v_{z_1}$, and $D \sin(\omega_e t' + \phi) + D_2 \sin \omega_\sim t' = L(t')$. In final form we have

$$D_1 = \left[(v_z - D_2 \omega_\sim \cos \omega_\sim t)^2 / \omega_e^2 + \left(L(t') - D_2 \sin \omega_\sim t' \right)^2 \right]^{1/2} \tag{11.116}$$

$$\phi = \arctan \left[\omega_e \left(L(t') - D_2 \sin \omega_\sim t \right) / (v_z - D_2 \omega_\sim \cos \omega_\sim t') \right] - \omega_\sim t'$$

For the potential regime there exists no analytic solution of the equation of motion of the electron, and therefore the trajectories were calculated using Störmer's formula [267] within automatic steep-width selection. An indicator of reliability was the fact that the functions $\overline{f_2}(\vec{v})$ in the current regime, which were obtained numerically and analytically, differed by less than 10%.

In calculating $\overline{f_2}(\vec{v})$, the time integration was performed by an approximate numerical method of rectangles on a uniform grid with an automatic pitch selection. The automatic steep-width selection was done by dividing the steep-width factor by two within an interval, until the relative error of integration $\delta = \left| (X_{h/2} - X_h) / X_{h/2} \right| / \left(M_{1/2-1} \right)$ became smaller than the given value of δ_0. Here X_h is the numerical value of the integral for a grid pitch h, and $M_{1/2}$ is the index of increase in integration accuracy when the pitch is divided by two. In our case, the values $M_{1/2} = 8$ and $\delta_0 = 0.01$ were selected.

For the most typical values of the parameters of the electrode sheath and electron temperatures, the calculated function $\overline{f_2}(\varepsilon)$ is characterized by the following features. For a small parameter $\beta = L_0 \omega_\sim / 2v_t$, the velocity distribution of the reflected electrons is similar to the distribution of the incident particles, which is taken as Maxwellian, for simplicity. This actually corresponds to reflection from a stationary potential barrier. In the range of values $\beta = 1$, the concentration of slow particles of the distribution with $v_{z_2} < v_t$ decreases effectively in comparison to its equilibrium value, but in addition beams of suprathermal electrons are formed [256, 257, 261], represented as the nonmonotonic regions of the velocity distributions or "humps" which shift to higher velocities when β increases [254]. When $\beta \leq 2$, the character of the distribution of the reflected particles depends slightly on the degree of rigidity of the boundary (parameter $4\pi e N_0$). The finite rigidity of the boundary begins to play a significant role mainly at values $\beta \geq 3$. Further on, for high values of β, this effect of the energy enrichment of reflected electrons disappears, and when $\beta \simeq 10$ the distribution function of the reflected electrons also tends to an equilibrium function coinciding with the distribution of the incident particles. This can be explained, since at high velocities of the sheath boundary an electron moves under the action of the repulsive field averaged over the period, and it corresponds to the case of reflection from a stationary potential barrier. As follows from direct calculations, the effect of acceleration in the near-electrode region is more effective at the current regime.

A numerical calculation of the degree of polarization in the vicinity of the near-electrode layer was conducted using the Fermi acceleration mechanism in [268, 269] for noble gas spectral lines. For the helium 492.2 nm line, the energy dependence of the degree of excitation was approximated by a semiempirical expression $\sigma(\varepsilon) = \sigma' \left[\varepsilon/\varepsilon_t^{-1}\right]^{1/2}$ from [270], in which $\sigma' = 2.93 \times 10^{-20}$ cm^2 was calculated using data of [271].

The threshold excitation energy of the upper level of the transition studied, $4\ {}^1D_2$ (23.72 eV), was taken from [270, 271], and the alignment cross section as a function of energy $\sigma^{(2)}(\varepsilon)$ was determined on the basis of its relationship to the dependence of the degree of polarization on energy during excitation by an electron beam, $P_0(\varepsilon)$ [272]: $\sigma^{(2)}(\varepsilon) = 0.017\,\sigma(\varepsilon)\left[P_0^{-1}(\varepsilon) - 1/3\right]^{-1}$.

With allowance for the actual accuracy of the experiments in determining $P_0(\varepsilon)$ [271], in calculations of the alignment cross section $\sigma^{(r)}(\varepsilon)$ the following piecewise approximation was used: $P_0 = 0.6$ for $23.72 \leq \varepsilon \leq 24.5$ eV, then a linear change $0.35 \leq P_0(\varepsilon) \leq 0.45$ eV in the range $24.5 \leq \varepsilon \leq 28.0$ eV, and $P_0 = 0.45$ eV for $28.10 \leq \varepsilon \leq 50$ eV. The calculated degree of polarization at the boundary of the electrode sheath for the potential regime of the discharge used, for a given value of the electron temperature, showed a maximum that corresponds to optimal conditions of the Fermi mechanism of energy transfer [263]. The degree of polarization decreases when T_e increases, since the degree of anisotropy of the motion of the electrons decreases.

In the calculations of transferred power W, the integration with respect to the velocity was performed by the trapezoidal method over a nonuniform grid:

$$W = \frac{mSe}{2T} \int_0^x dv_z \int_0^\infty dt_1 \left[v_{z_2}(v_{y_1}, t_1) - v_z^2\right] v_{z_1} f_1(v_{z_1}) \tag{11.117}$$

In the integration interval from 0 to ∞, we chose a node distribution law that provided a minimum number of nodes for a given precision of integration. Calculations of power from Eq. (11.117) confirmed the behavior noted above: At constant electron temperature, the power transferred through the near-electrode layers reaches a maximum at $u \sim v_e$, i.e., in the most efficient regime for the Fermi mechanism of electron acceleration. The transferred power exhibits a tendency to increase with the temperature of the plasma electrons, which is a secondary effect reflecting an increase of their flux to the layer.

As a result the spectropolarimetric determination of the power transferred to the plasma of a high-frequency capacitive discharge was performed in helium for 0.18 mbar and 55 V applied voltage and this value was found to be (0.4 ± 0.05) W.

11.3.2. Character of the Motion of Electrons in Electrode Regions of a Capacitive High-Frequency Discharge

Despite appreciable progress in research on the capacitive high-frequency discharge simulated by applications in optical engineering, the physical processes in electrode regions have not yet investigated sufficiently [134]. A simplified assumption of an oscillatory motion of electrons with amplitude determined by the intensity of the high-frequency field has been advanced [133]. In [273] the motion of the ions was examined. Duplication of electrons in the electrode layer was reported [274], and a numerical analysis of the temporary structure of the flow of charged particles on the electrodes was given [275]. If the amplitude of fluctuations in the electrons is smaller than the thickness of the electrode region, an oscillatory motion of electrons will predominate. If the time during which electrons remain within the electrode sheaths is smaller than the duration of the field half-cycle ($\tau < T/2$) and the electric field in the sheath is not sign-variable, a unidirectional movement from the electrodes to the central regions takes place. The latter situation can be realized in a capacitive hf discharge, if their significant own electrostatic fields are present due to the nonlinear voltage–current characteristics of the electrode sheath and acceleration of the electrons appearing on the electrodes as a result of γ-processes on electrodes [256–259, 261, 276]. For higher field frequencies a unidirectional motion is related to acceleration of plasma electrons by the electrode oscillatory potential barrier (Fermi acceleration).

The nature of electron motion in electrode regions was compared for a hf discharge and for the cathode fall of a dc glow discharge [277]. In both cases a unidirectional motion of electrons is realized, and a characteristic distribution of the luminescence intensity $I(z)$ is formed. Dc and hf discharges were excited in the same discharge tube of diameter 40 mm, with flat internal electrodes (Figure 11.2). When monotonically increasing the hf voltage ($V_\sim$), the structure of the luminescence distribution in the electrode region essentially changes

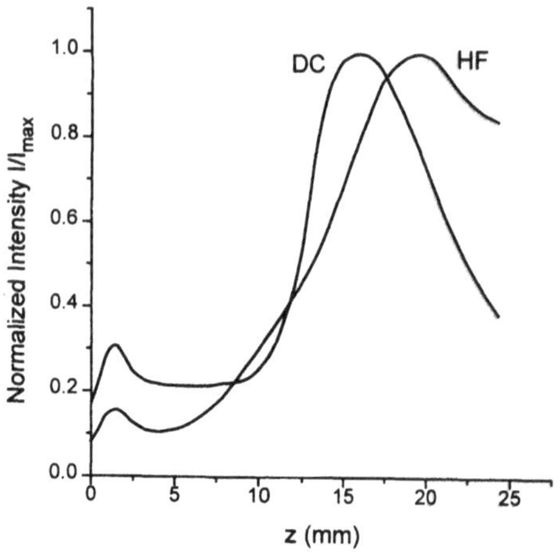

FIGURE 11.2. Spatial distribution of the total luminescence intensity in the electrode region of a capacitive high-frequency discharge and in the cathode fall region of a dc discharge in the same tube. Ne, $p = 0.033$ mbar. Capacitive high-frequency discharge: $f_{\sim} = 4$ MHz, $V_{\sim} = 500$ V. Dc discharge: $V = 700$ V.

(Figure 11.3). At low hf voltage the distribution $I(z)$ has a diffuse character, decreasing monotonically from the center of the discharge to the electrodes, which corresponds to a discharge, maintained exclusively by the hf-field (so called α-discharge).

When increasing $V_{\sim} > 100$ V in the electrode region, the value and role of the electrostatic fields in the electrode regions increases, and the distribution $I(z)$, averaged in time, becomes identical to the intensity distribution in the cathode fall of the dc discharge (Figure 11.2). The latter fact allows one to assume an analogous character of the electron motion in the electrode regions of the capacitive high-frequency and dc discharge, i.e., to assume a predominant unidirectional movement of electrons from the electrode to the center of the discharge. A confirmation of such an assumption could be an estimate of the average time of stay of electrons within the electrode sheath, being emitted from the electrode ($\tau \geq 10^{-8}$ s), which is less than a half-cycle of the hf field ($T/2 = 2.5 \times 10^{-7}$ s).

The difference between the hf and dc discharges in relation to the electron motion in the electrode region is that an oscillating electrostatic field [257] is present in the electrode region of the capacitive hf discharge. The nature of the intensity oscillations in the electrode regions of the capacitive high-frequency discharge was determined in [277] for the following points: in the cathode luminescence, in the dark cathode space, and in the region of a "negative luminescence" (points a-f, Figure 11.3a). At some points a significant asymmetry of oscillation of the luminescence intensity for the positive and negative half-periods of the hf field (Figure 11.3b) was observed. Such effects could be explained in the following manner. The large maximum in the negative half-cycle of the field was stipulated by an active excitation of gas by a flow of electrons from the electrode, accelerated by a maximum electrostatic field. In the other half-cycle the accelerating field was practically absent [257] and the luminescence was weak.

The second (which is counted from the electrode) bright region of the capacitive high-frequency discharge is analogous to the negative glow region of the dc discharge. It is possible to assume by analogy to the dc discharge that the border of the electrode sheath in the capacitive high-frequency discharge z is between the points d and e. The amplitude of displacement of the border of the sheath for a half-cycle of a field $\Delta z \sim (T/2)v_i$ reaches 2 mm, i.e., it is close to the value of the sheath thickness. The stationary distribution of radiation intensity $I_\lambda(z)$ of the spectral lines along an axis of the discharge was studied together with the degree of polarization $P(z)$ in the plasma of the high-frequency capacitive electrodeless discharge under pressure laws of noble gases for $f_\sim = 100$ MHz and a discharge cell dimension $l = 60$ mm and $d = 40$ mm.

For all studied lines $I_\lambda(z)$ was presented as a curve, increasing from an electrode to the center of discharge with a maximum at a certain distance from the electrode (Figure 11.4a). At decreased pressure the clearly expressed maximum disappeared, and the $I_\lambda(z)$ function was represented by a monotonic curve increasing to the center of the discharge.

Under the same cell conditions a sharp polarization maximum of the studied spectral lines was obtained, located closer to the electrode than the maximum of the radiation intensity (Figure 11.4 and 11.5). At the lowest pressure, when function $I_\lambda(z)$ did not exhibit any maximum, the electrode polarization maximum was still sharp and well pronounced.

A comparison of the coordinate polarization profiles $P(z)$ with the general nature of the discharge emission yielded some qualitative features of the electron motion. At pressures below 0.015 mbar a dark electrode space was observed near the electrode, the thickness of which was about several millimeters and increased with decreasing of pressure. The brightest region of the discharge emission was located directly at the border of this dark space. This bright region stretched to the center of the discharge with decreasing cross section (Figure 11.4b). Superposition of a weak magnetic field on the discharge volume has allowed us to find that the bright luminescence region was caused by excitation due to a beam of electrons, formed at the border of the electrode sheath and extending to the center of the discharge [261]. Localization of the near-electrode polarization maximum at the border of the dark space (Figure 11.4a) implies that the beam of electrons is formed at the boundary section of the electrode sheath.

Under conditions, when the radiation spectrum and its polarization are realized subject to direct excitation by the fast beam electrons, a comparison of functions $I_\lambda(z)$ and $P_\lambda(z)$ enables one to obtain information on the nature of the electron beam distribution in the plasma. Hence the function $I_\lambda(z)$ reflects energy relaxation of beam electrons during their movement out from the electrode sheath

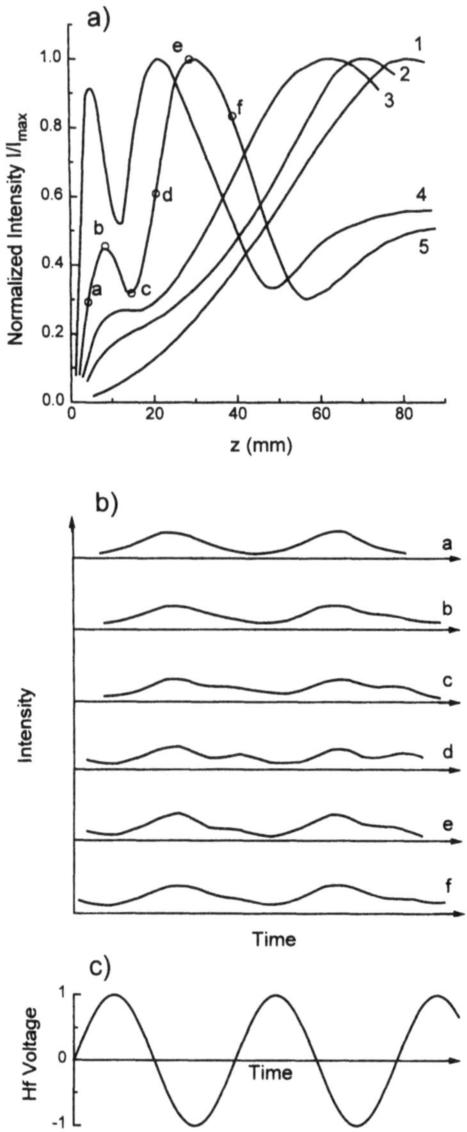

FIGURE 11.3. (a) Spatial distribution of the intensity $I(Z)/I_{max}$ in the electrode region of the capacitive hf discharge. Ne, $p = 0.33$ mbar, $f_{\sim} = 2$ MHz). $V_{\sim}$: 1, 100 V; 2, 300 V; 3, 600 V; 4, 900 V; 5, 800 V. a–f: points at which the luminescence oscillations of Figure 11.3b were observed. (b) Variation in the totally emitted intensity with time, for the conditions of curve 4. Curves a–f correspond to distances z from the electrode indicated in (a). (c) Time dependence of the hf voltage.

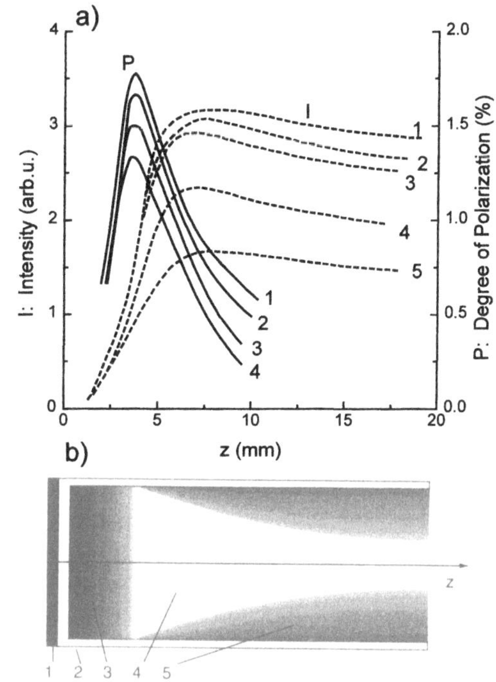

FIGURE 11.4. (a) Spatial distribution of the intensity $I(z)$ (dotted lines) of the Ar I-line $\lambda = 603.2$ nm and the polarization $P(z)$ (full lines). Ar, $p = 0.03$ mbar, $f_\sim = 100$ MHz. $V_\sim$: 1, 210 V; 2, 180 V; 3, 160 V; 4, 85 V; 5, 40 V. (b) General nature of the luminescence of capacitive high-frequency discharge under the same conditions at $V_\sim = 210$ V: 1, electrode outside the cylindrical tube; 2, glass tube; 3, dark space; 4, bright fluorescence zone; 5, decreasing cross section of the bright region.

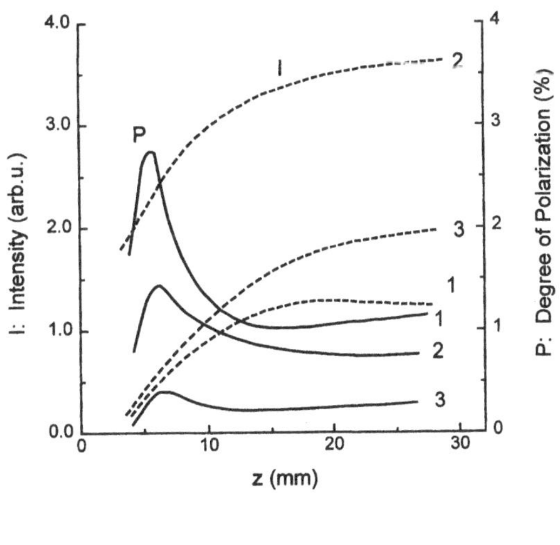

FIGURE 11.5. Spatial distribution of the intensity $I(z)$ (dotted lines) of the He I-line $\lambda = 492.2$ nm and the polarization $P(z)$ (full lines). $f = 190$ MHz. Helium pressure: 1, 0.027 mbar (mean free path of the electrons $\lambda_e = 0.04$ m), $V_\sim = 110$ V; 2, 0.053 mbar ($\lambda_e = 0.02$ m), $V_\sim = 110$ V; 3, 0.21 mbar ($\lambda_e = 0.05$ m), $V_\sim = 30$ V.

border. The decrease in $I_\lambda(z)$ after a maximum to the center of the discharge for all researched gases (see, for example, Figure 11.4a) is slower than expected, considering that it results in elastic scattering. Such behavior could be related to the influence of the beam formed by the opposite electrode sheath. The polarization profile reflects the momentum relaxation of electrons while the beam from the electrode sheath passes to the center of the discharge. Comparing $I_\lambda(z)$ and $P_\lambda(z)$, one can see that the momentum relaxation of the beam electrons proceeds essentially faster than the energy relaxation. The decrease in the $P_\lambda(z)$ recession in the vicinity of an electrode cannot be explained by elastic electron–atom collisions. The decreased length of $P(z)$ is smaller to an order of magnitude than the mean free path of the electrons under these experimental conditions. The beam–plasma instability effects, which provide effective momentum relaxation of the beam electrons, play an important role in this case.

Under the same discharge conditions, the measured average speed of the electrons beams formed in the electrode regions were measured by an optomagnetic method. For this purpose, the radiation of a small nonaxial volume of the discharge was projected to the entrance slit of a monochromator by an appropriate diaphragm. A magnetic field, oriented in the direction of observation perpendicular to the axis of the discharge tube, was imposed on the discharge. The intensity of radiation of the extracted spectral line was detected by a photomultiplier as a function of the magnetic field strength. The signal of the multiplier was amplified and recorded by a measuring computer. A sawtooth, steeply growing current was sent through the Helmholz coils, and the slope of the spectral line intensity was monitored as a function of the field strength. The maximum of this function corresponded to the value of the field, deflecting the beam at a chosen point of the volume. The nature of the displacement of the bright region by the magnetic field revealed that the beam electrons were rather monoenergetic.

Using the formula for the Larmor radius ρ, with known magnetic field intensity and coordinates of the selected observation location (x_0 is counted from the axis of the tube, z_0 from the electrode), one obtains the value of the average velocity of the directed fast electrons: $\overline{v_b} = 1.76 \times 10^8 \overline{B}(x_0^2 + z_0^2)/2z_0$ (B in mT). The results of the measurements, which depend on the amplitude of the high-frequency voltage, are shown in Figure 11.6.

The observed monotonic increase in velocities with increase in $V_\sim$ is explained for the well-investigated frequency range of the order of 1 MHz within frameworks defining the concept of electron acceleration in the near-electrode region by the electrostatic field in the electrode sheath V_0. For a frequency of 100 MHz at which the Fermi acceleration mechanism is realized [253, 254, 260], the $v_B(V_\sim)$ dependences were studied for different pressures of helium. When the pressure decreased the dependence became weaker, and disappeared at $p < 0.13$ mbar, confirming the Fermi mechanism of electron acceleration in the electrode sheath.

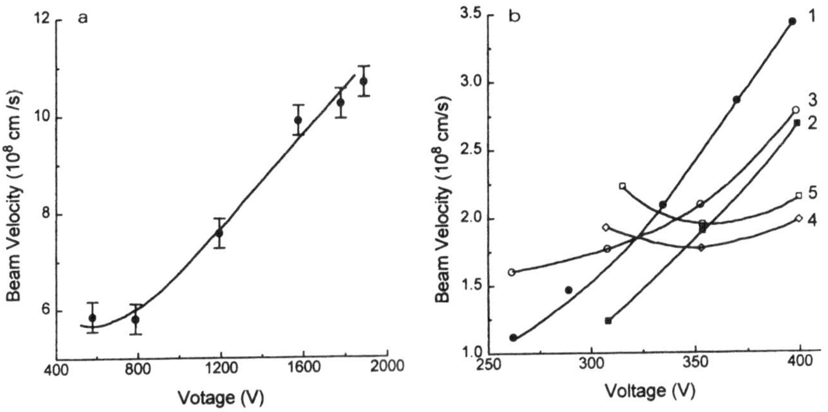

FIGURE 11.6. Dependence of the averaged velocity v_e of the electron beams on the hf field voltage. (a) $f_\sim = 1$ MHz, helium pressure 0.27 mbar; (b) $f_\sim = 100$ MHz, helium pressure: 1, 0.29 mbar; 2, 0.19 mbar; 3, 0.13 mbar; 4, 0.11 mbar; 5, 0.07 mbar.

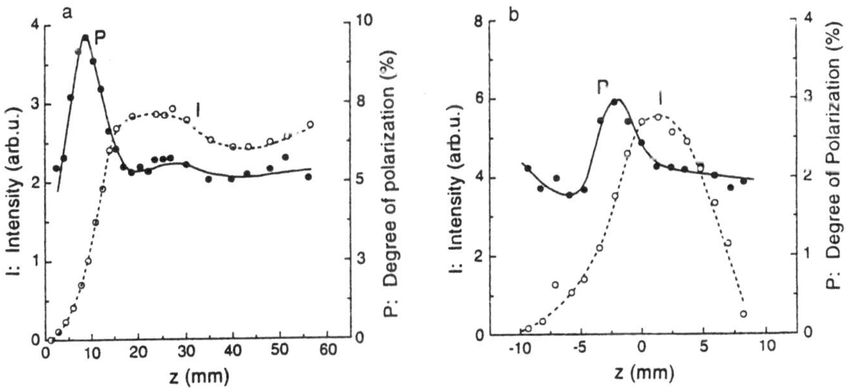

FIGURE 11.7. Spatial distribution of the intensity $I(z)$ (dotted line) of the He I-line $\lambda = 492.2$ nm and polarization $P(z)$ (full line) along the axis of the discharge for different discharge lengths. $f_\sim = 100$ MHz, $V_\sim = 190$ V. (a) Discharge length $L = 120$ mm, helium pressure 0.04 mbar; (b) $L = 20$ mm, helium pressure 0.09 mbar. The asymmetry results from accidental differences in the shape of the electrodes and the glass bulb.

11.4. Kinetics of Electrons in the Capacitive High-Frequency Discharge

Detailed spectropolarimeter measurements have been carried out to study the electron motion anisotropy in the capacitive high-frequency discharge. Observations on helium lines have been carried out in a cylindrical hf capacitive discharge maintained in a tube of 6 mm diameter with a variable length at $f_\sim = 100$ MHz [262].

The first group of experiments was devoted to polarization features of radiation at a large interelectrode distance, when the effect of one electrode region on the other could be neglected. The results on the He-line $\lambda = 492.2$ nm for a pressure of 0.04 mbar and with interelectrode distances of $L = 120$ mm are shown in Figure 11.7a.

As usual the dark space is observed in the vicinity of an electrode and at the border of the dark space a maximum of the degree of linear polarization is located. The degree of polarization decreases quickly when moving to the center of the discharge, reflecting a fast process of momentum relaxation of the beam electrons, and reached a constant value. This leads to the conclusion that in the central regions of the discharge there exists an additional anisotropy of the electron motion not connected with near-electrode processes. This polarization behavior is caused by penetration of the variable electrical field of the external generator through the near-electrode layers into the central region in order to maintain the discharge. The amplitude of the field can be evaluated from the criterion of the power balance [278] and

$$E_\sim^0 \sim \sqrt{m_e/M}\ \nu_{ea} v_T m_e/e \tag{11.118}$$

where M is the mass of neutral atoms, ν_{ea} is an effective frequency of the electron–atom collisions, and v_T is the thermal velocity of the electrons. Estimates of this value together with measurements in close conditions [279] give an $E_\sim^0$ value in the order of several V/cm. Nevertheless the electron motion anisotropy, which leads to the observed polarization in the central region of the plasma, is rather high. We think that besides the variable field, an additional reason for the fast electron anisotropy is connected with the formation of loss cones at the expense of a constant jump in potential at the lateral walls of the discharge chamber [255].

Qualitative interpretation of the obtained results in the electrode region was carried out with help of computer simulation using a model of a semi-infinite equilibrium and quasi-neutral plasma, located in the vicinity of an infinite flat electrode assuming the absence of collisions within the electrode sheath. The mean free path of the electrons was assumed to be larger than the characteristic length of the electrode sheath. The problem was considered as cylindrically symmetric with respect to an axis, perpendicular to the plane of the plates of electrodes, and one dimensional, as the forces acting on charged particles were directed orthogonally to the electrode surface, along the axis of the discharge. The frequency of the external field $f_{\sim}$ in the calculations was assumed to be much lower than the electronic Langmuir frequency, corresponding to the quasi-neutral plasma of the central part of the discharge with a Maxwellian distribution function.

The two-dimensional phase space z, v_z (where z was the distance up to the electrode, v_z the component of the electron velocity) was divided into 128×128 cells, and to each cell a certain value of the electron distribution function $f(v_z, z)$ was assigned. The algorithm realized in the program allowed the Liouville–Vlasov kinetic equation $df(v_z, z)/dt = 0$ to be solved. In this equation the differentiation was carried out along the trajectories of the electrons in phase space. At each step of the program, the shift of the phase space cells, proportional to their velocity and acceleration, was carried out. The acceleration was calculated by $a = -eE(z)/m_e$. The value of the field $E(z)$ was determined by the numerical solution of the Poisson equation $\partial E(z)/\partial z = \rho$, where $\rho = e(n_i - n)$ while n_i and n are the concentrations of ions and electrons, respectively. The concentration of ions was assumed to be constant and independent of time. The concentration of electrons was determined as the sum of the values of the distribution function in all cells of a line, corresponding to a given coordinate z. The initial value of the field on the electrode was determined from the condition $E(0) = \delta$, where δ was the charge density on the electrode. This charge density was modulated by the external current on the electrode with allowance for the current of the plasma ions and electrons, which penetrate the electrode potential barrier.

The results of this numerical simulation experiment show that while the boundary of the electrode sheath is moving to the center of the discharge, a pulsed reflected ejection beam is formed, moving from the electrode with an average velocity exceeding the thermal velocity of falling particles. This beam was observed directly [261] at low gas pressure and the polarization effect indicated rapid momentum relaxation of the beam electrons [254]. The velocity distribution function of the reflected electrons by the electrode sheath in the vicinity of its boundary, averaged over a period of the alternating electric field, is shown in Figure 11.8.

Polarimetric measurements for hf discharges with small interelectrode distance have been implemented to simulate the kinetic features of electrons in spectral lamps. The polarization maximum here was also observed, as can be seen in Figure 11.7b for the minimum interelectrode distance realized, 20 mm. Further reduction in interelectrode distance for this type of discharge down to values in the order of the double thickness of the dark electrode space leads to discharge extinction. The polarization effect shows that in order to study

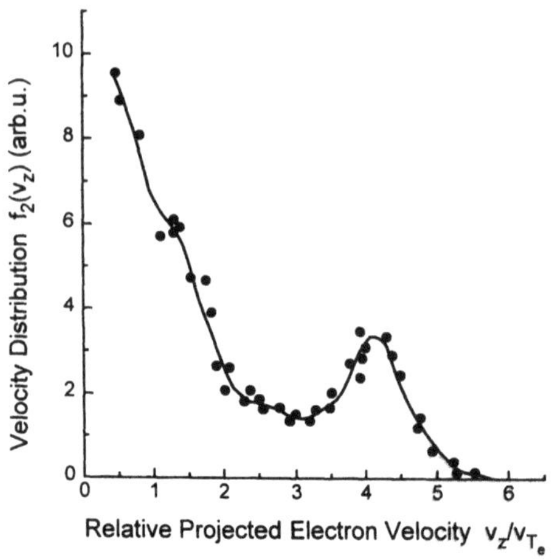

FIGURE 11.8. Velocity distribution function of the reflected electrons, averaged over one hf field period. The electrons are reflected from the oscillating boundary of the positively charged electrode sheath of a capacitive discharge. v_z is the velocity in the discharge axis direction, and v_{Te} is the mean velocity of thermal electrons.

processes near one electrode in the case of closely-located electrodes, it is impossible to neglect the influence of the opposite one. In the numerical model described above, a second electrode was introduced. The external current on this electrode was modulated in antiphase with respect to the current of the first one. A field of opposite direction from the opposite electrode, accelerating electrons from the second to the first electrode, was also taken into account. It was assumed that at the initial moment there exists a plasma with a Maxwellian distribution of electron velocities in the interelectrode interval. Results of the calculation show [265] that near the electrodes the clearly expressed oscillating electrode sheaths are formed. Between them there exists a region of the quasi-neutral plasma which oscillates with the frequency of the external electric field. The distribution of the z-projections of the electron velocities in various phases of the external field fluctuations are indicated in Figure 11.9. It has a quasi-Maxwellian nature, but displaced by the value of the velocity of the electrode sheath boundaries. In this case the velocity distribution of electrons does not possess the form of regular directed beams against the background of a general Maxwellian ensemble of thermal electrons; it is closer to the oscillating Maxwellian ensemble of electrons with the frequency of an external field.

The potential difference between the electrodes $V_\sim$ is obtained from the potential drops in the electrode sheaths, V_{S1} and V_{S2} and in the region of the quasilinear plasma, V_p:

$$V_\sim = V_{S1} + V_{S2} + V_p = 2\pi en\left(S_1^2 - S_2^2\right) - E_p\left(L - S_1 - S_2\right) \tag{11.119}$$

where E_p is the electric field strength at the center of the discharge, in the region of the quasi-neutral plasma: $E_p = 4\pi(\sigma_1 + neS_1)$ (σ_1 is the charge density on the first electrode). Taking into account, that $S_1 + S_2 = 2S_0 = \text{const}$ (S_0 is

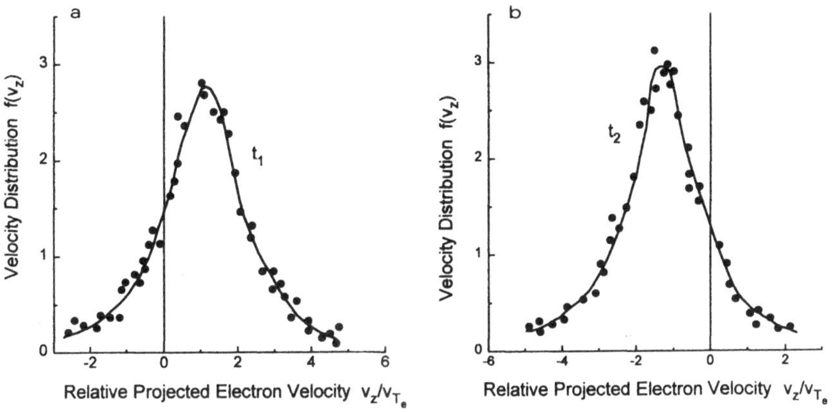

FIGURE 11.9. Velocity distribution function of the electrons at times. (a) $t_1 = 0.95T$, (b) $t_2 = 1.2T$ ($T = 1/f_{\sim}T$, hf field period) for small interelectrode distances. v_z is the velocity in the discharge axis direction and v_{Te} is the mean velocity of thermal electrons.

the mean thickness of the electrode layer), we obtain $V_{\sim} = 8\pi neS_0(S_1 - S_0) - 4\pi(\sigma_1 + neS_1)(L - 2S_0)$. Considering that the electronic cloud is oscillating as a whole under the effect of the internal field in the central region, we find $\ddot{S} = -eE_p/m_e = -4\pi\left(e\sigma_1 + ne^2S_1\right)/m_e$. If we set $S_1 = S_0 + S\exp(i\omega_{\sim}t)$ and $\sigma_1 = \sigma_0 + \sigma\exp(i\omega_{\sim}t)$, where σ_0 is the average value of the charge density on the electrode, we have $S_{\sim} = 4\pi\sigma e/m_e\left(\omega_{\sim}^2 - \omega_\ell^2\right)$ and $S_0 = -\sigma_0/ne$; here $\omega_\ell = \left(4\pi ne^2/m_e\right)^{1/2}$ is the Langmuir frequency. The expression for the impedance of the hf discharge ($Z = V_{\sim}/I, I = \sigma$) in the collisionless case assumes the following form:

$$Z = Z_s + Z_p = \frac{4\pi}{i\omega_{\sim}} \frac{2\omega_e^2 S_0 - \omega_{\sim}^2(L - 2S_0)}{\omega_{\sim}^2 - \omega_e^2} \tag{11.120}$$

where Z_s is the series impedance of the electrode layers and Z_p is the impedance of the central volume of the neutral plasma.

When $\omega_{\sim} \ll \omega_e$ and $S_0 \ll L$ this formula coincides with that obtained in [280]. The given expression specifies two types of resonances in a high-frequency discharge. The first of them, considered in detail in [280], takes place at frequency $\omega^2 = 2\omega_e^2 S_0/(L - 2S_0)$ and corresponds to zero impedance, $Z = 0$. The resonance of the second type takes place at $\omega = \omega_e (Z = \infty)$ or for an external source of the field with infinite internal resistance (ideal current source).

The electric field E_p inside the discharge can oscillate in antiphase to the external field $E_{out} = V_{\sim}/L$. In fact, the relationship between the potential drop in the region of the quasi-neutral plasma and the interelectrode potential difference is

$$V_p/V = Z_p/Z = -\omega^2 (L - 2S_0) / \left[2\omega e^2 S_0 - \omega^2 (L - 2S_0)\right] \tag{11.121}$$

In the considered case $\omega_e \ll \omega$, so we have $V_p/V_\sim < 0$.

This fact was checked by computing the electric field in the central part of the discharge interval. At some moments, the internal field in all cases appeared to be directed opposite to the external one. With allowance for electron–atom collisions, the phase shift between external and internal fields in the discharge can differ from π. Actually, the expression for the impedance in this case is

$$Z = \frac{4\pi}{i\omega_\sim} \frac{2S_0\omega_\ell^2 - (\omega^2 - i\nu\omega_\sim)(L - 2S_0)}{\omega_\ell^2 - \omega^2 + i\nu\omega_\sim} \tag{11.122}$$

Thus, the developed numerical model and the results of polarization experiments in the hf capacitive discharge for a small distance between the electrodes specifies that the anisotropy in the central part of the discharge is determined mainly by the oscillatory motion of the quasi-Maxwellian ensemble of electrons.

The degree of polarization of the plasma emission is determined by the quadrupole moment of the velocity distribution function [248], in the following way:

$$f^{(2)}(v) = A \exp\left(\frac{v^2 + \dot{S}^2}{v_T^2}\right) \int_{-1}^{1} d\alpha \left(3\alpha^2 - 1\right) \exp\left(-2\alpha v S / v_T^2\right) \tag{11.123}$$

Here A is a proportionality factor, v the speed of the electrons, $\dot{S}$ the velocity of the border of the electrode sheath of the space charge, while α is the cosine of the angle between the direction of the velocity vector and the OZ axis.

Analysis of the polarization effects together with the results of numerical model yield conclusions concerning the nature of processes in the high-frequency electrodeless discharge. The external field is almost completely shielded by the near-electrode sheath of the positive space charge. At rather large distances between electrodes, an effective transfer of energy from the external source occurs as a result of the inelastic interaction of the electrons with the oscillating electrode sheath. In the most optimum cases electronic quasi-beams, directed to the center of the discharge are formed. Electron beams, propagating to the center of the discharge, undergo effective momentum relaxation owing to collective plasma interactions. Outside the region of beam penetration (in the central part of the discharge) a region of quasi-neutral plasma exists, close to the positive column of a dc discharge. The internal alternative field in the central part of the discharge is significantly smaller than in the electrode region, but high enough to maintain the balance of the number of charged particles and energy in this region.

For distances between the electrodes smaller than the momentum relaxation length of the beam electrons, the velocity distribution of the electrons in the interelectrode space acquires a nonstationary quasi-Maxwellian form, shifted in velocity space by the value of the velocity of the electrode sheath borders.

12
Conclusions

Spectroscopy is a very wide field of physics, and it is very important to select an appropriate methodology when attempting to solve practical problems. This book addressed problems of designing, engineering, and improvement of sophisticated devices, using as their central part a high-frequency electrodeless gaseous discharge. The spectroscopic methods were used here not only for basic research such as precise determination of atomic constants, collisional parameters, investigations of electron and ion velocity distributions, and mechanisms leading to the frequency distribution of the emitted spectrum. In addition, practical spectroscopy was used to optimize spectral sources and manufacture with high reproducibility bulbs filled with exactly defined amounts of metals and noble gases.

The authors believe that the approach developed in this book, where basic scientific treatment proceeds in combination with very concrete and technologically oriented techniques and engineering problems, may be extremely fruitful in many other problems for which a real breakthrough in metrology and technology is sought.

References

[1] Ivanov, N. P. (1965): The atomic-absorption analysis, *Methods of analysis of chemical reactions and preparations* **10**, 21–23.

[2] Lvov, B. V. (1966): *The Atomic-Absorption Spectral Analysis*, Nauka, Moscow.

[3] Zaidel, A. N. (1985): *The Atomic Fluorescent Analysis*, Nauka, Moscow.

[4] Cohen-Tannoudji C. (1961): Quantum theory of optical pumping, in: *Advances in Quantum Electronics*, pp. 114–119, Columbia University Press, New York.

[5] Skudra, A. J. (1992): *Research and development of high-frequency electrodeless lamps of helium and mercury*, Thesis, University of Riga.

[6] Berzina, D. K., and Skudra, A. J. (1983): Research of optical parameters of high-frequency electrodeless lamps of helium, in: *Processes of Transfer of Energy in Vapor of Metals* (Kraulina, E., ed.), pp. 162–167, Latvian State University Publisher, Riga.

[7] Jackson, D. A. (1933): The hyperfine structure of the lines of the arc spectrum of rubidium, *Proc. Roy. Soc. (London) Ser. A* **139**, 673–682.

[8] Jackson, D. A. (1933): The hyperfine structure of the lines of the arc spectrum and the nuclear spin of indium, *Z. Phys.* **80**, 59.

[9] Jackson, D. A. (1930): The hyperfine structure of the lines of the arc spectrum of cesium and nuclear spin, *Proc. Roy. Soc. (London) Ser. A* **128**, 508–512.

[10] Tolanski, S. (1955): *High Resolution Spectroscopy*, Methuen, London.

[11] Meggers, W. F., and Westfall, F. O. (1950): Lamps and wave-lengths of mercury 198, *J. Res. NBS* **44**, 447–455 RP 2091.

[12] Kaptsov, N. A. (1950): *The Electrical Phenomena in Gases and in Vacuum*, Nauka, Moscow.

[13] Babat, G. I. (1942): The electrodeless discharges and some problems connected to them, *Vestnik Elektroprom.* **2**, 1–12.*

[14] Babat, G. I. (1942): The electrodeless discharges and some problems connected to them, *Vestnik Elektroprom.* **3**, 2–8.*

*The same material is presented (in English) in: Babat, G. I. (1947): Electrodeless discharges and some allied problems, *J. Institute of Electrical Enginners (London)* **84**, 27–37.

[15] Strauss, K. J. (1958): Untersuchungen über den Existenzbereich der elektrodenlosen Ringentladung, *Ann. Phys.* **1**, 281–300.

[16] Eidman, K. (1970): Gasaufheizung und Energiebilanz einer stationären hochfrequenten Ringentladung in Edelgasen, *Z. Phys.*, 99–100.

[17] Hsuchi, H., and Guongxiong, C. (1980): A study on the characteristics of radiofrequency H-discharges, in: *6th Int. Conf. Gas Discharges and their Applications 1, Edinburgh*. pp. 251–254, Heriot–Watt, London, New York.

[18] Le Bel (1938): *An electrical lamp*, Patent no. 2118452, USA.

[19] Corliss, C., Bozman, W., and Westfall, F. (1953): Electrodeless metal–halide lamps, *J. Opt. Soc. Amer.* **43**, 398–400.

[20] Zelikoff, M., Wyckoff, P., Ashenbrand, L., and Loomis, R. (1952): Electrodeless discharge lamps containing metallic vapor, *J. Opt. Soc. Amer.***42**, 818–819.

[21] Yacobsen, E., Harrison, W., and Westfall, F. (1949): Ultrafrequency excitation of Hg lamps for interferometer illumination, *J. Opt. Soc. Amer.* **39**, 1054.

[22] Bell, V., Bloom, A., and Linch, J. (1961): Spectral lamps of alkaline metals filled by vapor, *Rev. Sci. Instrum.* **32**, 588–583.

[23] Brewer, R. (1961): Powerful rubidium lamp with small noise, *Rev. Sci. Instrum.* **32**, 1356–1359.

[24] Franz, F. (1963): High intensity cesium lamp for optical pumping, *Rev. Sci. Instrum.* **34**, 589–590.

[25] Zherebenko, A. V. (1972): *Study of electrodeless high-frequency spectral lamps*, PhD Thesis, Alma-Ata.

[26] Zozulya, G. V. (1974): Spectroscopic research of rubidium light source, *Works of Metrological Institutes of USSR* **159** (219), 83–89.

[27] Zozulya, G. V., and Sivolova, O. V. (1971): Influence of frequency of excitation of the discharge to parameters of optical radiation of the high-frequency electrodeless discharge, *Works of Metrological Institutes of USSR* **7**, 375–381.

[28] Bazhov, A. S. (1973): *Definition of gold and elements accompanying it in mineral raw material*, PhD Thesis, Alma-Ata.

[29] Bazhov, A. S., and Zherebenko, A. W. (1971): High-frequency electrodeless spectral lamps with vapor of metals and their salts, *J. Appl. Spectros.* **16**, 1131–1141.

[30] Baranov, S. V. (1971): *Application of new sources of resonant radiation in the atomic-absorption analysis*, PhD Thesis, Moscow.

[31] Baranov, S. V., Baranova, I. V., Ivanov, N. P., Knyazev, V. V., Konstantinova, M. G., and Pyatova, V. N., (1977): Electrodeless lamps with high-frequency excitation of a spectrum as a source of light for atomic-absorption analysis. Part 3: High-frequency electrodeless lamps with thermostabilization of a discharge cylinder, *J. Anal. Chem.* **32**, 1524–1528.

[32] Baranov, S. V., Baranova, I. V., and Ivanov, N. P. (1982): Spectral lamps for atomic-absorption spectrometry (review), *J. Appl. Spectros.* **36**, 357–367.

[33] Baranov, S. V., Ivanov, N. P., and Pofralidi, L. G. (1969): The main characteristics of high-frequency electrodeless lamps, *J. Appl. Spectros.* **10**, 595–598.

[34] Kazantsev, S. A., and Subbotenko, A. V. (1994): *Spectropolarimetric Diagnostics of Gas Discharges*, St. Petersburg State University Publishing House, St. Petersburg (in Russian).

[35] Barvin, V. Q., Linker, B. Y., Samoilovich, V. I., and Fertik, N. S. (1974): A source of optical pumping for passive quantum standard of frequency, *Works of the Metrological Institute USSR 159 (219)*, 35.

[36] Bezlepkin, A. I., Khomyak, A. S., Aleksandrov, V. V., Voronina, T. N., and Mel'nikova, O. V. (1983): Serially delivered spectral lamps for atomic-absorption analysis, *J. Appl. Spectros. (USSR)* **39**, 367–373.

[37] Berzinya, D. K., Putnina, S. Q., Rewalde, G. V., and Skudra, A. J. (1991): The spectral characteristics of mercury isotopic lamps, *All-Union Conference on Physics of Low Temperature Plasma*, Minsk, pp. 151–152.

[38] Wyazovetskaya, N., Wyazovetskii, Y., Sechenkov, A., and Stankov, N. (1992): Application of gas absorbent at manufacturing of high-frequency electrodeless isotopic mercury lamps, *Scientific Works of Latvian University* **573**, Riga, pp. 87–93.

[39] Ginzburg, V. L., and Samarina, G. I. (1965): Application of various sources of light for the atomic-absorption analysis, *Industrial Laboratory* **31**, 249–250.

[40] Kurejchik, K. P., Bezlepkin, A. I., Khomyak, A. S., and Aleksandrov, V. B. (1987): *Discharge Light Sources for Spectral Measurements*, Universitetskoie, Minsk.

[41] Lezdin, A. I., Putnina, S. Q., and Skudra, A. J. (1985): Spectral parameters of helium in the high-frequency discharge, in: *Processes of Transfer of Energy in Vapor of Metal* (Kraulina E., ed.), pp. 99–105, Latvian State University Publisher, Riga.

[42] Lezdin, A. I., Putnina, S. Q., and Skudra, A. J. (1987): The high-frequency electrodeless spectral lamps, in: *Collision and Radiative Processes with Participation of Excited Particles* (Kraulina, E., ed.), pp. 133–140, Latvian State University Publisher, Riga.

[43] Lezdin, A. I., Putnina, S. Q., and Skudra, A. J. (1987): Population of excited levels in the high-frequency electrodeless discharge, in: *Elementary Processes at Collisions of Atoms and Molecules* pp. 31–38, Cheboxary State University Publisher, Cheboxary.

[44] Ubelis, A. P. (1992): High-frequency electrodeless lamps for the vacuum ultraviolet region of the spectrum, *Scientific Works of Latvian University* **573**, Riga, 87–93.

[45] Ubelis, A. P., Silinsh, Y. A., Berzinsh, U. V., and Rachko, Z. A. (1981): Spectra of high-frequency electrodeless lamps in vacuum ultraviolet (UV), *J. Appl. Spectros. (USSR)* **35**, 216–219.

[46] Atkinson, R. I., Chapman, G. D., and Krause, L. (1965): Light sources for the excitation of atomic resonance fluorescence in potassium and rubidium, *J. Opt. Soc. Amer.* **55**, 1269–1274.

[47] Berzina, D. K., Putnina, S. Y., Revalde, G. V., and Skudra, A. Ya. (1990): High-frequency electrodeless lamps with combined Hg–Cd filling, *Abstr. 1st Conf. on Analytical Atomic Spectroscopy*, Moscow, 89.

[48] Dagnall, R. M., and West, T. S. (1968): Some applications of microwave electrodeless discharge tubes in atomic spectroscopy, *Appl. Opt.* **7**, 1287–1294.

[49] Haarsmaa, J., and Jong, G. J. (1974): Preparation and operation of electrodeless discharge lamps — critical review, *Spectrochim. Acta.* **29B**, 1–18.

[50] Mansfield, J. M., Bratzel, M. P., Norgordon, H. O., Knapp, D. O., Zacha, K. E., and Winefordner, J. D., (1968): Experimental investigation of electrodeless discharge lamps as excitation sources for atomic fluorescence flame spectroscopy, *Spectrochim. Acta.* **23 B**, 389–402.

[51] Yacobson, E., Harrison, W., and Westfall, F. (1953): Electrodeless metal–halide lamps, *J. Opt. Soc. Amer.* **43**, 398–403.

[52] Pomerantsev, N. M., Ryzhkov, V. M., and Skrotskii, G. V. (1972): *Physical Principles of Quantum Magnetometry*, Nauka, Moscow, p. 448.

[53] Semenov, S. V., and Jacobson, N. N. (1968): Research of some parameters of optical pumping light in a rubidium gas cell frequency standard, *Radioelectronics. Ser. General Technical*, Leningrad, 138–143.

[54] Izotova, S. L., and Frish, M. S. (1972): Displacement of frequency of radiation of resonant lines of rubidium in gas filled electrodeless lamps, *Opt. Spectros.* **33**, 798–799.

[55] Izotova, S. L., Preobrashenski, N. G., Tambovtsev, B. Z., and Frish, M. S. (1975): About methods of correct processing of contours of spectral lines, *Opt. Spectros.* **38**, 842–847.

[56] Ionov, V. N. (1977): The small-size physics package, *Electronics* **10**, Moscow, pp. 73–75.

[57] Lopatin, W. M., and Terzeman, W. F. (1977): Change of the light flow from mercury lamps during long work, *J. Appl. Spectros.* **27**, 916–919.

[58] Rogov, V. I., Semenov, S. V., and Khutorshchikov, V. I., *A spectral source of light*, Patent no. 672512, USSR.

[59] Rogov, V. I., and Khutorshchikov, V. I., *A spectral source of light*, Patent no. 794397, USSR.

[60] Rogov, V. I., Khutorshchikov, V. I., Kalachnikova, A. I., and Nagapetyan, T. S., *A spectral source of light*, Patent no. 724504, USSR.

[61] Rozenblit, R. A., and Khutorshchikov, V. I., *A spectral source of light*, Patent no. 974142, USSR.

[62] Semenov, S. V., Smirnova, G. M., and Khutorshchikov, V. I. (1981): About ignition of the discharge in high-frequency electrodeless spectral lamps, *Radioelectronics. Ser. General Technical*, Leningrad, 90–94.

[63] Semenov, S. V., and Khutorshchikov, V. I. (1974): About durability of high-frequency electrodeless spectral lamps, *4th All-Union Conference on Discharge Devices*, Ryazan.

[64] Semenov, S. V., and Khutorshchikov, V. I. (1976): About a thermal mode of high-frequency electrodeless spectral lamps, *Reference Information about Transfers, Reviews, Deposited Manuscripts and Theme Indexes No. 3*, All Union Institute of Scientific and Technical Information.

[65] Semenov, S. V., and Khutorshchikov, V. I. (1977): About the discharge in high-frequency electrodeless spectral lamps, *Radioelectronics. Ser. General Problems* **12**, Leningrad, pp. 116–119.

[66] Semenov, S. V., and Khutorshchikov, V. I. (1984): About intensity of radiation of high-frequency electrodeless spectral lamps with vapor of metal, *Radioelectronics. Ser. General Problems of Radioelectronics* **8**, Leningrad, 120–125.

[67] Khutorshchikov, V. I. (1975): About measurement of pressure in high-frequency electrodeless spectral lamps, *Radioelectronics. Ser. General Technical Problems* **2**, Leningrad, pp. 110–114.

[68] Khutorshchikov, V. I. (1975): About durability of high-frequency electrodeless spectral lamps with Rb vapor, *Radioelectronics. Ser. General Technical Problems* **2**, Leningrad, pp. 115–118.

[69] Kazantsev, S. A., and Khutorshchikov, V. I. (1995): *A High Frequency Light Source of Line Spectra*, St. Petersburg State University Publisher, St. Petersburg (in Russian).

[70] Khutorshchikov, V. I., and Yushina, G. G. (1978): About measurement of conductivity of plasma of the discharge in high-frequency electrodeless spectral lamps, *Radioelectronics. Ser. General Problems* **4**, Leningrad, pp. 54–58.

[71] Clark, D., and Stanley, J. (1971): The manufacture of alkali vapor cells for optical pumping experiments, *J. Appl. Phys. E: Sci. Instrum.* **4**, 758–760.

[72] Fukuyo, H., Jga, N., Kuramoshi, N., and Taganawa, H. (1977): The dependence of the hyperfine structure of the Rb D_1-line on the temperature, *Jap. J. Appl. Phys.* **9**, 729–734.

[73] Gupta, R., Chang, S., and Happer, W. (1972): Cascade-decoupling measurements of excited S-states Na, Rb and Cs, *Phys. Rev. A* **6**, 529–544.

[74] Hirano, J. (1977): Spectre de la raie *D* du Rb dans le cas lampe spherique, *Rev. de Phys. Appl.* **12**, 1253–1262.

[75] Kuramoshi, N., Matsuda, J., Oura, N., and Fukuyo, H. (1980): Analysis of the temperature dependence of Rb lamp profiles, *J. Opt. Soc. Amer.* **70**, 1504–1507.

[76] Kuramoshi, N., Matsuo, T., Matsuda, J., and Fukuyo, H. (1977): Spectral profiles of the Rb *D* lines emitted from a spherical electrodeless lamp, *Jap. J. Appl. Phys.* **16**, 673–679.

[77] Kuramoshi, N., Naritsuka, S., and Oura, N. (1981): Combined Rb lamp for optical pumping, *Opt. Lett.* **6**, 73–78.

[78] Kuramoshi, N., Tanigava, H., Iga, K., and Fukuyo, H. (1970): Observation of the hyperfine spectrum of the Rb *D* line for optical pumping, *Bull. Tokyo Inst. Technol.* **101**, 63–70.

[79] Lennon, I., and Sexton, M. C. (1959): Recombination in xenon and krypton afterglow, *J. Electronics Control* **7**, 123–132.

[80] Oyamada, H., Takahashi, K., Sato, I., and Ushida, H. (1975): A consideration of rubidium lamp stability for rubidium frequency standards, *NEC Res. Dev.*, 340–343.

[81] Schmieder, R. W., Luri, A., Happer W., and Khadiavi, A. (1970): Level-crossing measurements of lifetime and hyperfine structure constants of the P states of the stable alkali atoms, *Phys. Rev. A* **2**, 1216–1228.

[82] Yamada, J., and Okuda, T. (1972): Impedance of a partially ionized cesium plasma, *Jap. J. Appl. Phys.* **11**, 1032–1038.

[83] Semenov, S. V., Smirnova, G. M., and Khutorshchikov, V. I. (1975): Physical processes in high-frequency electrodeless spectral lamps with vapor of metals, *Reference Information* **17**, Moscow.

[84] Tako, T., Koga, J., and Hirano, J. (1975): Spectral profiles of the Rb-*D* lines, *Jap. J. Appl. Phys.* **14**, 591–598.

[85] Kalachnikova, A. I., and Khutorshchikov, V. I. (1981): Problem of development of light source for optical pumping, in: *Processes of Transfer of Energy in Vapor of Metal* (Kraulina, E., ed.), pp. 164–173, Latvian State University Publisher, Riga.

[86] Luizova, L. A., Trukhacheva V. A., Semenov, S. W., and Khutorshchikov, V. I. (1985): A light source for quantum frequency standards, *All-Union symposium "Increase of Accuracy of Quantum Frequency Standards,"* Moscow, pp. 68–69.

[87] Guzhva, Y. G., and Semenov, S. V. (1967): Research of a rubidium light source for optical pumping, *Radioelectronics. Ser. General Problems* **8**, Leningrad, pp. 78–83.

[88] Knyazev, V. V., and Baranov, S. W., *An electrodeless spectral lamp*, Patent no. 396753, USSR.

[89] Rostov, Y. B., Ulyanov, A. A., Fateev, B. P., Sheronov, A. P., and Schurov, A.V., *A spectral source of light*, Patent no. 384158, USSR.

[90] Bezlepkin, A. I., and Shupeleva, N. M. (1980): Spectral lamps for quantum measures of frequency, *Electronic Engineering* **4**, Moscow, pp. 63–65.

[91] Grigoryanz, V. V., Zolin, N. N., and Zhabotinski, M. E. (1968): *Quantum Frequency Standards*, Nauka, Moscow.

[92] Ivanov, N. P., Minervina, L. V., and Baranov, S. V. (1965): Application of atomic absorption analysis of spectral lamps with high-frequency excitation, in: *Chemical Reactions and Preparations. Trudy instituta redkikh elementov* **27**, Moscow, pp. 1129–1131.

[93] Ivanov, N. P., Minervina, L. V., and Baranov, S. V. (1966): Electrodeless lamps with high-frequency excitation of the spectrum of In, Ga, Bi, Sb, Rb, Tl, Mg, Ca, Cu, *J. Anal. Chem.* **21**, 1129–1131.

[94] Agapov, A. S., Kalachnikova, A. I., and Khutorshchikov, V. I. (1983): Optimization of light sources for quantum gas cell frequency standards. Modeling of physical processes in high-frequency electrodeless spectral lamps with vapor of metals, *Thesis of All-Union conference "Application of Time–Frequency Methods in the National Economy,"* Moscow, pp. 193–194.

[95] Agapov, A. S., Kalachnikova, A. I., and Khutorshchikov, V. I. (1984): Optimization of light sources for optical pumping. Modeling of physical processes, *Radioelectronics. Ser. General Problems of Radioelectronics* **8**, Leningrad, pp. 91–98.

[96] Agapov, A. S., Matveev, A. A., and Khutorshchikov, V. I. (1984): Calculation of nucleus temperature in high-frequency electrodeless spectral lamps with vapor of metals, *Radioelectronics. Ser. General Problems of Radioelectronics* **2**, Leningrad, p. 114.

[97] Agapov, A. S., Matveev, A. A., and Khutorshchikov, V. I. (1985): Mathematical modeling of processes in high-frequency electrodeless spectral lamps with vapor of metals, in: *Processes of Energy Transfer in Vapor of Metals* (Kraulina, E., ed.), pp. 89–98, Latvian State University Publisher, Riga.

[98] Agapov, A. S., Smirnova, G. M., and Khutorshchikov, V. I. (1983): Formation of lines of radiation in electrodeless spectral lamps in modes of E- and H-discharges, in: *Processes of Transfer of Energy in Vapor of Metals* (Kraulina, E., ed.), pp. 152–161, Latvian State University Publisher, Riga.

[99] Agapov, A. S., and Khutorshchikov, V. I. (1984): High-frequency electrodeless spectral lamps with vapor of metals. Theory and experiment, *Electronics. Ser. General Problems of Radioelectronics* **9**, Leningrad, pp. 111–118.

[100] Bulychev, A. E., Denisova, N. V., Preobrazenski, N. G., and Suvorov, A. G. (1987): Mathematical modeling of the high-frequency electrodeless discharge, in: *Collisional and Radiating Processes with Participation of Excited Particles* (Kraulina, E., ed.), pp. 123–132, Latvian State University Publisher, Riga.

[101] Bulychev, A. E., Denisova, N. V., Preobrazenski, N. G., and Suvorov, A. G. (1988): Account of the characteristics of the electrodeless discharge, *Applied Mechanics and Technical Physics* **2**, 3–7.

[102] Bulychev, A. E., Denisova, N. V., and Skudra, A. J. (1989): The optical characteristics of an electrodeless RF discharge in argon and krypton, *Opt. Spectros.* **67**, 788–791.

[103] Putnina, S. (1992): Transfer of radiation in spectral lines of sources of radiation of low temperature plasma, *Proceedings of Latvian University*, **Vol. 573**, pp. 29–43, Riga.

[104] Revalde, G. (1992): Modeling of contours of spectral lines, *Proceedings of Latvian University*, **Vol. 573**, pp. 44–52, Riga.

[105] Skudra, A. J., and Khutorshchikov, V. I. (1992): High-frequency electrodeless spectral lamps with vapor of metals (review), in: *High-Frequency Electrodeless Sources of Light* (Kraulina, E., ed.), pp. 4–28, Latvian State University Publisher, Riga.

[106] Dindarov, V. E., Skudra, A. J., and Khutorshchikov, V. I., *Working substance for high frequency electrodeless lamps*, Patent no. 1170922, USSR.

[107] Dindarov, V. E., and Khutorshchikov, V. I. (1984): A rubidium source of light with a small level of radiation of background lines, *6th All-Union Conference "Metrology in Radioelectronics,"* Moscow, p. 324.

[108] Agapov, A. S., Kalachnikova, A. I., Khutorshchikov, V. I., and Smirnova, G. M. (1984): Accelerated tests of spectral lamps and their durability, *6th All-Union Conference "Metrology in Radioelectronics,"* Moscow, 201.

[109] Khutorshchikov, V. I. (1992): *Radiospectroscopic Method of Measurement of Gas Pressure*, Gosstandart, Moscow.

[110] Sheronov, A. P., Sapoznikov, Y. M., Selivanov, S. M., Treskov, A. V., and Eremina, N. M., (1976): Research of an aging of spectral lamps for rubidium frequency standards, *Communication Engineering. Ser. Radionavigation* **2**, 62–64.

[111] Nikolaeva, T. I., Semenov, S. V., and Khutorshchikov, V. I. (1974): Measurement of quantity of the metal in spectral devices, *Radioelectronics. Ser. General Problems* **4**, Leningrad, pp. 164–167.

[112] Khutorshchikov, V. I., Vidkovskii, O. O., and Smirnova, G. M., *A way of metal dosing in spectral lamps*, Patent no. 1141927, USSR.

[113] Khutorshchikov, V. I., and Luizova, L. A. (1984): Space distribution of atoms in high-frequency electrodeless spectral lamps with vapor of metals, *Opt. Spectros.* **56**, 238–243.

[114] Grigorev, M. B., Smirnova, G. M., and Khutorshchikov, V. I. (1989): About durability, long-term stability and reproducibility of spectral lamps with vapor of rubidium, *2nd All-Union Conference on Optical Orientation of Atoms and Molecules*, Leningrad, pp. 43–44.

[115] Gevorkyan, A. G., Khutorshchikov, V. I., and Yushina, G. G., *Way of stabilization of intensity of radiation of the working substance of a source of light of a quantum frequency standard*, Patent no. 130628, USSR.

[116] Skudra, A. J., Khutorshchikov, V. I., and Yushina, G. G. (1987): The spectral characteristics of high-frequency electrodeless spectral lamps, in: *Collisions and Radiating Processes with Participation of Excited Particles* (Kraulina, E., ed.), pp. 141–146, Latvian State University Publisher, Riga.

[117] Khutorshchikov, V. I., and Yushina, G. G. (1985): About the intensity of light of spectral lamps with metal vapors, *Radioelectronics. Ser. General Problems* **3**, Leningrad, p. 118.

[118] Khutorshchikov, V. I., and Yushina, G. G., *A way of creation of high-stability radiation in high-frequency electrodeless spectral lamps*, Patent no. 1321204, USSR.

[119] English, T., and Jechart, E. (1981): Development of a sapphire lamp for use in satellite-borne atomic Rb-clocks, *Proc. 35th Ann. Freq. Contr. Symp.*, pp. 637–645.

[120] Obukhova, E. S., Pikhtelev, A. I., and Rudnevskij, N. K. (1966): Spectral researches of a rubidium light source, *J. Appl. Spectros.* **5**, 793–794.

[121] Khutorshchikov, V. I., Rumyantsev, I. G., and Zholnerov, V. S. (1989): Flicker-noise in high-frequency lamps with vapor of metals, *2nd All-Union Symposium on Optical Orientation of Atoms and Molecules*, Leningrad, p. 45.

[122] Khutorshchikov, V. I., and Zemskova, M. V. (1990): About research of the spectral density of noise of electrodeless lamps with vapor of rubidium (f = 50–300 MHz.), *Radioelectronics. Ser. General Problems of Radioelectronics* **20**, Leningrad, pp. 96–99.

[123] Khutorshchikov, V. I., and Zemskova, M. V. (1992): Fluctuations of intensity of radiation of rubidium sources of light, in: *High-Frequency Electrodeless Sources of Light* (Kraulina, E., ed.), pp. 76–86, Latvian State University Publisher, Riga.

[124] Vanier, J., and Bernier, L. (1981): On the signal-to-noise ratio and short-term stability of passive rubidium frequency standards, *IEEE Transactions on Instrumentation and Measurement* **IM 30**, 277–282.

[125] Volk, C. N., and Frueholz, R. P. (1985): The role of long-term lamp fluctuations in the random walk of the frequency behavior of the rubidium frequency standard: A case study, *J. Appl. Phys.* **57**, 980–983.

[126] Godunov, A. M., Evlampiev, Y. K., Semenov, S. V., and Khutorshchikov, V. I. *A high-frequency electrodeless spectral lamp*, Patent no. 1124181, USSR.

[127] Volk, C. N., Frueholz, R. F., English, T. C., Linch, T. J., and Riley, W. J., (1984): Lifetime and reliability of rubidium discharge lamps for use in atomic frequency standards, *Proceedings of 38th Annual Control. Symp.*, pp. 387–400.

[128] Kazantsev, S. A. (1983): Astrophysics and laboratory application of the phenomenon of self-alignment, *Sov. Phys. Usp.* **26**, 328 (in Russian).

[129] Kazantsev, S. A., Petrashen, A. G., Polezhaeva, N. T., Rebane, V. N., and Rebane, T. K. (1990): Relaxation of collisional self-alignment of the drifting ions in the magnetic field, *Opt. Spectros.* **69**, 582–585.

[130] Aleksandrov, E. B., Bezuglov, N. I., and Jacobson, N. N. (1979): Optical self-pumping in Tl vapor, *Opt. Spectros.* **46**, 1061–1066.

[131] Aleksandrov, E. B., and Jacobson, N. N. (1980): Optical self-pumping in a superthin structure of main condition Rb, *Opt. Spectros.* **48**, 828–831.

[132] Altman, E. L., Ganeev, A. A., and Turkin, J. I. (1980): Redistribution of the relative intensities of the hyperfine structure components of a resonance line in a modulated discharge, *J. Appl. Spectros. (USSR)* **33**, 248–252.

[133] Levitskij, S. M. (1957): Research of potential of ignition of the high-frequency discharge, *J. Tech. Phy.* **27**, 970–977.

[134] Raizer, Y. P. (1987): *Physics of the Gas Discharge*, Nauka, Moscow.

[135] Yatsenko, N. A. (1981): About the existence of two modes of the high-frequency electrical discharge, *J. Tech. Phys.* **5**, 1195–1204.

[136] Bochkova, O. P., Razumovskaya, L. P. (1964): Spectroscopic research of the "weak" and "strong" hf-discharge in inert gases, *Opt. Spectros.* **17**, 16–23.

[137] Chandrakar, K. (1978): The transition from the first to the second stage of the ring discharge, *J. Phys. D.: Appl. Phys.* **11**, 1809–1813.

[138] Kolchin, E. A., Lisenkov, A. L., and Khutorshchikov, V. I. (1984): Research of parameters of the high-frequency discharge in a mixture Kr–Rb, in: *Vacuum and Discharge Electronics* (Potsar, A. A., ed.), pp. 9–12, Ryazan Radio-Technical Institute Publisher, Ryazan.

[139] Rovinski, R. E., and Sobolev, A. P. (1968): An optimum frequency range of the stationary induction discharge, *Thermophysics of High Temperatures* **6**, 219–223.

[140] Granovski, V. A. (1971): *Electrical Currents in Gases*, Nauka, Moscow, p. 543.

[141] Perrin-Lagarde, D. (1966): Perturbations pas des atomes isotopes du profil d'une composante spectrale de la raie $\lambda = 2537$ Å du mercure, *C. R. Acad. Sci. Paris* **263**, 1384–1386.

[142] Bogachev, V. M., Kunina, S. A., Petrov, B. E., and Popov, I. A. (1964): *Calculation of Semi-Conductor Transmitters*, Moscow Energetic Institute Publisher, Moscow.

[143] Borisova, Y. A., and Kozlov, L. N. (1977): Tests of spectral cesium lamps for the term of a service, *Geomagnitnoye priborostroeni*, Moscow, pp. 40–42.

[144] Shernoff, F. (1969): A mercury lamp for optical pumping, *Rev. Sci. Instrum.* **40**, 1418–1419.

[145] Silinsh, Y. A. (1992): Generators for high-frequency electrodeless lamps, *Proceedings of Latvian University*, **Vol. 573**, pp. 98–104, Riga.

[146] Silinsh, Y. A., and Ubelis, A. P. (1979): A generator with an oven for excitation of high-frequency electrodeless lamps, in: *Sensibilized Fluorescense of Vapor Mixes of Metals* (Kraulina, E., ed.), pp. 123–125, Latvian State University Publisher, Riga.

[147] Chelnokov, O. A. (1975): *Transistor Generators of Sine Wave Fluctuations*, Sovetskoie Radio Publisher, Moscow.

[148] MacDonald, A. D. (1969): *Microwave Breakdown in Gases*, Wiley, New York–London–Sydney.

[149] Nikolsky, P. P. (Ed.) (1952): *Handbook for Chemists 3*, Goskhimizdat, Moscow, Leningrad.

[150] Korchevoi, Y. P., Lukashenko, V. I., Lukashenko, S. N., and Khilko, I. N. (1977): Experimental determination of effective cross sections of ionization by electrons of atoms K, Rb, Cs, excited to the first resonance level, *High Temp.* **15**, 7–12.

[151] Frish, S. E. (1963): *Optical Spectra of Atoms*, State Publisher of Physical-Technical Literature, Moscow, Leningrad.

[152] Vainstein, L. A., Sobelman, I. I., and Yukov, E. A. (1973): *Cross Sections of Excitation of Atoms and Ions by Electrons*, Nauka, Moscow.

[153] Preobrazhenskii, N. G. (1971): *Spectroscopy of Optically Dense Plasma*, Nauka, Moscow.

[154] Golant, V. E., Zhilinskii, A. P., and Sakharov, S. A. (1977): *The Fundamentals of Plasma Physics*, Nauka, Moscow.

[155] Nikolaev, G. I., and Nemets, A. M. (1982): *Atomic-Absorption Spectroscopy in Research of Evaporation of Metals*, Metallurgiia, Moscow.

[156] Van Tongeren, H. (1975): Positive column of cesium– and sodium–noble-gas discharges, *Philips Res. Rept., Suppl.* **12**, 1–69.

[157] Fastovskii, V. T. (1972): *Inert Gases*, Atomizdat, Moscow.

[158] Radzig, A. A., Smirnov, B. M. (1986): *Parameters of Atoms and Nuclear Ions: Handbook*, Energoatomizdat, Moscow.

[159] Vriens, L. (1973): Energy balance in low-pressure gas discharges, *J. Appl. Phys.* **44**, 3980–3988.

[160] Zapesochny, I. P. (1967): Absolute cross-sections of excitation of levels of atoms of alkaline metals by electrons of small energy, *High Temp.* **5**, 7–13.

[161] Drukarev, G. D. (1992): *Collisions of Electrons with Atoms and Molecules*, Plenum Press, London.

[162] Biberman, A. M., Vorobieva, V. S., and Yakubov, I. G. (1979): Low temperature plasma with nonequilibrium ionization, *J. Phys. Usp. (USSR)* **128**, 233–271.

[163] Vasileva, I. A., and Asvadurov, K. D. (1979): Rules of similarity for energy distribution of electrons in mixes with alkaline additive, in: *Diagnostics of Low Temperature of Plasma*, pp. 179–187, Nauka, Moscow.

[164] Devyatov, A. M., Solovev, T. N., and Volkova, L. N. (1974): Research of a positive pole of discharge in vapor of alkaline metals. Probe measurements, *High Temp. Phys. (USSR)* **12**, 705–712.

[165] Zaidel, A. N. (1969): *The Tables of Spectral Lines*, Nauka, Moscow.

[166] Evlampiev, Y. K., Smirnova, G. M., and Khutorshchikov, V. I. *A spectral source of light*, Patent no. 1025280, USSR,

[167] Semenov, S. V., Smirnova, G. M., and Khutorshchikov, V. I. (1983): About intensity of radiation of high-frequency electrodeless spectral lamps, *Radioelectronics Ser. General Technical* **2**, Leningrad, pp. 95–99.

[168] Polushkina, I. N., Sorin, E. A., and Fedotov, J. F. (1974): Research on the hyperfine structure in a spectrum of rubidium radiation, *8th Siberian Conference on Spectroscopy*, Novosibirsk, p. 176.

[169] Kazantsev, S. A., Skvorzova, G. P., and Rish, O. M. (1989): A modern state of precise methods of spectrometry of colliding atoms, *State Standard Publisher*, USSR, Moscow, p. 49.

[170] Frish, M. S., Izotova, S. L., Khutorshchikov, V. I., and Zemskova, M. V. (1991): About the contour of radiation of resonance lines of rubidium in high-frequency electrodeless spectral lamps, *St. Petersburg State University Vestnik* **4**, 79–82.

[171] Mitchell, A., and Zemansky, M. (1961): *Resonance Radiation and Excited Atoms*, Cambridge University Press, London.

[172] Aleksandrov, E. B., and Prilipko, V. K. (1981): Optical pumping within the hyperfine structure of silver atoms in the ground state, *Opt. Spectros.* **51**, 218–221.

[173] Rytov, S. M. (1976): *An Introduction to Statistical Radiophysics. Part 1, Stochastic Processes*, Nauka, Moscow.

[174] Bogdanova, I. P., Molchanov, A. V., Semenov, R. I., and Chaika, M. P. (1991): Desorption of cluster by electrons in a pulse microwave discharge, *Opt. Spectros.* **71**, 477–481.

[175] Broussard, G., and Malnar, H., *Optical resonance cells*, Patent no. 3418565, USA,

[176] Gran, B. V., Kapustin, V K., and Poddubnyi, S. A., *Device for measurement of pressure*, Patent no. 238194, USSR.

[177] Pikhtelev, A. I., Saposhnikov, J. M., and Sheronov, A. P., *Manometer*, Patent no. 538251, USSR.

[178] Semenov, S. V., Khutorshchikov, V. I., *A way of measurement of partial pressure in a mixture of gases*, Patent no. 813191, USSR.

[179] Fateev, B. P. (1978): *Quantum Frequency Standards*, Sovetskoie Radio Publisher, Moscow.

[180] Matsuda, J., Kashivagi, Y., Kuramoshi, N., and Oura, N., (1987): Filtering effect of a Cs cell enclosing a foreign gas on the Cs lamp light, *Bull. Res. Lab. Precis. Mech. Electron.* **6**, 1–7.

[181] Mitin, Y. N., Smirnova, G. M., and Khutorshchikov, V. I. (1976): About hyperfine cell-filters on rubidium vapor, *Radioelectronics. Ser. General Technical* **7**, Leningrad, pp. 85–87.

[182] Batygin, V. V., Smirnova, G. M., and Khutorshchikov, V. I. (1984): Study of the double resonance in gas cells of small sizes, *Radioelectronics. Ser. General Problems of Radioelectronics 8*, Leningrad, 85–90.

[183] Gorny, M. B., and Matisov, B. G. (1983): Double resonance in gas cells of finite thickness, *J. Tech. Phys.* **53**, 44–52.

[184] Dmitriev, K. P., Stolyarov, O. E., and Khutorshchikov, V. I. (1988): Physics package on a gas cell-analyzer of a spectrum of microwave fluctuations, *7th All-Union Conference "Metrology in Radioelectronics,"* Moscow, pp. 264–265.

[185] Dmitriev, K. P., and Khutorshchikov, V. I. (1990): About restrictions of the limit sensitivity of double resonance signal by microwave fluctuations, *Radioelectronics. Ser. General Problems of Radioelectronics* **7**, Leningrad, pp. 92–96.

[186] Gorny, M. B., Matisov, B. G., Smirnova, G. M., and Khutorshchikov, V. I. (1987): About short-term stability of rubidium frequency standards, *J. Tech. Phys.* **57**, 740–746.

[187] Aleksandrov, E. B., Mamyrin, A. B., and Jacobson, N. N. (1981): Limiting sensitivity of a magnetometer, *J. Tech. Phys. (USSR)* **51**, 607–612.

[188] Smirnova, G. M., and Khutorshchikov, V. I. (1989): Limiting sensitivity of the double resonance, *All-Union Meeting on Optical Orientation of Atoms and Molecules*, Leningrad, pp. 46–47.

[189] Aleksandrov, E. B., Izotova, S. L., Mamyrin, B. A., and Frish, M. S. (1975): Formation of line profiles for optical pumping on potassium resonance lines, *Opt. Spectros.* **38**, 818–820.

[190] Robinson, N. G., and Johnson, A. E. (1983): A new heart for Rb-frequency standards: The evacuated, wall-coated sealed cell, *IEEE Trans. Instrum. Meas.* **32**, 198.

[191] Gershun, V. V., Khutorshchikov, V. I., and Jacobson, N. N. (1971): Shift of the rubidium line 794.8 nm by extraneous gases, *Opt. Spectros.* **31**, 866–869.

[192] Arditi, M., and Carver, T. R. (1958): Frequency shift of the zero-field hf splitting of Cs produced by various buffer gases, *Phys. Rev.* **112**, 449.

[193] Beaty, E. C., Bender, P. L., and Chi, A. R. (1958): Narrow hf absorption lines of Cs in various buffer gases, *Phys. Rev.* **112**, 450.

[194] Bender, P. L., Beaty, E. C., and Chi, A. R. (1958): Optical detection of narrow Rb hf absorption lines, *Phys. Rev. Lett.* **1**, 311–315.

[195] Margenau, H., Fontana, P., and Klein, L. (1959): Frequency shifts in hyperfine splittings of alkalis caused by foreign gases, *Phys. Rev.* **115**, 87–92.

[196] Ensberg, E. S., and Putlitz, G. (1969): Nonlinear hyperfine pressure shift by optical pumping with white light, *Phys. Rev. Lett.* **22**, 1349–1351.

[197] Ray, S., Das, G., Macdonald, P., and Wahl, A. C. (1970): Theoretical evaluation of the fractional hyperfine pressure shift of paramagnetic atoms in a noble gas, *Phys. Rev. A* **2**, 2196–2200.

[198] Khvostenko, G. I., Khutorshchikov, V. I., and Chaika, M. P. (1974): Stark constant in Cs, *Opt. Spectros.* **36**, 814–815.

[199] Kopfermann, H., and Krüger, H. (1936): Zur Hyperfeinstruktur der Resonanzlinien des Rubidiums, *Z. Phys.* **103**, 485–491.

[200] Vanier, J., Kunski, R., Cyr, N., Savard, J. J., and Tetu, M. (1982): On hyperfine frequency shifts caused by buffer gases: application to optically pumped passive rubidium frequency standards, *J. Appl. Phys.* **53**, 5387–5391.

[201] Batygin, V. V., and Zolnerov, V. S. (1975): Temperature dependence of the frequency of hfs transitions of Rb in buffer environment, *Opt. Spectros. (USSR)* **39**, 449–452.

[202] Semenov, S. V., Khutorshchikov, V. I., and Chalyi, V. V. (1979): About a method of accelerated determination of durability of gas cells, *Radioelectronics. Ser. General Technical* **6**, Leningrad, pp. 115–118.

[203] Gevorkyan, A. G., Smirnova, G. M., and Khutorshchikov, V. I. (1984): About manufacturing of spectral devices. Outgassing in the discharge, *Radioelectronics. Ser. General Problems of Radioelectronics* **2**, Leningrad, pp. 79–82.

[204] Eremina, N. M., Pikhtelev, A. I., and Sheronov, A. P. (1983): About the influence of physico-chemical processes of adsorption on the systematic drift of a rubidium frequency standard, *Communication Engineering Ser. Radiotechnical Measurement* **1**, Gorki, p. 35.

[205] Grodstein, A. E., and Kharitonova, L. S. (1969): Adsorption of gases in conditions of work discharge in electrovacuum devices, *Reviews on Electronic Engineering. Ser. Discharge Devices* **64** (133) pp. 3–58, Central Science-Research Institute "Electronica" Publisher, Moscow.

[206] Svetlov, V. A., Smirnova, G. M., and Khutorshchikov, V. I. (1984): Research of the surface of gas cells by a method of X-ray spectroscopy, *6th All-Union Conferences "Metrology in Radioelectronics,"* Moscow, pp. 325–326.

[207] Cook, R. A., and Frueholz, R. P. (1988): An improved Rb consumption model for discharge lamps used in Rb frequency standards, *Proceedings of 42nd Annual Control Symp.*, pp. 525–531.

[208] Gerard, V. B. (1962): Laboratory alkali metal vapor lamps for optical pumping experiments, *Rev. Sci. Instrum.* **39**, 217–218.

[209] Altman, E. L., Ganeev, A. A., Turkin, J. I., and Shalunov, S. G. (1977): Study of the width of the Hg 253.7 nm line in a source of light for sorption measurements, *J. Appl. Spectros.* **27**, 539.

[210] Khutorshchikov, V. I., and Rozenblit, R. A., *A way of thermoisolation of a Dewar vessel*, Patent no. 588447, USSR.

[211] Lagarde, D., and Lennuier, R. (1965): Effet de la pression sur le profil spectral de la raie de resonance $\lambda = 2537$ Å du mercure monobare 198, *C. R. Acad. Sci. Paris* **261**, 919.

[212] Rozenblit, R. A., and Khutorshchikov, V. I. (1986): Dewar vessel with variable thermoconductivity, *Radioelectronics. Ser. Thermophysics and Equipment* **6**, 103–110.

[213] Chidson, Y. S. (1973): A semi-conductor generator for excitation of a spectral lamp, *Instruments and Experimental Techniques* (Edelman, V. S., ed.), pp. 129–130, Academy of Sciences of the USSR, Moscow.

[214] Gilev, Y. V., and Konishchev, D. E. (1973): A UHF generator on transistors for a quantum magnetometer, *Geophysical Equipment* **53**, 35–37.

[215] Smirnov, B. I., and Kontorovich, V. L. (1976): A generator for excitation of the discharge, *Geophysical Equipment* **59**, 31–39.

[216] Laparde, D., Butaux, J., Lennier, R., and Prevot, J. Y., (1967): Determination de profils spectraux par la methode de balayage magnetique, *J. Phys.* **28**, 243.

[217] Butaux, J., and Lennuier, R. (1968): Comparative effects of different hydrogen molecules on position and profile of the line $\lambda = 2537$ Å of mercury, *C. R. Acad. Sci. Paris* **267**, 36–39.

[218] Butaux, J., and Lennuier, R. (1967): Effets comparés de l'hydrogène et du deuterium sur la position et le profil spectral de la raie $\lambda = 2537$ Å du mercure 198, *C. R. Acad. Sci. Paris* **265**, 43–47.

[219] Leboucher, E., Bousquet, C., and Bras, N. (1974): Measurement of isotope shift and hyperfine-structure of the Hg line $\lambda = 1849$ Å by means of the magnetic scanning method, *Nouv. Rev. Opt.* **5**, 121–128.

[220] Ben-Lakhdar-Akrout, Z., Butaux, J., and Lennuier, R. (1975): Experimental study of the mercury 4047 Å line ($6p\ ^3P_0 \to 7s\ ^3S_1$) absorption profile in presence of foreign gases, *J. Phys.* **36**, 625–629.

[221] Evdokimov, Y. V. (1970): *Experimental investigations of broadening and depolarizing collisions*, PhD Thesis, Leningrad, p. 276.

[222] Rish, O. M., and Chaika, M. P. (1975): Possibility of measuring spectral-line broadening by magnetic scanning, *Opt. Spectros.* **38**, 1035–1038.

[223] Kaliteevski, N. I., Rish, O. M., and Chaika, M. P. (1977): *Diagnostics of plasma*, pp. 74–79, Petrozavodsk State University Publisher, Petrozavodsk.

[224] Preobrazenski, N. G. (1976): *Ill-Posed Reverse Problem of Atomic Physics*, Nauka, Moscow.

[225] Chen, S., and Pao, C. (1940): Pressure effects of hydrogen and nitrogen on the second doublet of the Rb principal series, *Phys. Rev.* **58**, 884–889, 1058–1069.

[226] Granier, R., and Granier, J. (1965): Essai de determination des courbes d'energie potentielle du couple Rb–Ne. Relation avec une interpretation des satellites bleus, *C. R. Acad. Sci. Paris* **262**, 1502–1508.

[227] Granier, R. (1969): *Ann. Phys.* **4**, 383–422.

[228] Chen, S., and Fountain, C. W. (1964): The effect of xenon pressure on the red rubidium–xenon satellite bands, *J. Quant. Spectrosc. Radiat. Transfer* **4**, 323–330.

[229] Takeo, M., and Chen, S. (1964): Intensity analysis of the nature of Rb/Xe red satellite bands, *J. Quant. Spectrosc. Radiat. Transfer* **4**, 471–486.

[230] Ottinger, Ch., Scheps, R., York, G. W., and Gallagher, A. (1975): Broadening of the Rb resonance lines by the noble gases, *Phys. Rev. A* **11**, 1815–1826.

[231] Drummond, D. L., and Gallagher, A. (1974): Potentials and continuum spectra of Rb–noble gas molecules, *J. Chem. Phys.* **60**, 3426–3438.

[232] Carrington, C. G., and Gallagher, A. (1974): Blue satellite bands of Rb broadened by noble gases, *Phys. Rev. A* **10**, 1464–1472.

[233] Kaliteevskii, N. I., Rish, O. M., and Chaika, M. P. (1976): Measurement of the broadening and shift of the D_1 line of rubidium 87 due to collisions with atomic krypton, *Opt. Spectros.* **41**, 504–508.

[234] Gershun, V. V., Khutorshchikov, V. I., and Yakobson, N. N. (1971): Displacement of the 7947 Å line of rubidium by foreign gases, *Opt. Spectros.* **31**, 866–869.

[235] Kazantsev, S. A., Kaliteevski, N. I., and Rish, O. M. (1977): Magnetic scanning technique for measurement of broadening and shift of a spectral line, *Proceedings of 18th All-Union Meeting on Spectroscopy*, pp. 85–87, Gorki.

[236] Kazantsev, S. A., Kaliteevski, N. I., and Rish, O. M. (1978): Application of the magnetic scanning technique for measuring the broadening and shifts of the D_1 resonance line of rubidium caused by inert gases, *Opt. Spectros.* **44**, 638–642.

[237] Izotova, S. L., Kantserov, A. I., and Frish, M. S. (1977): Measurement of the shift and broadening of hfs components of a rubidium line in the presence of krypton, *Opt. Spectros.* **42**, 213–216.

[238] Izotova, S. L., Kantserov, A. I., and Frish, M. S. (1981): Constants for broadening and shift of rubidium 87 D_1 and D_2 lines by inert gases, *Opt. Spectros.* **51**, 196–199.

[239] Rouff, E. (1972): Broadening of alkali lines by atomic hydrogen, *Phys. Lett.* **38A**, 8–10.

[240] Rebane, V. N. (1976): Calculation of the broadening and shift of hyperfine components of rubidium and cesium D_1 lines in collisions with atoms of heavy inert gases, *Opt. Spectros.* **41**, 894–898.

[241] Granier, R., Granier, J., and Schuller, F. (1976): Calculations of impact broadening and shift of spectral lines for realistic interatomic potentials, *J. Quant. Spectrosc. Radiat. Transfer* **16**, 143–150.

[242] Rebane, V. N. (1978): Calculation of the collisional broadening and shift of the D_2 lines of rubidium and cesium during collisions with atoms of heavy inert gases, *Opt. Spectros.* **44**, 644–648.

[243] Rouff, E. (1970): A theoretical model for interaction between excited and ground state atoms. Application to pressure broadening, *Astron. Astrophys.* **7**, 4–15.

[244] Baylis, W. E. (1969): Semiempirical, pseudopotential calculation of alkali–noble-gas interatomic potentials, *J. Chem. Phys.* **51**, 2665–2678.

[245] Pascale, J., and Vanderplanque, J. (1974): Excited molecular terms of the alkali–rare gas atom pairs, *J. Chem. Phys.* **60**, 2278–2291.

[246] Chen, S., and Garret, R. O. (1966): Pressure effects of foreign gases on the absorption lines of cesium. I. The effects of argon on the first two members of the principal series, *Phys. Rev.* **144**, 59–71.

[247] Chen, S., and Garrety, R. O. (1967): Pressure effects of foreign gases on the absorption lines of cesium. IV. The effects of neon, *Phys. Rev.* **156**, 48–55.

[248] Kazantsev, S. A. (1983): Determination of the quadrupole-moment of the electron-distribution function in a plasma, *JETP Lett.* **37**, 159.

[249] Blum, K. (1981): *Density Matrix Theory and Applications*, Plenum Press, New York, London.

[250] Varshalovich, D. A., Moskalev, A. I., and Khersonsky, V. E. (1975): *Quantum Theory of Angular Momentum*, Science Publishers, St. Petersburg.

[251] Fano, U., and Macek, J. M. (1973): Impact excitation and polarization of the emitted light, *Rev. Mod. Phys.* **45**, 553.

[252] Kazantsev, S. A., and Hénoux, J.-C. (1995): *Polarization Spectroscopy of Ionized Gases*, Kluwer, Dordrecht, Boston, London.

[253] Kazantsev, S. A., and Subbotenko, A. V. (1987): Polarization spectroscopy of the hf discharge plasma, *J. Phys. D* **20**, 741–753.

[254] Kazantsev, S. A., Svelokuzov, A. E., and Subbotenko, A. V. (1986): Study of anisotropy of electron motion in the plasma of the capacitive high-frequency discharge of low pressure, *Sov. J. Tech. Phys.* **56**, 1091–1100.

[255] Kazantsev, S. A. (1982): Anisotropy of fast electron motion in plasma from the atomic state interference phenomena, *Opt. Spectros. (USA)* **52**, 559.

[256] Kuzovnikov, A. A., and Savinov, B. P. (1973): About the own stationary field's influence on the hf discharge properties, *Radiotechnics and Electronics* **18**, 816–822.

[257] Kuzovnikov, A. A., and Savinov, B. P. (1973): Space distribution of the parameters of the stationary hf discharge, *Moscow State University Vestnik, Ser. Physics Astronomy* **2**, 215–222.

[258] Kuzovnikov, A. A., Kovalevsky, V. L., Savinov, V. P., and Yakunin, V. G. (1979): The investigation of physical properties of the near electrode region of a hf discharge, *J. de Physique (Paris)* **40**, 459–460.

[259] Kuzovnikov, A. A., Kovalevsky, V. L., and Savinov, V. P. (1980): *8th All-Union Conference on Generators of Low Temperature Plasma* **3**, Novosibirsk, pp. 127–130.

[260] Kazantsev, S. A., and Subbotenko, A. V. (1984): Polarimetric diagnostics of the low temperature plasma, *Sov. J. Plasma Phys.* **10**, 78.

[261] Kazantsev, S. A., and Subbotenko, A. V. (1984): Observation of electron beams in the capacitive HF discharge, *Sov. Tech. Phys. Lett.* **10**, 1251–1255.

[262] Kazantsev, S. A., Kovalevski, V. L., and Kuzovnikov, A. A. (1990): Electron motion in the electrode regions of the capacitive high frequency discharge, *St. Petersburg State University Vestnik* **4**, 26.

[263] Fermi, E. (1949): On the origin of the cosmic radiation, *Phys. Rev.* **75**, 1169.

[264] Kazantsev, S. A., and Rys, A. G. (1989): Anisotropy of fast electron velocity distribution in the dc glow discharge, *St. Petersburg State University Vestnik* **4**, 24.

[265] Drachev, A. I., Kazantsev, S. A., Rys, A. G., and Subbotenko, A. V. (1988): Estimate of the power transmitted through electrode sheaths of the capacitive hf discharge, *Opt. Spectros.* **64**, 706–708.

[266] Kazantsev, S. A., Smirnova, O. M., and Subbotenko, A. V. (1988): On the kinetics of electrons in the capacitive hf discharge under low pressure, *St. Petersburg University Vestnik* **18**, 9–17.

[267] Korn, G. A., and Korn, T. M. (1974): *Mathematical Handbook for Scientists and Engineers*, McGraw-Hill, New York.

[268] Drachev, A. I., Kazantsev, S. A., and Subbotenko, A. V. (1990): On the possibility of determination of the alignment cross section of atomic states by electron impact, *Opt. Spectros.* **69**, 585–589.

[269] Drachev, A. I., Kazantsev, S. A., and Subbotenko, A. V. (1991): Spectropolarimetric determination of energy input into the near-electrode region of a low-pressure high-frequency capacitive discharge, *Opt. Spectros.* **71**, 527.

[270] DeHeer, F. J., Jansen, R. H. J., and Van der Kaay, W. (1979): Total cross sections for electron scattering by Ne, Ar, Kr, and Xe, *J. Phys. Ser. B* **12**, 979–1002.

[271] Smirnov, Y. M. (1984): The cross-section of electron impact excitation of He, *Astron. J.* **61**, 1087.

[272] McFarland, R. H. (1964): Cause of the observed polarization of electron-induced radiation from helium, *Phys. Rev. A* **133**, 986–990.

[273] Kovalev, A. S., Rakhimov, A. T., and Feoktistov, V. A. (1981): Possible mechanism for the instability of a glow-discharge which occurs after the pulse from an external ionizing agent, *Plasma Phys. (USSR)* **7**, 1411–1418.

[274] Smirnov, A. S. (1984): Near-electrode layers in the capacitive hf discharge, *J. Tech. Phys.* **54**, 61–65.

[275] Raizer, J. P., and Shnaider, M. N. (1987): The structure of the near-electrode layers of hf discharges and the transitions between their forms, *Plasma Phys. (USSR)* **13**, 471–479.

[276] Butler, H. S., and Kino, G. S. (1963): Plasma sheath formation by radio-frequency fields, *Phys. Fluids* **6**, 1346–1355.

[277] Kazantsev S. A., Rys A. G., Subbotenko, A. V. (1988) Polarization diagnostics of boundary regions of gas discharge plasma, *Sov. J. Plasma Phys.* **14**, 143–145.

[278] Livshitz, A. E., and Pitaevski, L. P. (1979): *Course of Theoretical Physics. Physical Kinetics*, Physics and Mathematics State Publisher, Moscow.

[279] Kazantsev, S. A., Polynovskaya, N. J., Pyatnatsky, L. N., and Edelman, S. A. (1985): Study of the anisotropy properties of the low temperature plasma by polarization spectroscopy, *Opt. Spectros.* **58**, 28.

[280] Godyak, V. A. (1976): Stationary high-frequency discharge of low pressure, *Plasma Phys. (USSR)* **2**, 1.

[281] Bloom, A. L., and Carr, J. B. (1960): Pressure shifts in the hyperfine structure constants of potassium, *Phys. Rev.* **119**, 1946–1947.

[282] Kosulin, G. M., Khutorshchikov, V. I., and Jacobson, N. N. (1971): Effect of the buffer gases on the radiooptical transition in the ground state of Rb, *Reference Information on Radioelectronics* **19**, 254.

[283] Anderson, L. W., and Ramsey, A. T. (1963): Study of the spin-relaxation times and effects of spin-exchange collisions in an optically oriented sodium vapor, *Phys. Rev.* **132**, 712–723.

[284] Dehmelt, H. (1957): Modulation of a light beam by precessing absorbing atoms, *Phys. Rev.* **112**, 1924–1925.

[285] Violino, P. (1968): Analisi dei metodi ottici per lo studio del rilassamento di spin an metallo alcalino per urto contro gas estranei, *Nuovo cimento, Suppl. 2*, 440–499.

[286] Aleksandrov, E. B., Popov, V. I., and Jacobson, N. N. (1985): A relaxation of orientation and alignment of Tl atoms in buffer gas, *Opt. Spectros.* **58**, 507–511.

[287] Arditi, M., and Carver, T. R. (1961): Light and temperature shifts in optical detection of 0–0 hyperfine resonance of alkali metals, *Phys. Rev.* **124**, 800–809.

[288] Strumia, F., Beverini, I., Moretti, A., and Rovera, G. (1976): Optimization of the buffer gas mixture for optically pumped Cs frequency standards, *Proc. 30th Annu. Symp. on Frequency Control*, pp. 468–472.

[289] Anashkin, V. I. (1970): Effects of space charge on the high-frequency electrodeless discharge, *Tech. Phys. J. (USSR)* **40**, 1262–1267.

[290] Cherepnin, N. V. (1973): *Absorption Phenomenon in Vacuum Engineering*, Radiotekhnika, Moscow.

[291] Birkhof, G. (1958): Messung der elektrischen Vorgänge innerhalb einer Hochfrequenz-Ringentladung, *Z. Angew. Phys.* **10**, 204–206.

[292] Antanevichus, O. L., Serapinas, P. D., and Shimkus, P. L. (1987): Effect of receipt of substance in plasma of the high-frequency discharge in noise of its radiation, *Opt. Spectros.* **63**, 224–225.

[293] Asmolov, E. S., and Nosik, V. I. (1979): About downbreak of high-frequency H-discharges, *Tech. Phys. J. (USSR)*, 647–658.

Index

Series Publications

Below is a chronological listing of all the published volumes in the *Physics of Atoms and Molecules* series.

ELECTRON AND PHOTON INTERACTIONS WITH ATOMS
Edited by H. Kleinpoppen and M. R. C. McDowell

ATOM–MOLECULE COLLISION THEORY: A Guide for the Experimentalist
Edited by Richard B. Bernstein

COHERENCE AND CORRELATION IN ATOMIC COLLISIONS
Edited by H. Kleinpoppen and J. F. Williams

VARIATIONAL METHODS IN ELECTRON–ATOM SCATTERING THEORY
R. K. Nesbet

DENSITY MATRIX THEORY AND APPLICATIONS
Karl Blum

INNER-SHELL AND X-RAY PHYSICS OF ATOMS AND SOLIDS
Edited by Derek J. Fabian, Hans Kleinpoppen, and Lewis M. Watson

INTRODUCTION TO THE THEORY OF LASER–ATOM INTERACTIONS
Marvin H. Mittleman

ATOMS IN ASTROPHYSICS
Edited by P. G. Burke, W. B. Eissner, D. G. Hummer, and I. C. Percival

ELECTRON–ATOM AND ELECTRON–MOLECULE COLLISIONS
Edited by Juergen Hinze

ELECTRON–MOLECULE COLLISIONS
Edited by Isao Shimamura and Kazuo Takayanagi

ISOTOPE SHIFTS IN ATOMIC SPECTRA
W. H. King

AUTOIONIZATION: Recent Developments and Applications
Edited by Aaron Temkin

ATOMIC INNER-SHELL PHYSICS
Edited by Bernd Crasemann

COLLISIONS OF ELECTRONS WITH ATOMS AND MOLECULES
G. F. Drukarev

THEORY OF MULTIPHOTON PROCESSES
Farhad H. M. Faisal

PROGRESS IN ATOMIC SPECTROSCOPY, Parts A, B, C, and D
Edited by W. Hanle, H. Kleinpoppen, and H. J. Beyer

RECENT STUDIES IN ATOMIC AND MOLECULAR PROCESSES
Edited by Arthur E. Kingston

QUANTUM MECHANICS VERSUS LOCAL REALISM: The Einstein-Podolsky-Rosen Paradox
Edited by Franco Selleri

ZERO-RANGE POTENTIALS AND THEIR APPLICATIONS IN ATOMIC PHYSICS
Yu. N. Demkov and V. N. Ostrovskii

COHERENCE IN ATOMIC COLLISION PHYSICS
Edited by H. J. Beyer, K. Blum, and R. Hippler

ELECTRON–MOLECULE SCATTERING AND PHOTOIONIZATION
Edited by P. G. Burke and J. B. West

ATOMIC SPECTRA AND COLLISIONS IN EXTERNAL FIELDS
Edited by K. T. Taylor, M. H. Nayfeh, and C. W. Clark

ATOMIC PHOTOEFFECT
M. Ya. Amusia

MOLECULAR PROCESSES IN SPACE
Edited by Tsutomu Watanabe, Isao Shimamura, Mikio Shimizu, and Yukikazu Itikawa

THE HANLE EFFECT AND LEVEL CROSSING SPECTROSCOPY
Edited by Giovanni Moruzzi and Franco Strumia

ATOMS AND LIGHT: INTERACTIONS
John N. Dodd

POLARIZATION BREMSSTRAHLUNG
Edited by V. N. Tsytovich and I. M. Ojringel

INTRODUCTION TO THE THEORY OF LASER–ATOM INTERACTIONS (Second Edition)
Marvin H. Mittleman

ELECTRON COLLISIONS WITH MOLECULES, CLUSTERS, AND SURFACES
Edited by H. Ehrhardt and L. A. Morgan

THEORY OF ELECTRON–ATOM COLLISIONS, Part 1: Potential Scattering
Philip G. Burke and Charles J. Joachain

POLARIZED ELECTRON/POLARIZED PHOTON PHYSICS
Edited by H. Kleinpoppen and W. R. Newell

INTRODUCTION TO THE THEORY OF X-RAY AND ELECTRONIC SPECTRA OF FREE ATOMS
Romas Karazija

VUV AND SOFT X-RAY PHOTOIONIZATION
Edited by Uwe Becker and David A. Shirley

DENSITY MATRIX THEORY AND APPLICATIONS (Second Edition)
Karl Blum

SELECTED TOPICS ON ELECTRON PHYSICS
Edited by D. Murray Campbell and Hans Kleinpoppen

PHOTON AND ELECTRON COLLISIONS WITH ATOMS AND MOLECULES
Edited by Philip G. Burke and Charles J. Joachain

COINCIDENCE STUDIES OF ELECTRON AND PHOTON IMPACT IONIZATION
Edited by Colm T. Whelan and H. R. J. Walters

PRACTICAL SPECTROSCOPY OF HIGH-FREQUENCY DISCHARGES
Sergei A. Kazantsev, Vyacheslav I. Khutorshchikov, Günter H. Guthöhrlein, and Laurentius Windholz

MIX
Papier aus verantwortungsvollen Quellen
Paper from responsible sources
FSC® C105338

If you have any concerns about our products,
you can contact us on
ProductSafety@springernature.com

In case Publisher is established outside the EU,
the EU authorized representative is:
Springer Nature Customer Service Center GmbH
Europaplatz 3, 69115 Heidelberg, Germany

Printed by Libri Plureos GmbH
in Hamburg, Germany